World Regional
GEOGRAPHY
HUMAN MOBILITIES, TOURISM DESTINATIONS, SUSTAINABLE ENVIRONMENTS
Second Edition

Alan A. Lew
C. Michael Hall
Dallen J. Timothy

Kendall Hunt
publishing company

Contents

CHAPTER 1

Introduction: Geography, Tourism & The World

CHAPTER 1 OVERVIEW QUESTIONS

HUMAN MOBILITIES

- In what ways does travel and migration interact with local geographies to create a sense of place in the communities that you know best?

TOURISM DESTINATIONS

- How have human and physical geography come together in a holistic manner to create the tourism destinations that you are familiar with?

SUSTAINABLE ENVIRONMENTS

- How does the knowledge of physical and human geography concepts enhance your understanding of current sustainability issues in your community and the world?

INTRODUCTION

This chapter introduces fundamental concepts in geography and tourism studies, along with an overview of the entire book. Geography is about places: why they exist where they are, why they look, feel and operate the way that they do, and how they interact with their larger social and natural environments, from local to global scales. Tourism is among the most common ways that people interact with and come to know both distant places and, increasingly, their own local communities. Both geography and tourism are, therefore, about the world we live in, how we come to know that world, how we move through that world, and how we value that world. This is the goal of this book: to inspire students toward a greater understanding, appreciation and a love of the world.

1.1 | WHAT IS GEOGRAPHY?

Current Geography Issues and
Resources (http://wrgeography.com/
geography.html)

Geography as a Holistic Science

Geographers are interested in the study of the earth as the home of humankind. It is a venerable discipline, recognized as a fundamental realm of knowledge by the early Greeks, and incorporating many of today's social sciences. As such, one aspect of geography that makes it distinct is the degree to which it is able to include disciplinary perspectives from across both the social and physical sciences in a comprehensive manner. A *geographic understanding* of the places, regions and continents that make up our planet requires this type of holistic approach, including knowledge of the physical processes that created the landforms, climates and vegetation as we see and experience them, and the historical and social processes that shape the diversity of human settlements and regional identities.

To achieve this, geographers divide their study of place and space into two general fields: *physical geography* and *human geography*. The *landscape* of a place, country or region reflects the complex interplay of human culture in a specific physical environment. Thus, knowledge of both the physical sciences (primarily the earth sciences) and the social sciences is required for a sound geographic education.

This chapter introduces many of the essential concepts of geography in general, with a special extended focus on the type of geography that most people encounter on a regular basis: *tourism geography*. The remaining chapters primarily focus on understanding the geography and tourism of the world's major regions and destinations.

Physical and Human Geography

Physical geography is divided into three basic areas of emphasis: *geomorphology* (land-forms and physiography), *climatology* (climates and weather patterns) and *biogeography* (flora and fauna distributions). Human geography, on the other hand, is not as easily broken down into subdisciplinary areas. In very

© Strejman, 2011. Used under license from Shutterstock, Inc.

World geography is all about an appreciation for the diversity of the world we live in.

general terms, however, it is possible to identify an *economic geography* branch (including population, urban, transportation, location, agricultural and industrial geographies) and a *cultural geography* branch (including historical geography, environmental perception, political geography and human ecology). In all areas of geography, there is considerable crossover with other disciplines (such as economics, history and biology). Geographers, however, tend to maintain a unique perspective in their emphasis on place and spatial phenomena. One way that this is apparent is in the geographer's use of maps and other representations of space.

Maps, Map Scale and GIS

Maps are generalized "representations" of places. An aerial photo shows all of the intricacies of what is actually on the ground in a place. By contrast, a map only shows the most important of those elements, based on the purpose of the particular map.

Map scale is important in identifying the place being represented. A *small-scale* map will show a large area (such as the entire continent of Australia), but everything will appear very small. A *large-scale* map will show a very small area (such as the Australian town of Alice Springs), but all the streets and buildings will appear large. *Small-scale* maps show much less detail, with many more objects deleted, than do *large-scale* maps, which can sometimes show every tree and sign on a city block.

Both physical and human geographers use maps, which are fundamental to geographic analysis. More sophisticated map analysis today is accomplished using *geographic information systems* (GIS), which combine computer maps with large databases, such as a census of population, to view spatial patterns and relationships and to perform spatial statistics and analysis.

Cartography and Satellite Images

Map making, referred to as cartography, and map reading are fundamental techniques common to both physical and human geography. Map reading is the ability to interpret processes by looking at the lines, dots and shaded areas that are drawn on most maps. If you can tell where the oldest part of a city is located by simply looking at the street and block patterns on a map, then you are reading a map as a geographer would. A related technique that geographers use is the interpretation of aerial photographs and satellite

Map of the known world in the 1700s. The Old World is in the right circle of this map and the New World is in the left circle.

images, also known as *remote sensing*. While cartography involves a generalization (or simplification) of selected elements (mostly roads and buildings) on the earth's surface, aerial images (from airplanes and satellites) show an exact replication of everything on the surface of a place. The high degree of detail that appears in many of these images requires special techniques, typically on a computer, to interpret their complex patterns.

Site

Geographers use the term site to refer to the immediate physical setting of a place. This includes the *topographic* or *physiographic* features that shape the visible surface of the earth (mountains, plains, rocks and soils), as well as the environmental opportunities of a particular location (such as access to a navigable river or a waterfall that can be used to generate electrical power). The site characteristics of San Francisco Bay, for example, include the fact that it is a very large body of water connected directly to the ocean, while also well protected from ocean storms. These characteristics have contributed to the development of major port facilities and accompanying international financial activities. Its site characteristics have also fostered a linear settlement pattern surrounding the bay. (The opposite of site is situation, which is discussed in more detail below.) One of a place's site characteristics is its absolute location.

Absolute Location—Latitude and Longitude

When we ask "where is it?" what we often want to know is "what is its absolute location?" The absolute location of a place never changes. Most absolute locations are based on some type of location system. Street addresses are one type of absolute location system. There are many different types of street addressing systems, but usually a specific house address and street number (accompanied with a community name) identifies only one specific location. Lines of latitude and longitude are another example of an absolute location system. Latitudes and longitudes are intersecting imaginary lines drawn on the surface of the earth to identify locations. There is one, and only one, specific location on the surface of the earth for each set of latitude and longitude coordinates. Latitude lines run parallel to one another in an east–west direction. The longest line of latitude is the *equator* (which is at 0 degrees latitude). As one moves north and south of the equator, the lines of latitude crossed get shorter and shorter, until the *north and south poles* are reached (at 90 degrees north and south latitude), where they have no length at all. Longitude lines run north–south. They all touch both the north and south poles, from which they spread out away from one another as they approach the equator. Unlike lines of latitude, longitude lines are all the same length.

Situation

Situation refers to the position of a place as it relates to other places. Much of the predominance of the city of New York within the United States, for example, can be attributed to its situation in relationship to the agricultural and industrial Midwest, as well as to the major sea routes from North America to Europe. When the Midwest was first being settled for agriculture, and in the 1800s for industry, the Hudson and Mohawk River Valleys, which empty into New York Harbor, were the most convenient way of moving goods from the Midwest through the Appalachian Mountains and to the East Coast. Ocean currents in the Atlantic Ocean circulate in a clockwise direction, coming up the East Coast of the United States and then shooting across the Atlantic from the New York area directly to France and England. Thus, these two situational characteristics (connections to the Midwest and to Europe) gave New York a tremendous locational advantage that is largely responsible for its prominent global position today. In summary, situational characteristics may be thought of as the external relationships that a place has with other places, while site characteristics are the internal geographies of a place.

Relative Location

Relative location is closely related to the concept of situation. It is the opposite of absolute location and can be defined as the location of a place in comparison with another place. As such, it can change when circumstances change. For example, as you travel through a city, the location of objects to where you are is constantly changing, and your relative location to them is constantly changing. Relative location can also change in response to innovations or disasters. For example, the construction of a new highway, a port facility or an airport can serve as a stimulus for economic growth in areas that were formerly on the periphery of a country. Even at the local level, the construction of a major new road or light rail system will increase property values in some locations and foster investment in residential, retail and office developments. In both of these examples, places improve their relative location by becoming more accessible and "closer" to other places with new infrastructure projects. Natural or human-made disasters (such as wars) that destroy infrastructure can, correspondingly, make places more isolated.

Form and Function

An example of how place (or site) and space (or situation) work together is in the shaping of the form and function of human settlements. The form of a city refers to the shape that it exhibits on the land, which can sometimes be seen on a simple road map. Some cities are very circular, while others are linear or rectangular. Most of the larger metropolitan areas are multi-nodal—they have many different commercial centers, each of which may have a form of its own. The location and distribution of different types of land uses (such as residential, commercial and industrial) are also part of the form of a city. The functions of a city are the predominant economic activities that take place there. Using indices of predominant functions, cities can be classified as being primarily agricultural, educational, industrial, recreational or commercial centers. For example, Beijing is China's premier administrative city, while Shanghai is its financial center, and Guangzhou is the country's manufacturing center.

Distance Decay and the Gravity Model

One of the major relative location relationships between places is measured by the degree to which two places interact with one another. In general, the closer two places are, the more they will interact. Such interactions include economic activities (people will shop in neighboring communities of the same size that are closer than those that are further away), recreation activities, social activities and political activities. The nearness attraction is known as the gravity model, while the fact that people are less likely to travel far for the same resources that are available nearby is known as the law of least effort and is related to the concept of distance decay.

Many geographic variations influence and distort simple gravity and distance decay modeling. For example, communities will leverage and promote their locational advantage, such as proximity to a waterway or other natural resources. They may have universities in them, or specialize in some activities that will give them a much broader market area than their population size might indicate. In addition, certain groups, such as ethnic groups, may have unique histories that draw them to some destinations over others, despite simple proximity measures. Changes in transportation and communication technology will also influence gravity and distance decay relationships between places. All of these variations are research areas for economic geographers.

Human Mobilities

Three themes dominate the overall topics covered in this book. The first is human mobilities. From a geographic perspective, mobilities refer to physical movements from one place to another. Humans are one of the most mobile large animals on planet Earth. They have migrated across all of the planet's land areas, and they have adapted their cultures to match the opportunities and to adjust to the challenges of a wide variety of environmental settings. Human mobilities can be *permanent* (usually referred to as a *migration*) or *temporary* (common examples are daily work commuting, business travel and tourism). They can be *voluntary* or *involuntary* (such as human trafficking and war refugees). They can be *domestic* (within a country where one is a citizen) or *international*. Other human mobility patterns include seasonal migrations (such as by agricultural workers), regular travel between a main residence and a holiday home or cottage and movements between urban and rural areas. All of the different forms of mobility involve push and pull factors. *Push factors* are aspects of a person's home place that push them to move to another place. These may include poor living conditions or lack of employment opportunities. *Pull factors* are those that attract a person to a specific destination, and they may include greater security, family relationships and opportunities for rest and entertainment.

Tourism Destinations

The second theme followed through this book is that of tourism destinations. Unlike more general regional geography books, this one has a strong focus on understanding places around the world through the lens of tourism. *Tourism is fundamentally a geographic phenomenon.* It involves mobility, the movement of people from one place to another. And it includes an appreciation of the many ways that places become significant and important, often as a result of visitation or at least of awareness by the outside world. Tourism destinations form in many ways, including: through their history; through their contemporary marketing, planning and design; and by virtue of their relationships to transport routes and networks. This is true of all places, but it is especially salient in those places that have become significant tourism destinations. In the end, most of the readers of this book will encounter the world beyond their home place by being a tourist. It is essential that tourists have an appreciation of the full human and physical geography of the places that they visit to ensure that those places are cared for in a holistic and sustainable manner.

Sustainable Environments and Places

The third theme underlying the approach to world regional geography this book takes is that of sustainable environments. Sustainability has provided an important guiding perspective on local and global development since it was first defined in the 1987 publication, *Our Common Future*, also known as the *Brundtland Report* (published by the United National World Commission on Environment and Development). The report's most commonly cited definition of sustainable development is *"development that meets the needs of the present without compromising the ability of future generations to meet their own needs."* The report identified five basic principles of sustainability:

1. The idea of holistic planning and strategy-making that links economic, environmental and social concerns
2. The importance of preserving essential ecological processes
3. The need to protect both biodiversity and human heritage
4. The need for development to occur in such a way that productivity can be sustained over the long term for future generations (the concept of intergenerational equity)
5. The goal of achieving a better balance of fairness and opportunity among nations

From this definition, way common way of looking at sustainability is that it seeks to balance the economic, social and environmental impacts of development and change in a community and the world. This is sometimes known as the *triple bottom line* and implies an economic accounting approach to understanding impacts. Whether such a balance is actually possible is, of course, highly contested. In

Photo: Alan A. Lew

Rural landscape in the KwaZulu Natal province of South Africa. A geographer can interpret this scene in terms of its physical geography, which includes the geology, vegetation and climate, or its human geography, which includes the types of buildings and their spatial distribution pattern.

practice, the concept of sustainability is open to many different interpretations and is mostly used to give a stronger voice to environmental and social concerns in debates over economic development initiatives.

1.1.1 Physical Geography and Our Environment

As noted above, physical geography is divided into three areas of study: climatology, geomorphology and biogeography. *Climate* is the annual pattern of weather in a place. By contrast, *weather* is the day-to-day condition of the atmosphere. Climate has a major impact in shaping the seasonal distribution of tourist travel to destinations and the activities they participate in when they are there.

Global *climate change* will have major impacts on the world's tourism industry, as many of the most tourism-dependent economies are located in coastal areas (where sea levels are rising) and in mountain regions (where snow is becoming less dependable). Sea level, for instance is rising significant. From 2005 to 2013, the overall global mean rise in sea level has been 2.78 mm (1/8 in.) per year. Most of this rise is from the melting of ice on land (2mm per year).

Geomorphology is the study of the shape of the earth's land surfaces and the spatial distribution of continents and islands. This includes plate tectonics, mountain building processes and surface erosion processes. These geologic processes are constantly, albeit slowly, changing the surface of the planet and creating some of the most spectacular scenic and recreation destinations in the world. Biogeography is the study of the geographic distribution of plants and animals. Climate and geomorphology define the type of vegetation that would be endemic to a place, although humans have often changed the natural vegetation over time. An understanding of biogeography is essential for the management of most eco-tourism destinations and outdoor recreation resources.

Geomorphology Concepts
Sedimentary, Igneous and Metamorphic Rock

The surface physiography of a region is shaped by its underlying rock structure. The three main types of rock are sedimentary, igneous and metamorphic. *Sedimentary* rock consists of successive layers of wind or waterborne materials. Under pressure, these deposits turn into rock, such as sandstone, shale, limestone and coal. Sedimentary rocks are the most common type found on the surface of the planet.

Igneous rock originates deep within the surface of the earth. High temperatures in the center of the earth cause rock to melt. Molten magma plutons push their way up through the earth's crust. If the plutons solidify before they reach the earth's surface, they become what are known as an intrusive igneous formation. A batholith is a massive intrusive feature. Batholiths exposed through the erosion of upper rock layers, such as California's Sierra Nevada Range, are usually composed of granite. Plutons that reach the surface of the earth extrude as lava.

Metamorphic rocks were once either sedimentary or igneous materials that have been changed through intense heat and pressure. This often happens where there is folding and crumpling of the earth's crust. Marble is a metamorphosed form of sedimentary limestone, while diamonds are a metamorphosed form of sedimentary coal.

Continental Drift

Although the earth seems to be a solid, it is actually much more of a liquid. Scientists believe that the very core of the earth consists of solid iron, measuring some 780 miles (1,250 km) from the center to its edge. The next 3,000 miles (4,800 km) out is all liquid. The first 1,380 miles (2,200 km) are liquid iron and nickel, while the remaining 1,550–1,750 miles (2,500–2,800 km), varying in different locations, are a mixture of molten and solid rocks known as the *mantel*. The outermost 5–40 miles (8–65 km) consist of the solid crust we stand on, most of which is igneous rock with a top layer of sedimentary rock.

In fact, the molten iron ore and rock deep beneath us is in constant motion, churning upwards in the same way the bubbles boil from the bottom of a boiling pot of water. This churning motion causes the hard crust to crack and shake (earthquakes). The mantel material pushes its way to the surface through volcanoes and undersea oceanic rift zones. The lightest material in the earth's crust has gradually come together to form the continents, which literally float on top of the heavier crustal material. The churning molten rock below pushes the continents around in a process known as *continental drift*. In addition to the continents, large blocks of heavier ocean floor are also being pushed about the surface of the earth. The study of this process is known as *plate tectonics*. Scientists have identified some 20 major ocean and continental plates on the surface of the earth today.

It is believed that the earth solidified as a planet approximately 4.7 billion years ago. For the first 1.5 billion years of its existence, there was no life on the planet. While some areas of the planet have exposed rock that dates back some 4 billion years, most of the surface rock is much younger. The first plant life, algae, probably came into existence 3.2 billion years ago. Two billion years of plant development passed before the first waterborne animal life came into existence. Trilobites dominated the seas 600 million years ago, and the first fish came into existence 100 million years later. About 440 million years ago, the first surface plant life appeared. Fish began to crawl from the sea 400 million years ago, leading to the development of the first amphibians.

Pangaea

About 360 million years ago, all of the continental land masses moved together to form a single land mass call Pangaea (meaning all lands). Many large mountain chains were created from the collision of the continents. After some 75 million years of evolution, the reptiles became what we now call dinosaurs and dominated animal life over the entire planet. Trilobites finally became extinct at this same time (225 million years ago). Geologic evidence indicates that, starting approximately 200 million years ago, Pangaea started to breakup into separate continents that eventually became what we know today. The older mountains also stopped growing at this time and have been eroding ever since, forming some of the lower mountain systems of the world now, including the Appalachians in North America and the Urals in Russia.

The Emergence of Life

The first birds came into existence some 160 million years ago. Most dinosaurs suddenly died out 90 million years later. It was at this time that the tallest mountains today first started forming, including the Himalayas (Asia), the Alps (Europe), the Andes (South America) and the Rocky Mountains (North America). All of these mountain systems were created from the collision of continental and ocean plates. Current evidence indicates that the first primates were also coming into existence some 65 million years ago, as mammals replaced the dinosaurs as the dominant species on all the continents, except Australia. Elephants appeared 45 million years ago.

After over 60 million years of evolution, primates eventually evolved human variations, a mere 2 million years ago (in the form of the *Homo habilis* species). Modern humans (*Homosapiens*) emerged 40,000 years ago, at about the same time that the earliest crossing of humans into North America from Asia may have occurred. (Most scholars accept that humans crossed into North America 12,000 years ago; evidence of crossing prior to that is somewhat controversial.) The earliest recorded history of any civilization dates back 5,000 years. Our modern calendar dates back 2,000 years. The modern period of European settlement in North America is only 500 years old. If a 100-story skyscraper were to represent the entire history of the planet earth, the top 12 in. would be the part in which the various human species have been present.

Glacial Till and Moraines

There are two types of glaciation: continental and alpine. *Alpine glaciers* are still seen in the highest mountains around the world. *Continental glaciers* only exist today in the Antarctic and Greenland, but they were significant in shaping the landscape of the northern hemisphere. The Pleistocene Epoch, (which was the last Great Ice Age), started about 2 million years ago and has been characterized by several periods in which glaciers advanced from the north to blanket large portions of North America and the Eurasian land mass. Up to a mile (5,280 ft; 1600 m) thick, these glaciers scoured the land and flattened the hills that they covered. The last of these glacial periods ended about 12,000 years ago. The glaciers left behind

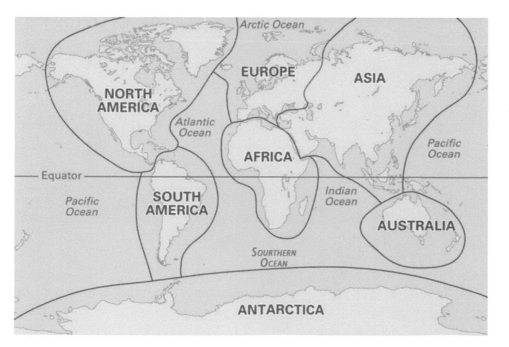

Continents and Oceans of the World.

many small lakes (kettles) and erratically meandering (or "lost") rivers. The Ohio and Missouri Rivers roughly mark the southern boundary of the glacial advances in North America, while most of the North European Plain was covered with ice at one point or another.

The scoured materials carried by the glaciers (both continental and alpine) were deposited in a variety of ways. This material is known as *glacial till*. The largest depositional features are called *moraines*, which define the edges of a glacier. Long Island (New York), Nantucket and Martha's Vineyard islands (in Massachusetts) are terminal or end moraines that marked the end of massive glaciers. They are both composed of material carried by glaciers from the interior of the continent.

Estuary Bays

Because a large amount of the earth's surface water was stored in glacial ice during the glacial advances, sea levels were about 450 ft below their current levels. Much more of the earth's continental shelves were exposed and a land bridge existed between Asia and North America. Because rivers seek to erode until they reach sea level, the lower sea levels allowed rivers to cut deeper valleys, especially in coastal areas. When sea levels rose, these valleys were filled in. Along the eastern seaboard of the U.S., Chesapeake Bay (Susquehanna River), Delaware Bay (Delaware River) and New York Harbor (Hudson River) were all created this way. Bays created by the filling in of river valleys are called *estuaries*. More estuaries are likely to form in the decades to come if ocean levels continue rising at the rate they have been since the early 1900s.

Factors in Weather and Climate

As noted above, weather refers to the day-to-day changes in the atmosphere of a place. Climate, on the other hand, refers to the annual pattern of atmospheric conditions in a place. Climate classifications are primarily based on (1) the seasonal variations in temperature and (2) the annual rainfall amount and seasonality. The major factors that influence climate are (1) latitude, (2) continentality, (3) air masses and (4) surface physiography.

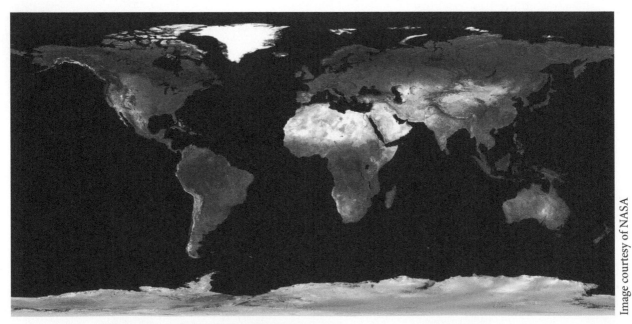

NASA composite satellite image of Earth with natural vegetation color.

Image courtesy of NASA

Because the earth moves around the sun on a tilted axis, the northern hemisphere is more directly under the sun in June and tilted away from the sun in December. These extremes become most pronounced as one gets closer to the poles. North of the Arctic Circle, Alaska's summer days have 24 hrs of sun, while winter days have 24 hrs of cold darkness. At the other extreme, the length of Florida's days and nights stays about the same all year round, and average winter temperatures are only 10°F less than average summer temperatures. Higher latitudes (closer to the poles), therefore, have greater annual temperature extremes and especially cold winters.

Continentality and Maritime Climates

Land (rocks and dirt) heats and cools faster than does water. Climatically, this means that the large continents of the world are warmer than the oceans in their summer months and colder than the oceans in their winter months. This further contributes to the intense winter cold experienced by the interior of the continents. Conversely, summer temperatures are as much as 50°F hotter in the interior of the continent than in coastal areas. Along the edges of most continents, the moderating influence of the oceans keeps summer temperatures down and winter temperatures relatively high. The onshore ocean winds hitting San Francisco, for example, give it almost the same temperature all year round. The maritime climate of San Francisco is the opposite of the continentality found in the interior of North America and Asia.

High and Low Pressure Air Masses

Large movements of air over the surface of the earth generally have either rising or falling patterns. Air that is rising from the earth's surface forms low pressure air masses. When the rising air falls, it forms a high pressure air mass (because the air is pushing down on the earth's surface). Rings of predominantly high- and low-pressure encircle the earth. Low pressure on the equator brings year-round rainfall, creating tropical rain forests. Descending dry air creates high pressure north and south of the equator. These are where the large deserts of the world are located. Humid low-pressure systems ring the globe north of here,

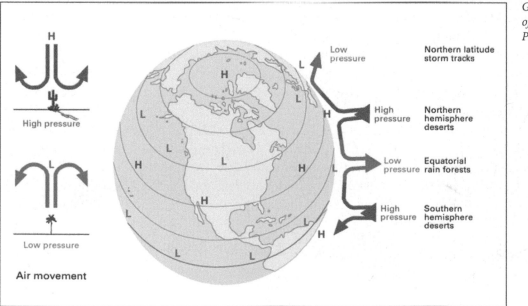

Global Systems of High and Low Pressure Belts.

*World Climates,
Wind Directions
and Ocean
Circulations*

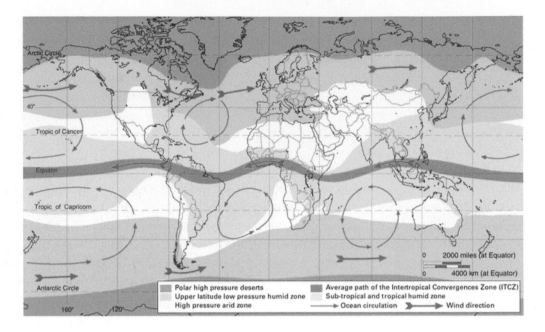

creating cool weather rain forests on the coasts of Washington and British Columbia in North America and in the British Isles in Europe. Finally, a dry high pressure system is situated over the North and South Poles, turning the north coast of Alaska and central Antarctica into a frozen desert.

Air Masses on the Western and Eastern Sides of Continents

Winds in a high-pressure air mass circulate in a clockwise direction. Winds in a low-pressure air mass circulate in a counter-clockwise direction. This results in air near the equator moving from east to west (e.g., from West Africa to Central America). Air masses at higher latitudes (closer to the poles) move in the opposite direction, from west to east. The direction that these air masses move has a major impact on the types of climate and weather that different parts of continents experience.

Large, and sometimes deadly, tropical storms along the equator all move from east to west and are generally known as hurricanes (North and South America), typhoons (Asia) and cyclones (Australia). But even when there are no major storms, warm and humid air moves in the same direction, impacting the eastern seaboards of all continents that are near the equator, including, for example, the southeastern United States and northeastern Australia. The opposite sides of these continents, on the other hand, are places of warm and dry air (as seen in the southwestern United States and in northwestern parts of Chile in South America).

Meanwhile, further north and south of the equator, cold and humid winter storms move off the oceans to hit the west coasts of the northern parts of North America and Eurasia. The opposite east coasts of these continents are affected more by continental air masses, which are drier, but also much colder in the winter and much hotter in the summer. Most of the very violent upper latitude storms in the southern hemisphere circulate in the oceans around the Antarctic and only affect the southernmost tips of Africa, South America and New Zealand.

Surface Physiography: Altitudinal Zonation and Rain Shadows

One of the two ways that surface physiography influences climate is through altitudinal zonation. Altitudinal zonation results from the fact that higher elevations are colder than lower elevations. As one moves up a high mountain slope, the climate and resulting vegetation change from warmer to colder

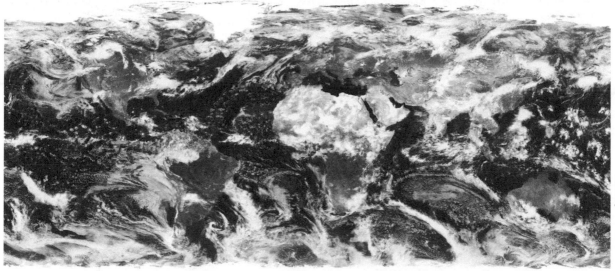

NASA composite satellite image of the earth with cloud cover. The Intertropical Convergence Zone is visible along the equator. High and low pressure desert areas are visible with little cloud cover, and upper latitude storm tracks are visible over northern North America and Eurasia and below Australia and South America.

Altitudinal Zonation

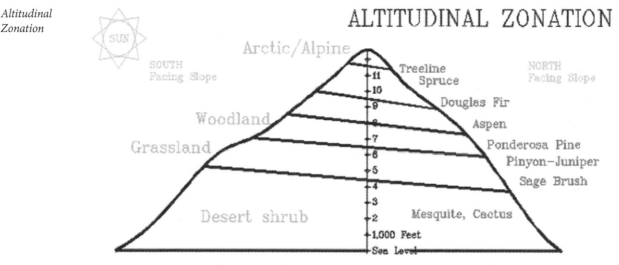

An example of Altitudinal Zonation on San Francisco Mountain in Arizona. Elevations for the zones shown here would be lowered as one moves closer to the N. & S. Poles and higher closer to the Equator.

(c) 1996, Alan A. Lew

weather types. The effect is similar to moving from lower to higher latitudes (i.e., from the equator to the poles). Altitudinal zonation is most pronounced in arid regions, like the southwestern United States, where mountain peaks stand out as lush islands in a sea of desert.

The second way that physiography influences climate is through the creation of a rain shadow. Water molecules slow down when the air gets colder, allowing condensation and clouds to form. When the temperature is warmer, the molecules warm up and move faster, and the clouds dissipate. As a mass of warm air rises into an area of colder air, it will eventually cool to a temperature at which condensation will occur and a cloud will form. This is known as the dew point temperature. If the air mass continues to rise and cool, condensation increases until precipitation (rain, snow, hail, etc.) occurs.

This process takes place regularly in mountain areas throughout the world. Air masses are pushed up the windward side of the mountains in a process known as *orographic* uplift. This cools the air and typically results in precipitation (rain or snow).

By the time the air mass reaches the top of these ranges, it has lost much of its original moisture. The air mass then starts down the back side of the mountain. This causes the air to warm up and the molecules to speed up, with the result that the clouds disappear and the little moisture that is left is now spread out over a wider area. The downwind (or leeward) side is in the rain shadow of the mountain.

Climate

Climate has a direct impact on the biogeographic distribution of the natural vegetation in a region. In fact, the two are often inseparable. The major factors of climate that affect vegetation patterns are humidity (precipitation) and temperature. High humidity and high temperature are associated with lush, green vegetation, such as that found in tropical rain forests. High temperature and low humidity are associated with arid, desert vegetation. Humidity becomes less important as temperatures decrease.

The lowest temperature regions are associated with spruce coniferous trees (found in most of Siberia and Canada) and tundra (a treeless type of grass and moss found in the northern extremes of the continents). Between these three extremes (high temperature and humidity, high temperature and low humidity, and low temperature) is a mixture of evergreen trees, deciduous trees (those that lose their leaves in winter) and grasslands.

Climate Change

Since the 1970s, the world has been warming at a rate faster than has been recorded in recent history. Of course, the recorded human history of global temperatures only dates back to the mid-1800s. Tree ring, coral growth and ice core data extend our temperature knowledge back about 2000 years, and they also show that the recent rise in temperature has been dramatic.

In fact, you would need to go back about 150,000 years ago to see a similar spike in global temperatures, based on Antarctic ice cores and geologic records. So there is no doubt that global warming, and accompanying climate changes, is taking place. More debatable has been the degree to which humans are contributing to the current global warming trend. The very close correlation between increased atmospheric carbon dioxide (CO_2) levels since the start of the Industrial Revolution (about 1800) and global temperature increases has convinced the vast majority of international climate scientists that humans have had a significant impact on the warming process. Human activities that increase atmospheric CO_2, as well as other greenhouse gases, include the burning of fossil fuels, the use of aerosols, the production of cement, deforestation, and beef and pork production.

Tourism and Climate Change

It is estimated that tourism is responsible for about 5 percent of total global CO_2 emissions if calculated on the basis of energy throughput. With respect specifically to tourism, most CO_2 emissions are associated with transport, with aviation accounting for 40 percent of tourism's overall carbon footprint, followed by cars (32%) and accommodation (21%). Cruise ships probably account for around 1.5 percent of global tourism emissions, although they also produce more localized air pollution as a result of the fuels they use.

In addition, as the construction of hotels, cars, airports and other infrastructure all consume considerable amounts of energy, a lifecycle perspective accounting for the energy embodied in tourism related

ISSUES AND INSIGHTS   Global Environmental Change

Although we often recognize change at a local level, it is becoming increasingly apparent that environmental change is occurring on a global scale and that tourism is involved in these change processes in various ways. Human impacts on the environment can have a global nature in two ways. First, global refers to the spatial scale, for example, the climate and the oceans have the characteristic of a global system and both influence and are influenced by tourism production and consumption.

A second kind of global environmental change (GEC) occurs if a change takes place on a worldwide scale, or if it represents a significant fraction of the total global resource, for example, biodiversity loss and urbanization. Tourism is significant for both types of change, as it encompasses several different areas of environmental impact, including:

- Climate change
- Land Use change
- Biological Mobility, Including Diseases, Pests and Weeds
- Water Availability

- Urbanization
- Biodiversity Loss

The news media focus is often on climate change, but it is important to recognize that this is only one, albeit extremely important, area of GEC. Furthermore, it must also be recognized that these changes are interrelated, although change is never uniform across time and space. Therefore, the changes you see in your neighborhood will often be different from those occurring in other parts of the world. How we react to those changes though will depend on such factors as our knowledge level, political institutions, culture and how we rank environmental risk in relation to the other factors that affect our lives and decision-making.

Sources

Gössling, S., and Hall, C. M. (eds.) 2006. Tourism and Global Environmental Change. London: Routledge. Hall, C. M. 2010. Tourism and biodiversity: More significant than climate change? Journal of Heritage Tourism. 5(4), 253–266.

construction would lead to significantly higher estimates. Tourism also leads to emissions in associated sectors, including tour operators and their offices and the commuting of employees to work, as well as the food requirements of tourists that appear to lead to higher emissions than those for food consumed at home. More recent broader assessments of tourism's contribution to global warming suggest that, particularly because of the effects of the transport component of the tourism industry, tourism's contribution may be on the order of 5.2–12.5 percent.

Natural Hazards

The earth's physical geography (geomorphology, climate and biological life) has been changing and evolving for over 4.5 billion years. And it will continue to change for many millennia to come as environmental conditions on the planet change. For humans, these changes can cause dramatic upheavals that are known collectively as *natural hazards*. These include earthquakes and volcanic eruptions caused by shifts in the earth's crust, landslides and floods caused by irregular rainfall patterns, wildfires made worse by forest management policies, and outbreaks of biological pests that attack crops and diseases that attack plants, animals and humans.

Recent rapid climate change has exacerbated weather-related hazards, including heat waves and droughts on the one hand, and hurricanes and blizzards on the other. All of these natural hazards are signs that we occupy a living and evolving planet, and most of them are not even considered hazardous when they occur in places uninhabited by humans. This is why the study of natural hazards focuses as much on human responses to the threat of natural hazards, as they do on how to manage and mitigate them.

1.1.2 Human Geography and Globalization

As noted above, human geography is a much more diverse field of study than is physical geography, and is not so easily broken down into clearly distinct subdisciplinary fields. Some feel that because of this, human geography lacks a sufficient focus of study. Most geographers, however, relish in the *eclectic*

The Western Wall in Jerusalem. Also known as the Wailing Wall, this site is actually an open air synagogue and is considered the holiest public praying place in Judaism. Religion is one of the many ways by which cultures create places of deep meaning in geographic space.

Photo: Dallen J. Timothy

nature of their profession, and see it as the only way to study and fully understand the nature of geographic space, place and regional phenomena.

Place and Space

Geography deals with two basic areas of inquiry about the world around us: place and space. Geography seeks to portray accurately the character of places. Place location (where is it?) is fundamental to understanding a place's characteristics. Place description (what is it like?) is part of the art of geography. These are the types of questions that most people would readily identify with geography. Geographers attempt to develop an awareness and understanding of the qualities of a place that make it special. We say that places that exhibit these special qualities have a strong *sense of place*—they are places that have a personality and significance and are often remembered long after we have left them.

Places are points of presence. A place exists and has a location. Geographic places exist in geographic space—typically some location on the surface of the earth. Other, non-geographic places also exist. These are mostly *fictional places*, but they can also be virtual places. In this book, the word place always refers to geographic places. All places, whether geographic or not, share in common (1) some means of distinguishing one place from another (for geographic places, this may simply be a location address, but typically also includes physical and cultural landscape features such as mountains and buildings), and (2) a relationship with other places within its spatial realm (for geographic places, you can at least measure a distance between any two places on the planet, although many other types of relationships also exist, such as economic and cultural).

Geographic Space

Space is another central or transcendent theme of geography. Besides knowing where a place is and what it is like, geography seeks to determine the reason places are located where they are and why they develop the characteristics that they have. The answer often involves understanding how places relate to one another over space. Examples of *spatial relationships* between places include transportation routes and communication linkages (both of which have changed over time with technological developments), ethnic ties (as people migrate from one place to another) and political associations (which often involve complex historical processes). The distribution of something over an area is called its *spatial pattern* or spatial organization. The word spatial here refers to geographic space, rather than to outer space. Geographic space is the three-dimensional space that encompasses the livable surface of the earth.

Geography is sometimes called a *spatial science* because of the importance of spatial relationships in a geographic understanding of the world. This spatial aspect of geography is most easily seen and perhaps understood in the production of maps and map-like diagrams, most of which show the distribution of one or more variables over space. Maps are also identified by the general public as being a fundamental aspect of geography. How these patterns came to be and how they function are key questions in geography. Geography can sometimes be defined as the study of the spatial organization of the world in which we all live.

Regional Geography and Regions

All of these various aspects of geography are brought together in the study of places and regions. For this reason, regional geography has been referred to as a virtuoso performance—it requires familiarity with the breadth of geography and competence in pulling together the diverse strands of knowledge to create a sense of place or regional character. The difference between a region and a place is dependent on scale and perception. A metropolitan area (comprising many cities that grow into one another) can be considered a place, at a national or international scale, or a region from a local scale perspective.

Another definition of a region is an area of land that is larger than a place and that contains a common characteristic, such as the growing of a particular crop or the market area of a product. There are many ways in which space can be regionalized in this way. Examples include political regions, economic regions, physical regions and cultural regions. Most regions are single, contiguous areas, although there are exceptions to this, such as the separation of Alaska from the rest of the United States by the country of Canada. Regions can be divided into smaller units, or subregions, if appropriate for analysis. Each region or unit, however, contains some collection of shared characteristics. Sometimes, these are more physical in nature; at other times, they are more social.

The geographic regions of the world, around which the chapters of this book are constructed, include both physical and human characteristics. For some regions, the human characteristics are emphasized more, while other regions are based more on physical characteristics. There is no fixed rule regarding this, and regional definitions can and do change through time as people's image and perception of them change. In general, regions with sparse populations are more likely to be defined in terms of their physical geographic features, while those with high population densities, such as Europe, are characterized more by their cultural and economic characteristics.

Core Area and Transition Zone

Most of the discussion of different regions in this text focuses on core area characteristics. The *core area* is the place where the shared characteristics used to define a region are most predominant. The core area of one region is distinctly different in some way from the core area of a neighboring region. *Transition zones* are areas between two regions that share characteristics of both to some extent. The characteristic(s) that defines each region are generally weaker in the transition zone. For example, the Mediterranean Sea clearly marks a distinct boundary between Europe and Africa, with almost no transition zone. You are either in Europe or you are in Africa. On the other hand, the boundary between Europe and Asia is a broad transition zone. Though the Ural Mountains provide a convenient demarcation between the two continents, there is actually nothing especially noticeable about the Urals that separates the European physical and cultural realm from that of Asia.

Similarly, the divisions between the United States South, the United States Midwest and the United States Great Plains subregions are not easily defined. The state of Missouri, for example, is included in the South by some authors, although it could justifiably be included in the Midwest or the Great Plains, because it shares core area characteristics with all three of these regions.

Homogeneous Regions

Homogeneous regions contain a common characteristic found throughout an area in equal degree. Island nations, such as Japan, tend to be more homogeneous than continent-based nations, though exceptions exist. The common shared feature may be one or a group of characteristics. A political entity, such as a country or state, which by definition encompasses its own citizens within its boundaries, is a *homogeneous region*. Also known as uniform und formal regions, homogeneous regions have more clearly defined boundaries (i.e., more narrow transition zones) than do nodal regions.

Nodal Regions

Also known as functional and focal regions, *nodal regions* have a central point at which the characteristic defining the region is most predominant. The farther away from this central point the characteristic is, the less predominant it is. A city's area of economic and social influence is usually nodal. The farther one moves away from a major metropolitan city, such as Sydney, Australia, the less its economic influence is, until one moves into the influence area of another major city, such as Melbourne, Australia. The area within the sphere of influence of a nodal center is known as its *hinterland* or *periphery*. Nodal regions have well-defined core areas and poorly defined boundaries, with broad transition zones. The core-periphery model of economic development, which assumes that the development of peripheral locations is controlled by decision makers in core locations, is fundamentally the description of a nodal region.

NASA simulation of nighttime light distribution over the surface of the earth, based on population density and level of economic activity. Notice that the population distribution for India is more uniform than that of Australia, which tightly hugs the coastlines. Geographers seek to understand these distributions.

Image courtesy of NASA

A Priori Regions

An a priori region is one that is arbitrarily drawn on the surface of the earth. Political units, such as states, provinces and countries, are the most common type of *a priori region*. Sometimes, the political boundary separating one country from another is not related to any natural or human geographical boundary but is instead decided by negotiation or expedience. Almost every straight line on a map is a sign of a prior negotiation that ignores cultures, mountains, valleys, and other terrain features. Many of the political boundaries in Africa and Southwest Asia (the Middle East), are the result of political processes that, either intentionally or out of ignorance, divide major cultural groups into more than one country.

A Sense of Place

Geographer Yi-Fu Tuan has suggested that a place comes into existence when humans give meaning to a part of the larger geographic space. Anytime we give a name to a location, we separate it from the undefined space around it. Some places, however, have more social meaning, or stronger definitions and names, than others. For geographers, and others, these places are said to have a strong *sense of place*.

Geographers are interested in why certain places hold special meaning to particular people. Places that have a strong sense of place have a strong identity and character that is deeply felt by both its inhabitants and visitors. It is a social phenomenon that exists independent of any one individual's perceptions or experiences. This feeling may be derived from the natural environment, but more often it consists of a mix of natural and cultural features in the landscape, including the people who occupy the landscape.

Places that lack a sense of place are referred to as placeless. Placeless landscapes are those that have no special relationship to the place in which they are located—they could be anywhere. Roadside strip shopping malls, gas stations, and convenience stores, fast food chains and national department stores are examples of placeless landscape elements.

Sense of Place Tourism Examples

Can you think of a place that has a strong sense of place? How about a placeless place? Most communities have examples of both, and leisure landscapes can be of both types, as well. Have you ever heard that Paris is a romantic place to visit? Does Paris have some magical air or water that transforms people into romantics? What Paris has developed is a sense of place that includes love and romance. The environment provides an image and supposed atmosphere of love based on its history, architecture, culture and mythology.

Many tourists are attracted to specific places because of an established sense of place expectation. For example, certain older retail districts provide particular fascination for visitors by offering experiences of the past or of other places. Thematic retail districts include historic preservation districts, waterfront fisherman's wharf districts and more playful cowboy towns and Scandinavian or German communities.

The importance of the sense of place and tourism is clear. Tourists go to places for specific reasons. Many of those reasons are because the destinations have a known *sense of place*. Whether the sense of place is based on religion, art, romance or food, or whether it is just for cultural information, people travel for the experience. Places all over the globe provide those experiences for different meanings and at different degrees of personal significance. This has relevance to the tourism-related concept of place branding, wherein some places become "branded" or known for being unique or special in one way or another.

Individual Sense of Place

Individuals also have a personal sense of place, which creates special existential relationships between them and places that they feel most attached to. The place where one lives often has this special meaning, even though most outsiders might not feel such a strong sense of place there. Places that have served as significant settings in one's life also may have a strong sense of place for an individual. Ancestral home-lands are often like this. Examples include Israel for Jewish people everywhere, China for overseas ethnic Chinese and Ireland for Irish–Americans. These special places often attract us again and again, as tour-ists, recreationists, and sometimes as genealogists.

People tend to remember places by their feelings. Sight, sound, smell, taste and hearing are all meth-ods that we use to remember particular places. Geographers, psychologists and other social scientists have only recently begun to explore the multiple ways that we experience and remember the places we have visited.

Environmental Perception and Behavior

Environmental perception is an area of study that overlaps geography and psychology, and sometimes philosophy. It deals with how people perceive and, as a result, behave in the environment where they live. For example, the northern Europeans initially perceived the Great Plains as a useless environment, referring to it as the *Great American Desert*. Alternatively, the Spanish coming from Mexico recognized the region as having great potential for cattle raising, similar to what they were used to in central Spain. In another example, many more traditional societies today consider communal ownership of land as far more appropriate than private land ownership, which has long been predominant among those originally from European and East Asian cultures. The way people perceive and behave in the world reflects many aspects of their culture, experiences, socioeconomic background and value system. All of these have a lot to tell us about the cultural landscapes they create.

Globalization and Localization

The world has become a much smaller place in recent decades due to advances in technology and the opening up of global trade, both of which have contributed to an increase in *placelessness* worldwide. Through television and movies, values and experiences are shared across cultures and continents. Through the Internet, people today are able to communicate closely with friends and colleagues on op-posite sides of the globe just as if they were in the building next door. *Transnational corporations* (TNCs, operating in two or three countries) and *multinational companies* (MNCs, with offices in three or more countries) are often cited as one of the new forces bringing about this global transformation by operating beyond the political and geographic confines of traditional countries. While there is some truth to this, it is also true that countries are as important as they have ever been, and a trend that is directly opposed to globalization seems to have become particularly significant in recent years.

Increasing *localization* is emerging at the same time as globalization, as communities seek to express their individuality and local autonomy. Localization even becomes a strategy to compete more effectively in the global economy because localization emphasizes the difference of "your place" from "other places." In some places, this has even resulted in demands for greater political autonomy or even independence. In other cases, it can be seen in such things as the rapid rise of local microbreweries across North America in the 1990s, as well as the celebration of local foods in the growth of farmers' markets.

Postindustrial and Post-Fordist Society

The trend toward localization has been attributed to a fundamental change in North American and European society—the transition from an industrial economy (in which more people work in industries than in any other economic sector) to a postindustrial economy (with the majority of people employed in service jobs). Accompanying this *economic transition* has been a shift from an emphasis on assembly line, mass-produced merchandise (which Henry Ford first perfected in building the model-T automobile, and is therefore known as *Fordism*) to an emphasis on more personalized and individualized products, is therefore known as *post-Fordism*. This is reflected in an increasingly diverse array of products (including landscape experiences) designed to match the interests of smaller niche markets of people who are willing to pay more for personal and customized services. Specialty travel, such as culinary and bicycle tours, are examples of post-Fordist tourism.

Postmodernism and Historic Preservation

Another concept that is related to these economic trends is *postmodernism*, which came out of architecture but has since been expanded to encompass a broad realm of values in contemporary society. Modernism was the total rejection of historical approaches to architectural design and was an architectural trend that dominated much of the late 19th and early 20th century building construction in North America. Skyscrapers are typical of the modernist approach. Postmodernism was a rejection of modernism and is sometimes viewed as a return to, or an embracing of, a more historical approach to building design, as well as to social values. This is most clearly seen in the historic preservation movements that first became popular in North America in the 1960s.

On a broader social scale, a postmodern world is also more *relativist* (everything can and should be judged on its own merit, rather than based on universal moral values), and thus we can select from not only the past, but also from an eclectic and diverse realm of elements of different cultures around the world. Thus, postmodernism brings together both the local and the global and, hopefully, does not create something that is placeless in the process.

McDisneyization

McDisneyization is a postmodern opposite of sense of place. This term was coined by George Ritzer and Allan Liska to describe the *McDonaldization* (another Ritzer term) of service industries (fast and mass produced) and the *Disneyfication* of tourism (the epitome of which includes Disneyland and Las Vegas). An important element of such developments is the integration of place with media promotion—the two are often seemingly inseparable, with place being a product placement in films and TV programs. The McDonalds restaurants and the Disney theme parks are considered hypermodern models of:

1. *Efficiency*: Getting the most for one's money, which usually means seeing, doing and eating as much as possible
2. *Predictability*: Safety, known cleanliness and service standards, plus the ability to communicate in a common language
3. *Calculability*: Precisely defined itineraries, with no unexpected costs or other surprises
4. *Control*: Service employees whose behavior is tightly controlled by scripts (telling them what to say and how to behave), and the preferred use of advanced technologies to control employees and clients/guests).

These models now influence many aspects of the contemporary modern landscape and lifestyle. Examples of McDisneyfied places include: theme amusement parks, cruise ships, Las Vegas hotels and casinos, themed shopping malls and strip malls, some chain restaurants (e.g., the Rainforest Cafes) and a variety of public entertainment spaces, including historic preservation commercial districts. The very success of these environments seems to indicate that this is what many people want. The holiday life style is becoming omnipresent in the global landscape. Yet, these McDisneyfied places can be as placeless as the less entertaining mass shopping and work environments that are even more widespread in globalizing international urban landscapes.

Creek Street in Ketchikan, Alaska. The street dates back to the early 1900s when brothels were forced out of town to this less desirable location. Today, its dramatic location, built over Ketchikan Creek, is one of the most popular tourist sites in town, within easy walking distance of the large cruise ships that ply the inside passage to Glacier Bay National Park. All of these factors make it an example of historic preservation, postmodernism and McDisneyization.

Photo: Alan A. Lew

Environmental Impacts

Globalization encompasses many of the major economic and social impacts that international tourism has on destinations. Tourism, however, also impacts the physical environment. The natural environment is the destination's major attraction in many places. Ocean beaches, high mountains and tropical and temperate forests are all popular tourist environments. Many of these natural environments, however, are sensitive to overuse and misuse.

In Northern Arizona the major tourist attraction is the Grand Canyon, which is also one of the best known international attractions in the United States. However, with over 3 million tourists a year, mostly concentrated in the summer months and mostly arriving by private automobile, the Grand Canyon National Park has experienced significant urban-like congestion and pollution. Haze has increased in the canyon, and photographers are finding it more difficult to get a good picture because of a decrease in the number of very clear days. While tourists themselves contribute to the haze problem, it mostly comes from urban air pollution from the metropolitan areas of Las Vegas and Los Angeles. Will people want to continue visiting the Grand Canyon if air pollution masks its beauty and magnificence? What would the Grand Canyon experience be if visitors could not see it or if most of their experience was of traffic congestion?

Community Planning

Globalization and the social and environmental impacts of development present challenges to local communities in terms of how they will respond to create places that reflect their values and history, while at the same time enhancing their quality of life. These community goals are typically achieved through city and regional planning departments within the local government. Planning is also one of the major career fields for students who major in human geography.

Planners seek to increase all aspects of the quality of life in a community, including social interactions, environmental conservation and economic success. These are the same triple bottom line goals of sustainable development, and they ultimately require a broad geographic understanding of the physical, cultural and institutional character of a place. That includes, for example, the local vegetation and climate, sources of food supply, energy demands and resources, public health issues, natural habitat protection issues and public safety needs.

Sustainable Development

The planning process typically starts with the identification of a community's desires, needs and long-term goals. This is followed by the development and official adoption of a plan to address the needs and long-term goals of the community, which local and regional governments then implement. This is a highly political process, as different people have different opinions on what a community's goals should be and how they should achieve them. Sustainable development principles, such as those cited previously in the *Brundtland Report*, are intended to help guide that political process by setting local decisions in the context of larger, global goals.

Interpretation

Interpretation, or education, about the natural environment is a key component of any tourism experience. Without it, tourists might not be inspired to engage in an activity, might not be able to find their way to locations, and they might not understand the full significance of the areas they are visiting. Subliminal forms of interpretation tell us about places in indirect ways, such as through music, news stories, fiction, movies and architectural design. More direct and intentional forms of interpretation include maps, information boards and human guides. How places are interpreted in these many different ways influences people's perception and experience of places and environments. Interpretation serves as a key variable in creating satisfactory visitor experiences, though many places are subject to different and conflicting interpretations that represent divergent views within a society. Interpretation is also related to the concept of sense of place and how people interpret places and place experiences.

Summary

Geography is important to the study of tourism because tourism is geographical in nature. It occurs in places, it involves movement and activities across space (between places), and it includes activities in which place character and personal self-identities are formed through the relationships created among places, landscapes and people. Physical geography provides the essential background against which tourism places are created. Environmental impacts and concerns are major issues that must be considered in managing the development of tourism places. Human geography provides an understanding of the social and economic relationships that exist in providing tourism opportunities and activities, as well as the special meaning that these places have to individuals. An area's sense of place, in many circumstances, is the driving force behind tourism development. After all, without the uniqueness of places, tourism would be mundane and uninteresting. A goal of this book is to provide that sense of place for destinations around the globe through an understanding of the physical and human geography of the earth.

1.2 | GEOGRAPHY AND THE STUDY OF TOURISM

Current Tourism Issues and Resources (http://wrgeography.com/tourism.html)

As mentioned above, recreation (including tourism) is one of the major functions of some cities. The number of communities that promote tourism and recreation as important economic activities has increased in recent decades, as demand for these activities has steadily grown worldwide. The world's developed economies have seen this as they have transitioned from manufacturing employment to service and amenity industries. This has prompted communities to develop their leisure resources to build that portion of their economies. However, even in the developing world, the demand for recreation and tourism has grown at a phenomenal rate, along with a corresponding growth in the middle classes. As a result, every year more tourists are traveling domestically (within their home country) and internationally across the globe.

The old adage of *needing a vacation to recover from a vacation* reflects the often strenuous nature of many holiday travels, as we try to squeeze in as many activities as possible within the short period that we are free from work. Travel and tourism are typically considered leisure activities and a form of recreation that takes place away from the home place. The fact that tourism involves travel from one place to another, and that it occurs in places often shaped intentionally by the tourism industry, also makes it inherently very geographical.

Leisure Studies
Leisure as Non-Work

Most people have an inherent sense of what they consider to be a leisure activity. We generally know that it is not work and not something that you must do. In fact, it is easier to define what leisure is not than what it is. In part, this is because leisure is very subjective—what one person considers a leisure activity, another person may not consider leisure at all. A hike in the woods might be considered leisure by some, and work by someone else. Furthermore, the same person can hold different views from one day to the next, depending on the social context and the individual's attitude. Despite the subjective complexities of leisure, we can generally say that leisure occurs when an individual is undertaking an activity that he or she wants to do and enjoys doing.

Leisure, therefore, can be defined as time spent free of obligation and necessity, when one is in control of one's own destiny. During pure leisure, there are no overt outside factors or forces that determine what we do and how we spend our time. While we have each been socialized to consider some forms of leisure activities as more acceptable than others, we feel that during our leisure time we are free from society's expectations and demands of us.

This definition of leisure raises some intriguing questions about *free will* and social expectations. However, it is also limited in that it is not easily quantifiable because it does not specifically address different types of leisure (for counting purposes), nor does it describe to what extent one is free from social obligation (such as when a business trip becomes a tourist holiday). Because defining leisure as the opposite of work and obligation is so vague, we will move on to a second definition of leisure that focuses on activities.

Leisure Activities

Leisure can also be defined as a specific activity that results in the physical or mental relaxation and rejuvenation of an individual. Some popular activities normally associated with these results include watching television, participating in sports and other outdoor recreation activities, reading books and magazines, going to the movies and listening to music. These activities bring relaxation to the mind, body and soul, for most people. As mentioned above, individuals can vary a lot in the types of activities they find relaxing and stressful. For some, work may actually be considered a leisure activity, while for others a leisure activity may be considered work. Defining leisure as a set of specific activities makes leisure quantifiable, but it may not address an individual's motivations and goals for leisure. Therefore, a third definition has also been proposed.

Leisure as a State of Mind

Leisure can also be defined as a state of mind. This idea was expounded by the Greek philosopher Aristotle (384–322 BCE), who saw leisure as a form of a self-meditation and self-improvement. He saw leisure as the most essential element of humankind because it allows for *self-development, creativity,* and *self-actualization.* It enables us to step back from the constant struggles of daily life to consider what is really important in life. Only then will we know what areas of our life need improvement. While this view of leisure offers many avenues for *humanistic* understanding, it shares the challenges of the non-work time definition by leaving a lot of room for subjective interpretation.

Each of these definitions of leisure, above, has its flaws. However, a combination of these views offers a fuller definition of leisure. As such, leisure could be defined as an activity that spans a period of time and is chosen of one's own free will instead of a work or a social obligation, and that allows for personal relaxation, contemplation and rejuvenation.

Geography and Leisure

Almost every aspect of life is interwoven with leisure (including both recreation and tourism), either as a leisure activity or as its opposite. Because of this, the study of leisure is undertaken from a wide range of disciplinary perspectives, the most prominent being sociology and economics. Geographers have also made major contributions to *leisure studies*, especially in the area of tourism, because of their interests in the nature and development of places, how people use and behave in places and the varied relationships that exist between places. As noted above, the study of geography is distinct in that it encompasses both the physical sciences (physical geography) and the social sciences (human or cultural geography).

Sustainable Tourism

As noted previously, sustainable development is defined as development that meets the needs of the present without compromising the ability of future generations to meet their own needs. The application of

Photo: Alan A. Lew

The Golden Triangle is where the geographic borders of Myanmar, Thailand, and Laos meet on the Mekong River. In this photo, the tourists are in Thailand; Myanmar is seen below the sign, separated from Thailand by a small river; and a sandbar in the Mekong River is in Laos to the right. In the past, this was a major heroin production and transportation area due to its climate, isolation and poverty. Today, however, its unique geographic features form the basis of a significant tourism economy.

sustainable development to tourism is known as *sustainable tourism*, which has become an important part of planning for the conservation and development of natural and cultural resources. Sustainable tourism often includes an emphasis on mass transit, renewable resources, leave-no-trace activities and community development. The challenge of the sustainable development concept is how to balance *environmental conservation* (the sustainable part) with *economic development* (the development part). This is typically simplified into a political dichotomy between the environmental left and the pro-development right. Successful sustainable development efforts have overcome these political divisions to find and work toward a common future.

1.2.1 The Tourism Industry

Tourism is a global phenomenon that includes a multitude of different types of businesses and government entities. It is a source of substantial economic and social consequence that requires objective and scientific understanding. However, just as the definition of leisure presents a challenge, so too does the definition of the tourism industry.

Defining Tourists

Without tourists there would be no tourism industry. Tourists travel to attractions and find new attractions even when developers and locals have not yet found or promoted them. It is more common, though, that the tourist (or potential tourist) is the one who is bombarded with advertisements and hawkers trying to get them to spend their money.

We usually think of a tourist as someone who is vacationing somewhere far from home. But the technical definition of a tourist is a little more specific. Most of the definitions used for statistical purposes consider a tourist as anyone who travels to a place outside of his or her usual residential environment and stays away for at least one night, but no more than one year. The motivation for the trip is irrelevant—the trip could be for business reasons, to visit friends and family (known as *VFR travel*), for education or simply for pleasure. This definition is used by organizations such as the *World Tourism Organization* (UNWTO) to compare tourist arrivals among countries internationally.

Different definitions are sometimes used within a country and within local destinations. China, for example, counts day visitors (those not staying overnight) in its total tourist numbers, although they also provide standardized numbers to the UNWTO. Many countries separate business travelers from leisure travelers. For international travel overall, leisure travelers generally outnumber business travelers two to one. This, however, varies considerably among destinations. Business travelers tend to pay more for their airline tickets and hotels than do vacation travelers, but their stay at a destination is usually shorter.

Recreationists

At the opposite end of the tourist trip is the recreation activity. The clearest form of a non-tourism recreation activity is any leisure activity that occurs within someone's residential home place. (The exception would be planning a vacation trip.) As recreationists travel further away from home, the same activity might be transformed into tourism. In general, recreation activities become tourist activities (or tourism products) when provided by private sector companies that mostly service non-resident visitors. They are considered recreation products when visiting tourists comprise a minority of the clients served.

Day Trippers

Day trips (also known as *excursions*) are trips that take people away from their homes but do not involve an overnight stay. A day excursion can be considered a recreation activity though because it normally occurs well beyond the home place, and the activities and services normally used are the same as those provided to tourists by the tourism industry. At many tourist destinations separating those who are tourists from those who are day-tripping recreationists is difficult. One way of doing this is to compare overnight tourist numbers estimated from hotel taxes and total visitor numbers estimated from attraction receipts. The methodology is not perfect, but it can be an important first step in segmenting visitors into major types for marketing purposes. While cruise passengers stay away from home for extended periods of time (e.g., a few days or a few weeks), they are considered day-trippers from the destination's perspective, because they arrive and depart on the same day, and they typically spend the night onboard the ship.

Tourism as an Industry

An industry is defined as a set of businesses that share in the production of a common product. For example, the auto industry consists of businesses that manufacture, sell and service automobiles; the gaming industry includes businesses that are associated with the activity of gambling. A set of businesses must meet three criteria to be considered an industry:

1. They produce essentially the same product.
2. They use essentially the same technology.
3. The product output is large enough to warrant data collection and reporting.

By most measures, tourism is the largest *service industry* in the world. International travel expenditures and receipts account for 26 percent of all international trade in services (2009 data, World Trade Organization). This definition of tourism includes a large part of the recreation activities that occur on the planet, especially anything considered an attraction, as well as almost all of the hospitality industry (hotels, restaurants and related businesses) and a good chunk of the transportation industry (especially airplanes, intercity buses and trains). However, international definitions of what constitutes particular sectors are important in determining estimates of their relative size.

Demand for travel and tourism continues to increase, despite terrorist threats, and especially as leisure time and economic well-being increase in the newly developing economies of Asia, Latin America, Eastern Europe and elsewhere. In fact, according to the UNWTO and aircraft manufacturers, such as Airbus and Boeing, the number of international travelers and installed seats in aircraft is expected to double between 2000 and 2020. As a result of growth in the number of aircraft worldwide and demand for travel, the overall distances of flights are predicted to increase, with the average distance flown growing from 1,437 km in 2002 to 1,516 km in 2022.

International Standards

Some accounts claim that tourism is the world's largest industry overall, not just the largest service industry. These claims are subject to debate. More conservative estimates place tourism as the world's fifth or sixth largest industry when compared to heavy industrial sectors, accounting for 7 percent of global trade, and behind fossil fuels, telecommunications and computer equipment, automotive products, and agriculture.

To make such a claim requires that tourism be defined and measured in a way that is consistent with the conventions and tools used in macroeconomics. These include the *International Standard Industrial Classification* (ISIC), the *Central Product Classification* (CPC) and *Systems of National Accounts* (SNA). (The SNA is the analytical frame work used by most countries to collect, order and analyze macroeconomic performance.) Tourism is not normally a part of such standard industry classifications and requires its own set of accounts.

Tourism Industry Elements

However tourism is defined, most people would include the elements of movement (transportation), of remaining temporarily in one place (accommodations), of consuming food and drink (which could be an attraction) and of participating in activities (attractions). The transportation sector generally includes airlines and airports, trains, buses, taxis, private automobiles, boats and ferries, the servicing and repair of these transportation modes and travel agents and tour companies that facilitate transportation. The accommodations sector can include hotels, motels, resorts, campgrounds, rural bed and breakfasts, the homes of friends and relatives, cruise ships, dude ranches, accommodation booking agencies and businesses that service these different accommodations.

Food and Beverage

Food and beverage consumption is often closely related to accommodations, though not always. The food and beverage industry includes all forms of restaurants and eateries (including those in hotels and at attraction sites), other providers of food and drink to tourists (including grocery stores), wholesalers that sell food and drink to restaurants and businesses that provide other services to restaurants and food providers. Food and beverage, as a sector of the tourism industry, is especially challenging to measure because most of these businesses serve local residents as much, or more, than tourists. The balance between tourists served and residents served varies considerably from one establishment to the next.

Attractions

Attractions are extremely diverse, varying by scale and interest. The city of San Francisco is considered an attraction, as is the city's Fisherman's Wharf district, and individual restaurants and amusements within it. Attraction types can be categorized based on their form (e.g., a famous building, mountain or theme park) and their experience (e.g., recreation, relaxation, entertainment and education). The challenge in identifying the attraction sector further complicates definitions of the tourism industry.

These four sectors (transportation, accommodations, food and beverage and attractions) contribute to providing a product called *tourism*. They are all service industries, which they market to a common

The Geographer Cafe in the historic port city of Melaka, Malaysia. This postmodern themed cafe brings together a contemporary coffee and tea experience with the historical tradition of geographic explorations in a former shophouse situated in one of the oldest European colonial port cities in Asia. As such, it becomes an example of eatertainment—the combination of eating and entertainment.

Photo: Alan A. Lew

clientele (tourists). Even though they are among the largest industries in the world, there is no clear and simple way to aggregate their diverse products meaningfully into a single generic product that can be compared to more traditional industries. This is because tourism is not a material product, but an experience that is purchased and kept more in your memory than in your living room.

Tourism Economics

Anything that can be purchased is a commodity. By creating tourist attractions, the tourism industry tends to commodify almost all aspects of the contemporary world. This is known as *commodification*. The UNWTO has defined a tourism commodity as any good or service for which a significant portion of demand comes from persons engaged in tourism as consumers. Significant portion is not defined, however. If we assume that this is more than 50 percent, then a tourism commodity is any product or service for which at least 50 percent of the buyers are tourists. (Depending on the product, service and place, the tourist demand could be less than 50%.) Furthermore, almost anything can become a tourist attraction, with the right marketing. Attractions can include forms of transportation (e.g., San Francisco's cable cars), food and beverage (e.g., wine tourism in California) or even sewage tunnels (Paris).

The magnitude of tourism as an economic activity in a destination is usually measured primarily by estimating the total expenditures made by visitors in the course of a trip (or on behalf of a visitor, such as contracts with a hotel made by a tour operator in assembling a tour for sale). A few purchases before and after a trip can be considered part of the trip's expenses, such as the purchase of consumables made immediately before departure, gas for the car and such expenditures as the dry cleaning of travel clothes immediately after the trip.

Tourism and Non-Tourism Commodities

Four dilemmas sum up the major challenge in measuring the economic magnitude of tourism:

1. Visitors consume both tourism and non-tourism commodities.
2. Residents consume both tourism and non-tourism commodities.
3. Tourism industries produce tourism and non-tourism commodities.
4. Non-tourism industries produce tourism and non-tourism commodities

Non-tourists also purchase many commodities that tourists purchase. Food and drink, for example, are products that both tourists and local residents use widely. They are a tourism commodity, they are a recreation–leisure–entertainment commodity, and they are a basic needs commodity. In addition, the balance in the proportion of tourist and non-tourist consumption of food and drink will often vary among communities and neighborhoods, from one season to another, and on different days of the week.

Input–Output Models

In general, only a small number of commodities accounts for the bulk of tourism spending. Identifying the portion of demand for a commodity that is directly attributable to tourism requires significant data about *tourism supply and demand*. In the past, this was done with extremely complicated and time consuming *input–output models*. These models measure the monetary flows into and out of a community, and between every business and business sector within a local economy, including both tourism and non-tourism businesses.

While the resulting model provides a comprehensive snapshot of a local economy, it is too difficult to apply to regional and national economies. Measuring tourism's contribution to a nation's economy is a particular challenge because, for reasons noted earlier, tourism is not an industry in the SNA.

Tourism Satellite Accounts

National accountants in France were the first to explore ways to analyze aspects of a nation's economy that are not adequately represented within the SNA. To do this, they developed a concept called *comptes satellites* (satellite accounts), which is known as *Tourism Satellite Accounts* (TSAs) when applied to the tourism industry. The TSA approach identifies the tourism percentage of each traditional industry (account) in a country's overall economy (SNA). (For example, tourism may account for 15% of the food and beverage industry profits in one country, but 50% in another.) The tourism portions can then be added up to determine tourism's total contribution to a country's economic performance.

Tourism Satellite Accounts are usually constructed for a national economy, so national averages guide the selection of commodities included in the analysis. The identification of tourism commodities will, however, vary among nations and the results cannot be compared among different countries. The arts, textiles, agricultural products and transportation products may receive a significant portion of their demand from visitors to some countries, but much less in others.

1.2.2 Tourist Demand and Mobility

A common approach to studying tourism is to divide it into the broad areas of demand and supply, and then to examine each separately. The demand side is the market for tourism attractions and facilities. It includes the reasons why people choose to travel and why they prefer some activities to others. Looking at it from a particular destination's point of view, it is knowing who the client or market is for a place. The supply side of tourism refers to the destination resources that are available for the tourist and recreationist. These include facilities and attractions of all kinds (such as sports fields, parks, beaches and entertainment), as well as supporting infrastructure (such as transportation, hotels and restaurants) and services (such as travel agents and recreation programs and activities).

Migration studies have long categorized the reasons for choosing to move from one place to another into push-factors and pull-factors. Push factors encourage an individual to leave a place (or to take a trip away from home), while pull factors affect where an individual goes after establishing the desire to leave. This is similar to tourism demand. A traveler is pulled to the amenities advertised for a given resort after a propensity toward travel has been established by push factors (such as stress at one's work). This is similar to tourism supply.

Demand can create supply. In a liberal market economy, commodities will be produced to meet a market demand. When people want to consume something (including recreation and tourism), the market will respond by creating more goods for them. Economic success comes to those who are best able to meet a market demand.

Knowing the *motivations* and needs that drive people to consume (participate in) a leisure activity is essential to understanding patterns of demand for a leisure product. Motivations must be understood to successfully design, develop and promote a product. The location, design, amenities, and special features of a development must be attractive to clients and visitors who comprise its principal markets. And because markets are constantly changing, adjustments need to be made to address new desires and expectations. To market effectively, the needs of the consumer must be identified.

In a demand-driven model, tourist motivations influence the development of a destination. Conversely, a supply-driven model creates its own market demand. Some natural features (such as the Grand Canyon, the Himalayas and Ayres Rock) and some special cultural sites (such as the Taj Mahal, the Pyramids at Giza and the city of Venice) are seen by most people to be of inherent interest to recreationists and tourists. While there may be a natural market for a site, growing that market will also require knowledge of tourist motivations and market segments.

It has been argued that the supply and demand dualism is superficial and that supply and demand are actually different perspectives on the same leisure phenomenon. What we know of as tourism, for example, would not exist if we did not have both tourists and destinations or attractions. The tourist shapes the destination because the destination must respond to the tourist's presence in some way. And the destination is shaping the tourist by generating expectations, motivations and experiences that would not exist if that particular destination, and its interpreters, did not exist.

Tourist Motivation

There is no one approach or model that is widely accepted as the standard for assessing tourist motivation. This is due to the diverse combination of products and experiences that comprises the tourism industry, and the associated wide range of tourist interests and needs. Motivations and needs also change over time, such as over the course of a lifetime, from one trip to the next, and from one activity to the next on the same trip. In addition, many tourists are not aware of what motivates them to take a certain trip or to visit a particular attraction, except in the most general of terms, which typically include recreation, relaxation, education and escape from their work-a-day world.

One example of a tourist motivation model that incorporates significant elements of supply is that suggested by Mayo and Jarvis (1981). They proposed that travel motivation can be divided into the following four types, based on what the tourists most wish to gain, see or experience in the trip. More than one of these can occur at the same time, though one tends to be more prominent than the others for any single trip.

1. *Physical motivators*: The desire for physical rest, sports participation, beach recreation, relaxing entertainment and health considerations.
2. *Cultural motivators*: The desire for knowledge of other countries, including their music, art, folklore, dances, paintings and religion.
3. *Interpersonal motivators*: The desire to meet new people; to visit friends or relatives; to escape from routine, family or neighbors; to make new friendships away from the home setting; or to experience anomie (which refers to social interactions in an anonymous setting).
4. *Status and prestige motivators*: The desire for recognition, attention, appreciation and a good reputation among family, friends and acquaintances in the home setting. (This is also referred to as ego enhancement.)

The advantage of this approach is that participation in specific activities can be used to classify motivations. Visitation to a museum is, therefore, an indication of a cultural motivation, while going to a beach would be a type of physical motivation. The attachment of motivation to activities facilitates the easy collection of data on tourist behavior at its most apparent level, making this a preferred approach by local and national tourism boards.

Maslow's Hierarchy of Needs

The psychologist Abraham Maslow proposed a *Hierarchy of Needs* (1954) that is widely used in explaining motivations for human behavior. According to the original model, individuals first and foremost focus on satisfying their physiological needs for survival. Once these have been satisfactorily met, then the next level becomes the primary motivation of behavior. Belonging and love are generally achieved through the family unit, while esteem is typically achieved through work accomplishments. The ultimate goal of human existence is the top level need of self-actualization, where one takes part in activities or a lifestyle that is rewarding, fulfilling and allows for the full expression of a person's individuality, creativity and purpose for living. This is similar to Aristotle's definition of leisure. Many see tourism as a way

Maslow's Hierarchy of Needs

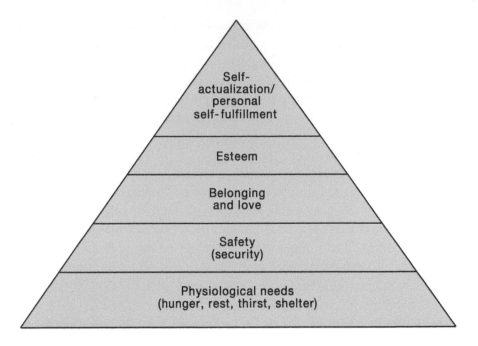

of achieving some degree of this level of need, if only for a transitory period and via a commodified and packaged product.

The major criticism of Maslow's hierarchy is that human needs occur simultaneously, and research indicates that even those who are struggling to meet basic physiological needs are looking for opportunities for self-actualization at the same time. Such a mix of motivations is also common in tourism and recreation. Maslow added several additional layers to the pyramid in later years, including a *Need to Understand* and an *Aesthetic Need* above the *Esteem level*, and a *Transcendence level* above the *Self-Actualization Need*. All have considerable implications for tourism and recreation. Maslow further identified deficiency or tension-reduction motivations (push factors) and inductive or arousal-seeking motives (pull factors) as supplements to his hierarchy of needs model.

The Leisure Ladder

Building on Maslow's hierarchy, Philip Pearce (1982) proposed a Leisure Ladder, with five motivational levels. The *leisure ladder* exists entirely within the realm of recreation and tourism. It defines the most basic level as relaxation and bodily comfort, which is probably somewhere above Maslow's basic need of food and shelter for survival. At this level, recreation and tourism are a recuperative therapy for dealing with the stress of work. The second level is leisure as stimulation and play, which can be physically and mentally as hard as work. The third level is leisure as a social experience where interpersonal relationships are a key component, while the fourth level is leisure as a mastery of skills and educational experience. The highest level is similar to Maslow's self-actualization peak fulfillment experience. According to Pearce, one of these levels dominates the motivation for any leisure activity. Lower levels must occur before one can experience a higher level, and lower levels can occur simultaneously with higher levels. Increased age and life experiences are associated with higher levels of leisure.

Psychocentric—Allocentric

Plog created a model in 1972 to analyze destination choices made by different types of tourists. He did this by classifying northeastern U.S. travelers on a scale of psychological types and then determining the destination that each type of tourist tended to choose. Travelers ranged on a continuum from *psychocentrism* (a focus on one's self) to *allocentrism* (a focus on other people). Psychocentrics are characterized as being conservative, inhibited and more safety conscious when they travel. They usually spend less money and travel to familiar places. Allocentrics are more outgoing, willing to take risks, open to new experiences and interested in travel to exotic places. He suggests that allocentrics would more likely travel to Africa or Asia, while psychocentrics would prefer travel to nearby destinations and theme parks (Table 1.1).

Table 1.1   Plog's Psychocentric-Allocentric Motivation Model

Psychocentric and Near-Psychocentric Motivations

- Travel for acceptance within a social group.
- Ego enhancement and an increase in social status.
- Travel as a cultural norm, including paid vacations required by law (more common in European countries).
- Visits to places seen or read about in the news, especially travel news.
- Visits to amusement parks.
- Pleasure, fun, enjoyment, parties.

Midcentric Motivations

- Rest and relaxation.
- Visiting close friends and relatives.
- Health, including a change in climate, search for sunshine, visits to spas and medical treatment.
- Change of pace or "escape" for a short period of time.
- Perceived glamour or prestige of a destination.
- Appreciation of nature, including trips to national and state parks, forests, lakes, canoe trips, ocean coasts.
- Sensual indulgence, including food (gourmet meals, buffets), comforts and luxuries for the body, romance.
- Shopping for souvenirs, gifts, luxury and specialty products, jewelry, art, cars, antiques.
- Transportation experiences, including cruise ships, comfortable trains, buses, airplanes, and cars.
- Pre- and post-travel experiences, including trip planning (anticipation, learning, dreaming) and sharing pictures and describing the trip upon return.
- Family or personal matters.

Near-Allocentric Motivations

- Religious pilgrimage or inspiration.
- Gaming, including casinos and horse races.
- Participation in sports events and sports activities.
- Travel as a challenge or test of endurance, such as rock climbing, mountaineering, sky and underwater diving.
- Business travel, conferences, meetings, conventions.
- Theater and special entertainment.
- Temporarily trying out a new lifestyle or home.

Allocentric Motivations

- Educational and cultural motives, including learning and appreciation, and scientific or purposeful trips with expert leaders.
- Genealogy research, study and exploration.
- Search for the exotic (from a Western perspective, major destinations include the Pacific islands. Southeast Asia, and West Africa).
- Satisfaction from a sense of freedom, including anonymity through travel, air and sea travel experiences, and fast trains and cars.
- Meeting people and making friendships in foreign places.
- Sharpening perspectives, such as an awakening of senses or a heighten awareness of the world.

Source: Adapted from Plog, S. 1987. Understanding psychographics in tourism research. In *Travel Tourism and Hospitality Research*, eds. J. Ritchie and C. Goldner. New York: Wiley.

Geographers have suggested that destinations evolve and change in the types of tourists they attract. New destinations tend to be discovered by adventurous allocentrics, with near-allocentrics soon following them. Over time, it may become a psychocentric destination. Palma de Mallorca (Spain) was a near-allocentric or midcentric destination for British tourists in the 1950s, and today it is considered a purely psychocentric destination. Phuket Island, off the Indian Ocean coast of Thailand, was an allocentric destination in the early 1970s, visited by alternative Western backpackers and hippies. It became a midcentric destination by the 1990s, and it is probably a near-psychocentric destination today (at least for Europeans and Australians).

A *model* is a simplification of reality that tries to eliminate all but the most important factors to explain how something works. Because human behavior is so complex, no single model will ever be able to address all of the variations in tourist motivations. The psychographic model does not consider the fact that people travel with different motivations on different occasions, and even at different times on a single trip. A wealthy allocentric may travel to Africa on an annual vacation, but may also take weekend trips for a psychocentric experience. Changes in life cycle, health and financial resources can all lead to fluctuations in travel behavior and psychographic levels. Finally, just as individuals are complex, so are destinations. Any large metropolitan area, for example, contains a full range of psychocentric to allocentric neighborhoods and attractions.

Market Segmentation

Once a tourist's motivation has been established, then the potential to match that person with an appropriate destination is increased. Such a match will most likely maximize satisfaction by producing the kind of experience the traveler seeks. Travel agents, tour wholesalers, destination marketers and tourism product developers can use this. Understanding the tourist's desire provides a basis for decisions about the types of environments and services that should be provided at a destination now and in the future, and it dictates the message content of promotional campaigns. *Advertising* can be more effective in targeting interested consumers. Advertising campaigns can be designed to dispel potential visitor worries, to create or confirm a desired image of a destination or simply to create an awareness of a destination for a new, potential market population.

As mentioned above, most destinations offer a range of different types of accommodations and attractions suited to the interests of different types of tourists. *Market segmentation* is the process of identifying the range of different tourist markets most suited to the range of sites and experiences available in a destination. Rather than dissipating promotional resources by trying to attract all potential travelers, or a very narrow portion of them, efforts can be specifically aimed towards the most likely prospects.

Understanding motivations for leisure, travel and recreation is essential to the tourism industry. To understand tourist motivations, researchers have created numerous models to simplify the complexity of travel behaviors. Several examples have been described above. With a better understanding of leisure motivations, services can be better catered to meet user needs, and advertising campaigns can be directed to the highest potential market. Travel agents can also better direct travelers to destinations and experiences that suit their wishes and temperament. The tourism industry, in general, can be strengthened with a fuller understanding of tourist motivations and destination offerings.

Tourism, Mobility and Migration

Another way of conceptualizing tourism is to view it as a form of *voluntary temporary mobility* by which people travel to another location, often for leisure, educational purposes or *visiting friends and relations* (VFR) or for business reasons. Both VFR and business travel are often more important for some destinations than travel for which leisure is the primary motivation. Although this is not to suggest, of course, that visiting friends and relations is something that may be totally without leisure! This way of thinking about tourism helps differentiate it from forced mobility, as in the case of political or environmental refugees, for example, people having to move because of a major flood, or permanent migration. The concept of temporary mobility therefore, includes a wide range of tourism-related phenomena, such as leisure travel, health tourism, volunteer tourism, educational travel, travel to second homes, pilgrimages, working holidays and business travel.

From a geographical point of view, tourism studies require an understanding of the places of tourist origin (or tourist generating areas), of tourist destinations and the relationship between the place of origin and the destination, which includes transportation routes, business and marketing relationships, and travelers' motivations.

Barriers and Constraints to Travel

As much as people want to travel, they often do not realize that most of the time and cost involved in touring is devoted to finding food and drinks, getting to lodging and restrooms, and sitting on some form of uncomfortable transportation. Many destinations are much dirtier and involve more hassling with local business people than their promotional literature makes out. Poor ground transportation systems are the leading cause of death for tourists around the world. Concern over the comfort and safety of a destination's facilities and accommodations, along with the distance required to get to a destination, will keep some people from taking a trip. In addition to these comfort concerns, there are a lot of other barriers to travel in general, as well as to travel to specific destinations. These include:

- Lack of personal time can prevent travel, including a lack of adequate free time and vacation time, and the need to stay home to care for children, the elderly or the ill.
- Institutional barriers, including legal restrictions on travel within one's own country (such as restrictions by the United States on travel to Cuba)
- The general difficulty of obtaining a passport and visa in a timely manner prior to a trip.
- Personal health and disability-related issues can prevent participation in some or all forms of travel and recreation activities.
- Crime concerns, including fear of financial scams and concern over being injured, robbed or kidnapped will prevent people from traveling to some destinations.
- Civil unrest, terrorism and war are very strong barriers to travel, due to the threat of being captured, injured or killed.
- Disease concerns can have a major impact on travel, including fear of infection from a specific disease (such as the SARS (severe acute respiratory syndrome) outbreak in 2003–2004, which temporarily decimated travel to and within Asia and the outbreak of Ebola in parts of West Africa in 2014); or there may be a more general concern, such as traveler's diarrhea from unhygienic food, which has become an increasing problem on Caribbean cruise ships.
- Personal phobias, with fear of flying (aerophobia) being a common affliction, prevent many people from traveling to some types of destinations.
- Weather and disaster concerns will keep people away from some destinations, including such undesirable weather as hurricanes, or natural disasters, such as earthquakes, tsunamis (tidal waves) and fires.
- Negative local attitudes toward tourists and other outsiders can reduce travel to a destination and especially prevent repeat visits (people visiting more than once).
- Lack of money to pay for travel expenses keeps some people from traveling, although there are many forms of travel that cost very little.

- Lack of information about a destination will keep people from going there.
- Lack of interest or desire to travel, which might be the most pervasive of all travel constraints, keeping many people close to home.

Travel Visa Issues

Visas are a form of documentation that are often required for the citizen of one country to enter and stay in another country. There are many different types of visas, including tourist visas, business visas, student visas, work visas and diplomatic visas. The visa limits the amount of time a visitor can stay in a country and stipulates what types of activities he or she can do in the country. Some people are required to apply for an exit visa to leave their home country. The visa is usually pasted or stamped in the traveler's passport, although a few are electronic. Some countries require that visas be obtained prior to arrival (from an embassy or consulate), while others issue visas upon arrival in the country.

Visas can be very expensive. Citizens of the United States, for example, must pay close to $150–250 to obtain a tourist visa to visit China, Brazil, Chile or Turkey (as of 2014). This includes the cost of the visa charged by the country, as well as the cost of mailing and paying a private company to take the traveler's passport to the embassy or consulate. These four countries have a reciprocal visa policy, which matches the price and requirements that the United States charges their citizens to visit the United States. Those requirements increased significantly following the September 11, 2001, terrorist attacks in the United States. For other countries, however, obtaining a visa can be inexpensive, if somewhat time consuming.

Special agreements between countries allow that the citizens of each not need a visa for entry, though a passport is usually still required. United States citizens, for example, can visit 160 countries (as of 2014) without getting a visa prior to arrival at a border. The leader, however, is the U.K., whose citizens can visit 174 countries without a visa, and who can often stay in those countries much longer than can visitors from the U.S. or elsewhere. This is a result of special arrangements between the U.K. and its former colonial territories, many of which are part of the 53 member Commonwealth of Nations (formerly known as the British Commonwealth), with some that still recognize the Queen of England as their titular head of state (including Canada, Australia and Belize).

Visa-free travel is not a standard policy within the Commonwealth of Nations. However, it is for several other country groupings, most of which are based on economic agreements. These include:

- The 28 countries of the EU (European Union), plus Switzerland, for which no visas are required for travel within the EU, though border formalities are still maintained for certain countries
- The 15 member countries of ECOWAS (Economic Community of West African States) which allow most of their citizens to travel freely among them without a visa
- Most of the ten Southeast Asian countries of ASEAN (Association of Southeast Asian Nations), for which visa free travel exists for all citizens, except for those of Myanmar
- The nine countries of the former Soviet Union that are members of the CIS (Commonwealth of Independent States)
- The six countries of the GCC (Gulf Cooperation Council) on the Arabian Peninsula
- The five countries of the EAC (East African Community).

Travel Advisories

Many countries issue official government advisories that they use to keep their citizens from getting into trouble overseas. In addition, many other organizations post travel warnings and other information to raise tourist awareness about the challenges of travel to certain destinations. Few countries are totally absent of warnings, and nowhere are tourists completely immune to predators. On any given week, civil

ISSUES AND INSIGHTS   The End of Oil—A Potential Future Constraint on Travel?

One of the most significant issues facing tourism is energy and issues of availability. According to the Energy Committee at the Royal Swedish Academy of Sciences (2005: 1), "It is very likely that the world is now entering a challenging period for energy supply, due to the limited resources and production problems now facing conventional (easily accessible) oil.

Nearly 40 percent of the world's energy is provided by oil, and over 50 percent of the latter is used in the transport sector." According to the Committee, mitigation measures must be initiated in the next few years to secure a continued adequate supply of liquid fuels, especially for the transport sector. Aviation will be more fuel efficient in the future as a result of technological innovations, but possibilities are limited, and although energy use per person per km will fall, the overall predicted increase in the number and distance of people flying will mean that overall fuel use and amount of emissions will continue to grow. However, over the longer term, completely new energy solutions are required given the decline of cheap oil. The key issues with respect to oil supply are outlined below.

Shortage of Oil
The global demand for oil is presently growing by almost 2 percent per year, with consumption at the end of 2011 set at 89 million barrels (including other liquid fuels) per day (1 barrel = 159 liters) or 30 billion barrels per year. Finding additional supplies is increasingly problematical because most major oil fields are well matured. Already 54 of the 65 most important oil-producing countries have declining production, and, one study found that by 2005, the rate of discovery of new reserves is less than a third of the rate of consumption. In 2009, the United Kingdom Energy Research Centre concluded in a review of studies that a global peak in oil production is likely before 2030, with a significant risk of a peak before 2020. Like climate change modeling, different models of the forecast of the demand and supply of oil use different parameters. Nevertheless, as in the case of climate change, although forecasts of when peak demand will actually occur do differ with respect to the exact date, there is nevertheless broad agreement that it will happen eventually, with the exception of some international oil companies and energy organizations.

Reserves of Conventional Oil
Since 1990 two-thirds of the increases in reserves of conventional oil have been based on increased estimates of recovery from existing fields and only one-third on the discovery of new fields. A conservative estimate of discovered oil reserves and undiscovered recoverable oil resources is about 1,200 billion barrels, according to the United States Geological Survey. This includes 300 billion barrels in the world's sedimentary basins, which have not yet been explored.

The Key Role of the Middle East
Only in the Middle East and possibly the countries of the former Soviet Union is there potential (proven reserves of 130 billion barrels) to significantly increase production rates to compensate for decreasing rates in other countries. The supply of Middle Eastern oil is often affected by political instability even while the cost of oil increases because of market perceptions of stability.

Unconventional Oil Reserves
There are very large hydrocarbon resources, so-called unconventional oil, which includes gas (c. 1,000 billion barrels of oil equivalent, much of which could be converted to liquid fuels), heavy oil and tar sands (c. 800 billion barrels), oil shales (c. 2,700 billion barrels), and coal. Problems with unconventional resources include long lead times in development, environmental impacts and the availability of water and natural gas for the production process.

Immediate Action on Supplies
Improvements in the search for and recovery of conventional oil as well as the production rate of unconventional oil are required to avoid price spikes, which would lead to instability of the world economy over the next few decades.

Liquid Fuels and the Transport System
Oil supply is a severe liquid fuels problem and less of a general energy supply problem. The transportation sector, for example, consumes 57 percent of the world's oil. Alternatives need to be developed to oil for the transport sector, otherwise there will not only be increased oil prices but also increased competition between transport and other oil users.

Economic Considerations
In the long run, the price of crude oil will be determined by the price of substitutes. Continued high oil prices are anticipated as long as the pressure from the expanding Asian economies such as China and India is maintained, as well as political instability in the Middle East.

Environmental Concerns
Unconventional oil will significantly extend the length of the hydrocarbon era and its subsequent contributions to greenhouse gas (GHG) emissions. Constraints similar to those imposed on other fossil fuels (e.g., emission controls and CO_2 sequestration) will be necessary and provide major challenges for industry. Unconventional oil and energy sources themselves are also sources of environmental concern. For example in the United States there has been a substantial concerns over the environmental effects of *fracking* and the potential impacts of the development of long-distance oil pipelines,

(continued)

ISSUES AND INSIGHTS Continued

especially from the Canadian tar sands. From a global point of view, transport is the most relevant sector in terms of the long-range environmental sustainability of tourism, accounting for an estimated 75–90 percent of all GHG emissions caused by tourism. Within tourism transport, the share of aviation-related GHG emissions corresponds to about 75 percent of all transport emissions. However, the contribution of GHG emissions from aviation is obviously not spread evenly across the world and instead is concentrated in the wealthier countries where the vast majority of tourists actually come from.

Sources

Royal Swedish Academy of Sciences (Energy Committee). (2005). Statements on Oil, Royal Swedish Academy of Sciences, Stockholm.

Scott, D., Hall, C.M. and Gössling, S. (2012). *Tourism and Climate Change*, Routledge, London.

UK Energy Research Centre (UKERC) (2009). *An Assessment of the Evidence for a Near-term Peak in Global Oil Production.* A report produced by the Technology and Policy Assessment Function of the UK Energy Research Centre. UKERC, London.

unrest, natural disasters, or diseases in many countries results in advisories urging people not to travel there. In some cases, those who are in a country are urged to leave, if they can do so safely. Official advisories can have a major impact on travel to a destination. Government advisories about the SARS (severe acute respiratory syndrome) outbreak in China and Southeast Asia in 2003 contributed to a huge decline in travel (2.8 million tourism-related jobs were lost in China alone), and governments and industry representatives in the region highly criticized them.

Conclusions: Tourism Demand

The desire to participate in tourism and recreation is strong and deeply rooted in the human psyche. However, motivations can vary considerably among individuals. How people participate in travel and tourism, in particular, is largely guided by social norms and the larger socioeconomic context of the travel industry. Numerous individual and societal barriers and limitations further shape the pattern of participation in tourism. Why people travel and what they seek from travel have major implications for people in local destinations, as well as for the global economy.

1.2.3 Tourism Supply: Attractions, Destinations and Landscapes

One way of looking at tourism supply is to focus on tourist attractions. Tourist attractions are fundamental to the very existence of tourism. Without attractions there would be no tourism. Without tourism there would be no tourist attractions. But what is a tourist attraction? Several definitions have been proposed, including:

- An exploitable resource.
- A marketable product and image.
- A place attribute or feature.

All of these definitions are simple and beg for additional definitions of what a resource, a product, an image, an attribute and a feature are.

The Tourist Attraction System

Another definition of tourist attractions is all those sites in a non-home place that are of interest to tourists. There are three essential elements in this definition:

1. a tourist;
2. a sight or site in a non-home place; and
3. an interest or reason for the tourist to want to view the sight or experience the site.

These three elements comprise a model of tourist attractions that was first suggested by MacCannell (1976), and later defined by Leiper (1990), as the *Tourist Attraction System*. Their argument is that you cannot have a tourist attraction unless all three of these features are present. If there is a site and a tourist, but no reason for the tourist to take an interest in the site, then the site cannot be considered an attraction. It is like Yi-fu Tuan's difference between *space and place*—the attraction emerges as a special place from the surrounding undifferentiated space through the meaning given to it by the tourist (or the tourism industry).

Similarly, if there is a site and a reason for tourists to be interested in it, but there are no tourists, then the site is not a functioning tourist attraction. And, finally, if there is a tourist who has a motivating interest to see or experience a certain type of attraction, but there is no site that offers that experience, then there is no tourist attraction.

In practice, however, the tourist attraction system has some limitations. The first limitation is the problem that almost anything can become a tourist attraction. Historic sites, amusement parks and spectacular scenery are almost always attractions. Food is often an attraction, accommodations (resorts) can be an attraction, and sometimes transportation (cruise ships) and the tourists themselves (Spring Break parties) are attractions. The diversity of potential tourist attractions is phenomenal. For example, alternative tours have been developed to take people into sewer systems, to homeless neighborhoods and slums, to cemeteries, to disaster locations and into working mines.

The second problem is that the tourist attraction system is not how most people think about tourist attractions. For most people tourist attractions include landscapes to observe, activities to participate in and experiences to remember. The role of the tourist is seldom considered because he or she is visually separated from the attraction. The tourist is, therefore, taken for granted in a more common definition of attractions. The meaning or purpose of the attraction is less taken for granted by most people. But instead of meaning, most people view this as the attraction's importance. Attractions are perceived by tourists to be ranked above or below each other in importance. The hierarchy of attractions will vary from one person to the next based on the set of available attractions that a tourist is aware of, the depth of knowledge that the tourist has of each attraction and the tourist's personal interests.

The Tourist Attraction Site

It is the attraction site that most people think of when they hear the word tourist attraction. Lew (1987) examined the different ways that researchers defined attractions when they conducted attraction inventories and found three distinct approaches. Although they were often mixed together in a single list of attractions, each is actually a characteristic that all attractions share. These characteristics are:

1. a form, identity, or name;
2. an organizational and developmental structure; and
3. a perceived experience.

At their most basic level, attractions are identified by their individual names. Lists of attractions by name are most often used in inventories of small areas, such as towns or small cities. At the other end of the form approach is the use of *Standard Industrial Codes* (SICs) to identify general types of attractions, such as accommodations, restaurants and retail stores. SIC-based attractions can be useful to determine monetary flows because governments use them to summarize economic census data. Schmidt (1979) suggested that attractions are best divided into geographical, social, cultural, technological and religious categories. Schmidt's approach is typical of that used in attraction inventories to assess the tourism supply in a destination, though these vary considerably in details from one place to the next. Tourist guidebooks usually classify attractions under a combination of both specific names for the most important sites and

Yorktown, Virginia, battlefield where United States and French forces defeated the British in 1781. The importance of many historic sites is often not apparent in their physical landscape alone. In this example, the visitor would be at a loss about the significance of this otherwise empty space without the site interpretation provided by signage and tour guides.

Photo: Alan A. Lew

more general categories. Guidebook categories vary from one destination to another, though they typically include the general categories of retail stores, restaurants, accommodations, museums, amusements, sports and recreation.

The two basic types of formal attractions are nature-based attractions and human-based attractions, although there is a transition between them, which results in a third type of attractions. Examples of these are shown in Table 1.2. In practice, these types of formal attractions are divided up among categories that are most relevant to a specific destination.

Tourist Attraction Structure

The structural approach to tourist attractions does not examine the individual attraction's form or identity, but instead it focuses on its spatial capacity and temporal nature. Geographic size is the simplest basis for categorizing the spatial character of an attraction. A spatial hierarchy of attraction scale would progress from the smallest specific object of interest in a site, then to cities, then regions, and then an entire country and continent. Asia, for example, is a destination, as are China, Beijing and Tiananmen Square. Scale considerations can provide insights into the geographic grouping or clustering of attractions, which is important in the planning and marketing of tourism. Tourism marketers promote the images of specific, small-scale attractions (that are easier to sell) to create identifiers for larger attraction complexes. An example of this is the Eiffel Tower as a *marketing icon* for France. Planners, however, are then faced with the problem of an over-concentration of demand at some tourist sites and the under-utilization of others.

In addition to the spatial structure of an attraction (or group of attractions), the temporal or time nature of attractions is important. Special events (such as the Olympic Games) can be huge attractions, but only for short periods of time. Outdoor recreational activities (such as snow skiing) are similarly time dependent. This can have a major impact on the seasonality of visitors to a destination, which must find ways to attract visitors during off-season periods. A third type of structural characteristic is capacity. This is similar to spatial scale, but it is internal to the attraction. Some attractions can hold a lot more visitors at one time than others can. This, also, presents challenges to tourism and recreation planners.

In describing attractions based on their structural characteristics, some of the terminology that has been used by tourism researchers includes:

- Attractions nucleus, inviolate belt and zone of enclosure (around the attraction).
- Remote or isolated attractions.

Table 1.2   Types of Formal Attractions

Nature-based Attractions

- Panoramas: mountains, sea coast, plains and valleys, arid lands, islands.
- Ecosystems: climate, sanctuaries (national parks nature reserve).
- Landmarks: geological, biological (flora and fauna), hydrological.

Mixed Nature-based and Human-based Attractions

- Observational nature: rural and agriculture landscapes, scientific gardens (animals zoos, arboretums trees), and archeology.
- Participatory nature: mountain activities (summer and winter), water activities, other outdoor activities.
- Leisure nature: trails, parks (beach, urban, other), resorts.

Human-based Attractions

- Settlement infrastructure: utilities (e.g., dams), settlement activities and morphology (retaining, finance, government, education and science, religion), people (way of life, ethnicity).
- Tourist infrastructure: forms of access (to and from a destination, tour routes), information and reception (welcome centers), basic needs (accommodations, meals).
- Leisure infrastructure: recreation and entertainment (performances, sports related, amusements), culture, history and art (museums, monuments, performances, festivals, cuisine).

Source: Lew, A.A. 1987.

- Structured or unstructured development.
- Planned or unplanned infrastructure.
- Accessible or inaccessible location.
- Touring, clustered or destination attractions.
- Rural, suburban or urban.
- Local, regional, national or international scale.
- Craft or industrial tourism.
- Slow or rapid growth.
- Small, medium or large capacity.
- Event, transitory, seasonal or permanent attraction.
- Single or multiple visitation.

Tourist Attraction Perception

Studies of tourist perceptions and experiences of attractions comprise the third major approach to the study of tourist attractions. Pearce defined a tourist place as "any place that fosters the feeling of being a tourist" (1982:98). One way this experience is felt is in the degree of *outsideness* and *insideness* that one feels when visiting a destination. Some have argued that a major goal of the tourist is to penetrate into the inside (or back region) of the attraction to experience its authenticity. For the tourist, some risk is required to take this leap into authenticity.

Approaches to tourist attraction perceptions and experiences are shown in Table 1.3 and are closely related to tourist motivation, as discussed above. This reflects the inseparable relationship between the site and the visitor in the tourist attraction system. Because most tourists are outsiders and newcomers to a place, many studies of attractions perception focus on security and the minimization of risk. To do this, they typically take place in staged, inauthentic and highly structured environments.

Tourists also mostly interact with the advertised *images*, rather than with a direct and unmediated experience of a site. MacCannell (1976) referred to this as *marker involvement* because the tourist is more

Table 1.3   Cognitive Approaches to Tourist Attractions

Tourist Activities

- Education/exercise or exploration
- Place to contemplate or to talk
- Guided tours or unguided touring

Passive or Active

- Attraction character.
- Touristy/contrived/staged or authentic.
- Especially animated or normal daily life.
- Evoked set (know/n attractions) or inert set (knowable attractions) or inept set (unknowable attractions).
- Tourism oriented or non-tourism oriented.
- Structured and organized or unstructured.
- Front region or back regions.
- Modern or traditional/historic.
- Heard a lot about/important place or absence of other tourists.

Tourist Experiences

- Expensive/luxury or economy/values or inexpensive/cheap.
- Safe/sanitary or different/getting aw/ay or escapism/freedom.
- Pleasant/friendly or companionship or novelty.
- Leisurely/restful or fun/swinging or adventurous/exciting.
- Mass produced or limited experience or individual experience.
- Common/ordinary or interesting or unique.
- Recreational or diversionary or experiential or experimental or existential experience.
- No role transformation or role transformation.
- Marker involvement or sight involvements.
- Familiar or exotic.
- Easy to tour or effort to tour.

Source: Lew, A.A. 1987.

focused on the label attached to the attraction than the attraction itself. Most historic sites, such as empty battlefields, are marker involvement attractions, as opposed to sight involvement attractions that would not require any interpretation. Outstanding natural landscapes and culturally unique places are examples where sight involvement often predominates over marker involvement.

The *perceptual* or *cognitive* approaches can be applied to any of the formal attraction types in Table 1.3. In fact, many researchers freely mix formal, structural and perceptual approaches in their attraction inventories, without thinking about the differences. For example, one study of attractions in Greece classified them into three formal types (ancient Greece, picturesque villages and sun and sea), two structural types (clustered-isolated and accessible-inaccessible) and one perceptual type (unique-common). In general, the selection of one or more approaches depends on the goals of the research. A comprehensive study would want to use all three major approaches, whereas a more narrowly focused study might only use one approach.

Tourism Places and Landscapes

Places are a major focus of geographic study because they are the focal point of relationships among local residents, outside visitors, economic processes, the physical environment and culture. The visual landscapes of places are readily accessible for anyone to experience. Landscapes have strong representational and symbolic meaning, and they form a veritable stage for play and recreation. Landscapes are often used to promote a destination through television and print advertisements, and through these media landscapes become the first and most basic setting for the tourist experience.

Landscapes are also captured by tourists through photographs that become lasting memories. However, because they are so omnipresent, forming the basis for everyday life, landscapes can be challenging to interpret and understand. Furthermore, different cultures have different understandings of what may constitute an ideal or even an attractive landscape, while landscape appreciation may change over time. For example, until the mid-18th century in European culture, the ideal landscape was an ordered urban or even garden landscape. In contrast, from the early 19th century on, natural landscapes and particularly those presented as Wild and Romantic started to figure prominently as the ideal, a change that was also important for the development of the national park concept in the United States and elsewhere.

Landscapes are often an integral part of the image of a destination. The New York's City skyline and Hong Kong's harbor are among the more recognized landscape icons. Tourism marketers turn landscapes into representations of destinations. Tourists use these representations to decide whether or not to visit a place, and, then once there, to assess and validate their experiences. At the same time, the destination manages its landscape to meet the needs of both locals (hosts) and visitors (guests). Some elements of the landscape are either transformed or intentionally preserved for the guest, while others are oriented more toward local needs. As tourism and leisure become more pervasive elements in daily life, it is becoming more difficult to separate host and guest landscape elements.

Landscape Studies

Geography and other social sciences have developed an increasing interest in tourism landscapes since the 1990s when the tourism economy became a dominant global service industry. This interest reflected a widespread realization of the degree to which tourism has been responsible for modifications to many landscapes around the world, from the revitalization of older retail districts and the building of suburban *mega-malls* (the Mall of America in Minnesota is one of the top attractions in the United States), to the conservation of nature areas and the rise in ecotourism. The bigger the alterations in land use and the larger the visitor flows, the more such development can result in environmental degradation and incompatible construction, carrying capacity issues and social change to the pre-existing landscape.

At the same time, there has been a growing interest in the landscape in general, including landscape policy, landscape values and landscape assessment. New analytical methods, some using geographic information systems (GIS), have skyrocketed. Many have had a focus on evaluating, protecting and enhancing the quality and use of the landscape at the local, regional, national and international levels.

Classifying Landscapes

The earliest approach to the geographic study of leisure and tourism landscapes was the classification of scenic quality and land use. This approach was most popular in the first half of the 20th century, though it is still widely used today. It involves classifying landscapes into different types (such as retail districts, public parks, open space, amusement parks and wilderness) and then assessing and evaluating the condition of each landscape type, as well as the overall situation. The major tools used in this approach are *land use maps* and demographic other social and economic data that lend them to quantitative and spatial analysis. This approach to landscape research continues to be dominant in environmental studies and in the design professions (urban planning and landscape architecture). They reflect an assumption that tourism and leisure management involve hard facts and clear typologies and follow predictable economic models, which is not always the case.

Resort Morphology

Geographers have long been interested in the distribution of land use in a community and what that can tell us about the past and future development of a place. Standard land uses include industrial, retail, residential and open space, although these are often further divided into many additional types. Urban morphology refers to the distribution or pattern of land use in a community, including which types of land uses are located where. This approach is also quantitative in nature, but it is more interpretive in application. When applied to tourism and recreation, it has mostly been used in the study of resort communities, islands and coastal landscapes. How retail, attractions, accommodations and residential land uses have changed over time in these areas has given rise to the concept of the *Recreation Business District* (also known as the *Tourism Business District*) as a distinct land use type in many coastal resort communities.

Structuralism

Another way that social scientists study tourism landscapes is from the perspective of *structuralism*. Structuralism focuses on larger social norms, belief systems and societal practices that shape and force certain forms of behavior and decision-making (and resulting landscapes) on society. Economic, social and environmental policies established by national governments, as well as decisions made by multinational corporations, are examined as the primary actors in local landscape modification.

Structuralists have focused on the nationalization and globalization of cultural and social norms, through mediums such as Hollywood movies and pop music videos, and how they have contributed to a homogenization of landscapes (*placelessness*). They examine how big money interests override local small money interests, commodifying local uniqueness and making destination landscapes more alike than different. For the structuralists (of which there are several different approaches, including *Marxism*), the principle struggle is between the individual and society, and between local culture and the dominant political and economic structure.

In tourism this is often seen in stories of the struggle between authentic villagers and multinational hotel developers, or between local mom and pop hotel and restaurant owners and outsider fast food and chain store companies. In most places the latter groups have been gaining ground over time, with changes in the landscape that are both quantifiable and visible.

Cultural Landscapes

Geographic research on tourism landscapes also can be shifted in scope and scale from the macro-analysis of symbolic landscapes to the micro-analysis of town and resort landscapes. The latter is somewhat related to classification and *morphology studies*, but with a greater emphasis on the symbolic nature of place features in the local communities. This cultural perspective stresses the distinctive physical and material nature of places, along with their social and cultural meaning and symbolism. One outcome of this shift is that many of the new landscape geography studies in recent years have discovered the unexpected importance of leisure and tourism in the everyday landscape.

Cultural meanings are inherent in the landscapes of our *built environments*, and in how we value and treat our natural surroundings. They visually illustrate the relationships between social structures and local cultures, as well as between the larger society and the individual. Various approaches have been used to interpret tourism development, tourism economics and tourism psychology through their imprint on the landscape. There is no question that the landscapes of tourism in any place will change as tourist markets change, creating new opportunities for entrepreneurs to pursue, and for students of tourism to study. The speed and form of how local communities respond, and how they manage their political economy and long-term cultural interests, are a major focus of interdisciplinary research on the landscapes of tourism, recreation and leisure.

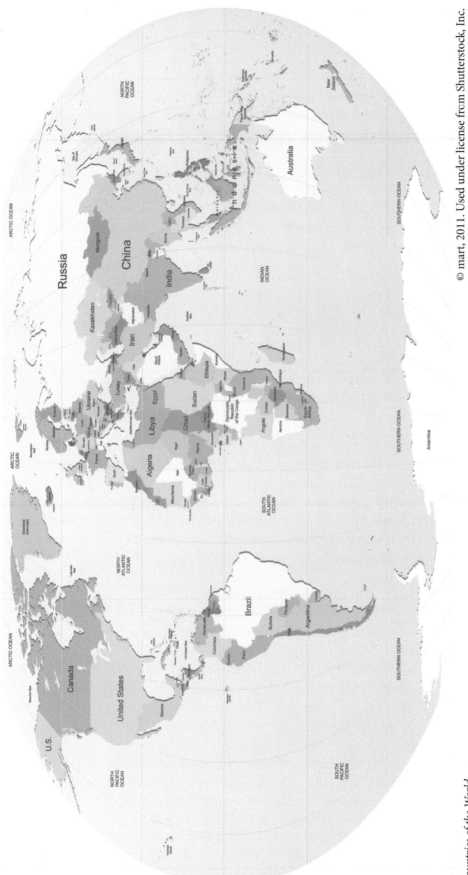

Countries of the World.

© mart, 2011. Used under license from Shutterstock, Inc.

1.3 | OVERVIEW OF THIS BOOK

This chapter introduced basic geography and tourism concepts, and themes that relate to discussions in the regional chapters of this book. The regional chapters are covered in four chapters, each of which includes several regional and subregional sections. The traditional approach to regional geography begins with physical geography (geomorphology, climatology and biogeography) and then covers human geography (historical, cultural, economic and political).

This book takes this basic regional geography approach. Each region begins with a section describing the natural features and their geographic variations. These discussions form the basis for describing the supply of nature-based tourist attractions and considerations, including opportunities and challenges for their development and visitation. The same approach is taken with the human geography of each region. Each region section identifies major cultural subregions, their historical formation and their contemporary development issues, particularly as these are related to human mobility, tourism destination development and sustainability issues.

In addition to the basic foundations of physical and human geography, and nature-based and culture-based tourism supply, special issue topics are discussed in each of the sections. These topics are related to the major issues that each region faces. The approach used in this book provides readers with a fundamental understanding of the diversity of our contemporary world, the uniqueness of the landscapes and human experiences in different regions and places in the world, and the importance of geography, history, sustainability and contemporary tourism and travel around the world.

1.3.1 World Regions

Traditionally, the continents of the world include Europe, Asia, Africa, North America, South America, Australia and Antarctica. With the exception of Europe and Asia, each of these is a large land mass completely surrounded by oceans and seas. Europe is actually a peninsula of the Eurasian land mass, and is treated separately from the rest of Asia because of its cultural difference and global economic importance. In addition, geography (and history) has mostly been interpreted from a European perspective. The world, however, is changing, which creates increasing challenges for geographers trying to divide the world into meaningful regions.

The problem is that, while geographic regionalization requires a broad, holistic view of both physical and cultural phenomena, at some point a decision must be made to emphasize one basic characteristic over another. This decision is sometimes easy, and other times not. Different decisions made by different geographers will result in different regionalizations. In this book, the regions of the world are grouped into four large chapters. The contents of each of these chapters are listed below. The chapters and their sections vary in size because the world is a complex place and the size (land area and population) and global importance of each region varies considerably. This book is primarily written for English-speaking students in North America, Europe and the Australian region. As such, there are some biases toward the way those regions view the world.

One example of this is in the naming of countries. Countries around the world are known by different names in different languages, as well as by their home country language. For example, the country known as *Germany* by the English-speaking world is called *Deutschland* by German speakers, *Allemagne* by French speakers, *Saksa* by Finnish speakers, *Niemcy* by Polish speakers, and *Doitsu* by Japanese speakers. In this book, we use country names that appear, in the opinion of the authors, to be in common use in the English media at the time the book was written. Such conventions change over time, and names will be updated periodically in future editions of the book. Other common names that appear in English for countries are also indicated parenthetically, where appropriate.

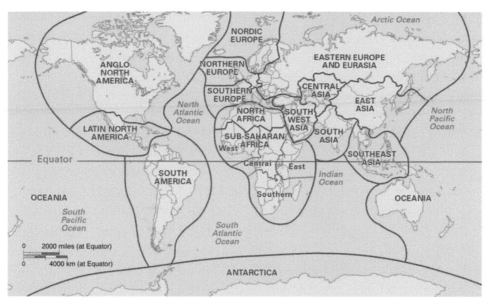

Geographic Regions and Subregions in this Book

Chapter 2—Europe, East and West

In terms of global historical importance, economic impact and international travel, Europe probably continues to be the most important region in the world. For these reasons, all of Chapter 2 is devoted to Europe and its subregions. The major subregions in Europe are (1) Western Europe and (2) Eastern Europe. These are sometimes treated as two separate world regions, though the transition between them has become increasingly difficult to define since the fall of the Soviet Union in 1991. Western Europe is further divided into the Nordic countries (mostly Scandinavia), the Northern European countries (excluding Scandinavia) and the Southern European countries on the Mediterranean Sea. In addition, the European Union (EU) has become a major political and social force in the countries that are its members. Because of this, EU and non-EU countries are discussed separately in Chapter 2. The most important non-EU country is Russia, which extends to the Pacific Ocean and includes areas that are part of the continent of Asia (everything east of the Ural Mountains). Eastern Europe also includes the *Transcaucasia* portion of Russia, where the Caucuses Mountain Range forms a border between Europe and Asia.

Chapter 3—The Old World: Central Asia through Africa

This chapter covers a vast territory that includes three major world regions: (1) Central Asia, (2) Southwest Asia and North Africa (which includes the Middle East) and (3) Sub-Saharan Africa. In contrast to Europe, the regions and countries of Chapter 2 have some of the lowest population densities and are among the least visited in the world. They include two major cultural realms. The area from Central Asia through North Africa is one of the two mostly Muslim regions of the world. A major difference is that Central Asia is a mostly Turkic speaking region that was formerly part of the Soviet Union, while Southwest Asia and North Africa were more influenced by the Arabic language and Western European colonial powers. Sub-Saharan Africa, which is all of Africa south of the Sahara Desert, is comprised of mostly black racial groups that are divided into a wide array of language and cultural traditions.

Chapter 4—Asia and Oceania

In terms of total population and the total surface area of the globe, Asia and Oceania are the largest of the four regional chapters in this book. This chapter covers five major global subregions: (1) South Asia, (2) Southeast Asia, (3) East Asia, (4) Oceania, and (5) Antarctica and the Oceans. Each of these could be treated as a separate chapter in other regional geography books, although it is also common for Antarctica and the Oceans to be omitted completely. However, these subregions share important historical and cultural traditions, and they are all related to the idea of Asia in important ways. Asia is the fastest developing region of the world in terms of economic importance, average income levels, and in travel and tourism development. It is also the most populated region of the world.

Oceania and the Antarctica and Oceans regions, which extend off the southeast edge of Asia, are the least populated of the world's regions. They include some very different subregions: Australia and New Zealand, the South Pacific Islands, and the continent of Antarctica. Because of the importance of the Pacific and Southern Oceans in these subregions, the chapter also discusses ocean geography and issues from a global perspective.

Chapter 5—The Americas, North and South

The New World of the Americas is one of the easier regions to define, since it is relatively isolated from the Old World of Africa and Eurasia. The subregions of the Americas are also widely accepted. These include (1) Anglo North America (Canada and the United States), (2) Latin North America, consisting of the Caribbean Islands and Central America (including Mexico), and (3) South America. North America is another area of considerable economic and cultural influence on a global scale. It has some of the highest international travel rates, and the United States' borders with Mexico and Canada are among those crossed the most in the world. In addition, the countries of the Caribbean Sea are among those most dependent on tourism in the world. South America, by contrast, has been one of the least visited regions, though its greater political stability and increasing economic importance are leading to changes in that situation.

Chapter 6—Geographic Futures and Opportunities

The final chapter of this book provides an overview of some of the major challenges and issues facing the tourism industry. We pose some questions as to what the future(s) of tourism may look like both on Earth and, possibly, in space. We will also return to issues such as climate and environmental change and the development of more sustain able forms of travel and tourism. In addition, we will also look at some of the potential opportunities in tourism at both an industry and a collective level.

Follow Current World Regional Geography Issues

World Regional Geography Issues and Resources Website (http://WRGeography.com)

Current World Regional Geography Issues on Facebook (http://on.fb.me/WRGeography) – case sensitive

Current World Regional Geography Issues on Twitter (http://twitter.com/WRGeography)

REFERENCES CITED

Lew, A.A. 1987. A framework of tourist attraction research. *Annals of Tourism Research* 14(4): 553–575.

Leiper, N. 1990. Tourist attraction systems. *Annals of Tourism Research* 17(3): 367–384.

MaCannell, D. 1976. *The tourist: A new theory of the leisure class.* New York: Schocken Books.

Mayo, E.J. and Jarvis, L.P. 1981. *The psychology of leisure travel, effective marketing and selling of travel services.* Boston: CBI Publishing.

Pearce, P. 1982. *The social psychology of tourist behavior.* Oxford: Pergamon.

Schmidt, C.J. 1979. The guided tour, insulated adventure. *Urban Life* 7: 441–466.

World Commission for Environment and Development. 1987. *Our common future: The Brundtland report.* Oxford: Oxford University Press.

Name _____

Human Mobilities

1. Identify three examples of human mobility that shape the community you live in. Describe the impact of each of these on your community.

_____

_____

_____

2. Identify and describe three examples of human mobility that have global impacts. Describe the impact of each of these on your community.

_____

_____

_____

Tourism Destinations

3. In what ways is the community that you live in a tourism and recreation destination? How does tourism influence (or not) the image of your community? What role does it play in your community's local economy?

_____

_____

_____

4. How is a tourism perspective useful in understanding the geography of the world and places in it? Cite three specific examples or concepts in geography (from this chapter) that could help you, personally, to enhance your understanding of the places that you visit.

_____

_____

_____

Sustainable Environments

5. In what ways is a geographic understanding of our environments and communities necessary to create a more sustainable world? Cite three specific examples or concepts in geography (from this chapter) that apply directly to contemporary sustainability issues.

_____

_____

_____

6. What is the relationship between economic and cultural globalization and local sustainability? How might globalization help local sustainability efforts? How might globalization hurt local sustainability efforts?

_____

_____

_____

Instructor Questions (answer any additional questions from your course instructor in the space below)

_____

_____

_____

CHAPTER | 2

Europe, West & East

2.1 THE GEOGRAPHY OF EUROPE

2.2 WESTERN EUROPE TODAY

2.3 EASTERN EUROPE TODAY

CHAPTER 2 OVERVIEW QUESTIONS

HUMAN MOBILITIES

- How has the long history of human movement across Europe shaped its contemporary cultural landscape?

TOURISM DESTINATIONS

- Why is Europe the biggest international tourism region in the world today?

SUSTAINABLE ENVIRONMENTS

- How has modern Europe, both west and east, overcome its past national and ethnic tensions to become the mostly peaceful and prosperous place that exists today?

INTRODUCTION

In terms of numbers of international tourist arrivals, Europe is the most visited region in the world accounting for just over half of all international visitor arrivals. Although much of this is *intra-regional* travel (between the many countries of the region), Europe is also a major inter-continental destination, especially from Asia and North America. France and Spain are two of the most visited countries in the world, while Italy, the United Kingdom, Germany and Russia are also usually in the top ten most visited. For this reason, this chapter is devoted entirely to Europe, along with its eastern and southeastern extensions into the Asian continent. Section 2.1 covers the basic physical geography and human geography of Europe. Section 2.2 discusses tourism in *Western Europe*, while Section 2.3 covers *Eastern Europe*. The division between Eastern and Western Europe is based on historical divisions that are increasingly less meaningful. As such, in both of these sections, the emphasis is initially on European Union (EU) countries, and then on non-EU countries, which reflects today's most important political and economic division within Europe.

2.1 | THE GEOGRAPHY OF EUROPE

Europe is typically considered one of the world's continents. However, it is unlike the other continents in that it is not completely surrounded by water. Instead, Europe is really a *peninsula*, jutting out from the much larger Eurasian land mass. Just where this peninsula begins and ends is not obvious on a map. For the most part, we accept the Arctic Ocean, the Atlantic Ocean and the Mediterranean Sea as the northern, western and southern sides of the European continent. The eastern edge of the continent is typically defined by the Ural Mountains in Russia.

Europe, Asia and Africa

The large *Eurasian* continent has several other peninsula-type regions that may be equally as deserving as Europe of the status of a continent (including South Asia and Southeast Asia). We will be discussing each of these later. However, only Europe is regularly designated as a separate continent. The reason for this is primarily due to cultural perceptions. One of those cultural perceptions relates back to the ancient Greeks. *Greek geographers* divided the world into three continents: Europe, Asia and Africa (which they called *Libya*). Greece was part of Europe. Africa was on the other side of the Mediterranean Sea, and Asia was on the other side of the Aegean Sea. Perceiving themselves to be culturally different from Asia and Africa has been one of the major factors in Europe's aspiration to be a separate continent.

Another reason for Europe's special status is that for most of the past 500 years, Europe has been one of the most important regions of the world economically, politically and militarily. Starting in the 15th century, European explorers ventured out to other parts of the world seeking to expand their economic and political influence. They established colonies in Asia, the Americas, Africa and elsewhere. Very few parts of the world were not subject to *European colonization* or the territorial claims of European countries. Through their colonies, they introduced new models of economic and political development. Today, those colonial ties still influence non-European lands, in the form of special trade relationships, cultural ties such as language, as tourist destinations and as sources of immigrants to European countries.

For all these cultural and historical reasons, most geographers consider Europe a separate continent. It is also the most popular tourist destination region in the world and, given its high level of economic development and relative ease of outbound travel, the most important source of tourists to other parts of the world.

European Union

Today, much of Europe is part of the European Union (EU). It is sometimes euphemistically referred to as the *United States of Europe*, although this description is not completely accurate. Like the United States, travel between most EU countries does not require a separate visa and seldom requires a stop at a border. This is quite an achievement for a continent that was the home of both World War I and World War II, two of the most devastating conflicts of the 20th century. However, the member nations of the EU hold considerably more power than individual states in the United States, with only a limited amount of authority having been passed on to EU central authorities and the European Parliament. Unlike many other *supranational* bodies, the European Parliament is democratically elected and its activities cover many aspects that affect the lives of European citizens. However, as in the United States where there is often tension between state and federal powers, in Europe there are often disagreements over the interaction between national and European powers.

Major Regions of Europe

Despite the creation of the EU, Europe remains one of the most diverse regions in the world both culturally and physically. Europe has one of the world's highest population densities, squeezed into one of the smallest major geographic realms that we will be discussing. Within the small land area of Europe, numerous languages and cultural traditions exist, as does a great variety of landform and climatic conditions, all of which serve to enhance its tourism attractiveness. The major geographic regions of Europe, as defined in this textbook, are based on a combination of culture and language, on political boundaries and alliances, and on climate and terrain. They include the following, though keep in mind that there is some overlap in these definitions.

Western Europe consists of the following areas:

- *Nordic Western Europe* consists of Denmark, Greenland, Norway, Sweden, Finland and Iceland. Greenland (in North America) and the Faroe Islands are autonomous territories of Denmark that are included in the *Nordic Group*. The Åland Islands situated between Sweden and Finland are the other territory recognized as part of the Nordic grouping. Although a part of Finland, this Swedish-speaking territory has considerable political independence. Denmark, Norway and Sweden are collectively referred to as Scandinavia, with the Scandinavian Peninsula consisting of Norway and Sweden.

- *Northern Western Europe* includes the British Isles; the continental European countries of Belgium, the Netherlands and Luxembourg (known as the Benelux countries); and the German-speaking countries of Germany, Switzerland and Austria. Switzerland, Liechtenstein and Austria are sometimes broken out to form a separate Alpine Europe region. The *British Isles* consist of the United Kingdom (England, Scotland, Wales and Northern Ireland), the Isle of Man (a self-governing British Crown dependency in the Irish Sea), and the Republic of Ireland. While typically considered part of Western Europe, the British Isles are sometimes treated as a separate region. It is also important to note that the Republic of Ireland does not usually support the term British Isles, as they feel that it implies that Ireland remains subject to British control. Nordic and Northern Western Europe are mostly *Germanic-speaking* countries, though notable exceptions include Finnish in Finland; Celtic areas in the British Isles (Scotland, Wales, Ireland, Isle of Man, Cornwall) Brittany in France and Galicia in Spain; and French in southern Belgium.

- *Southern Western Europe*, also known as *Mediterranean Europe*, consists of the following: France, which almost completely surrounds the coastal microstate of Monaco; the Iberian Peninsula countries of Spain, Portugal and the microstate of Andorra; Italy, which also contains the microstates of the Vatican City and San Marino; Greece; and the island states of Malta and Cyprus. Most of the people of southern Europe speak Romance (or Latin-based) languages, with the major exceptions of Greece, the Basque area of Spain and the Celtic Breton region in westernmost France. Turkey is sometimes considered part of southern Europe as well.

Eastern Europe is generally divided into two political and geographical groupings. *EU Accession States* have either joined the EU since 2000 or have applied to join, and *EU Non-Accession States* have not applied to join or are only at the very earliest stages of discussion with the EU. Because the status of EU accession is subject to change, the discussion of Eastern Europe in this chapter is valid only to the current edition of this textbook.

- *The EU Accession States* consist of Bulgaria, the Czech Republic, Hungary, Poland, Romania, Slovakia, Croatia and Slovenia, and the three *Baltic* states of Latvia, Lithuania and Estonia. The Baltic countries regained their independence in 1991, after the formal collapse of the Soviet Union in 1990. Estonia,

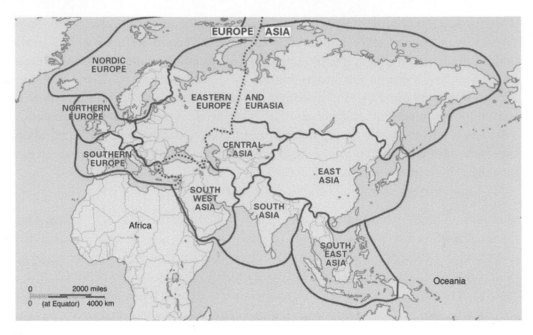

▶ *The Eurasian Land Mass and its Subregions. Western Europe includes the three regions of Nordic Europe, Northern Europe and Southern Europe.*

which has strong cultural, language and economic ties to Finland, has also attempted to portray itself as a Nordic country, although it is not formally a part of the Nordic group and the Nordic Council, which is the leading political body. The Czech Republic and Slovakia were formerly the country of Czechoslovakia, but they separated peacefully in 1993. Historically, the accession countries have had closer ties to Western Europe than have other Eastern European countries.

Bulgaria, Romania and Slovenia are also a part of the *Balkan Peninsula*, along with Albania, Serbia, Montenegro, Croatia, the Republic of Macedonia, Kosovo and Bosnia-Herzegovina. These countries comprise the southern part of Eastern Europe. Greece is also physically part of the Balkans but was never a State Communist-ruled country as were the others. So for cultural and political reasons, including its earlier membership in the EU, Greece is normally discussed separately from the other Balkan countries.

■ *The EU Non-Accession States* consist of Albania, Armenia, Belarus, Bosnia and Herzegovina, Georgia, Kosovo, the Republic of Macedonia, Moldova, Montenegro, Russia, Serbia and Ukraine. These countries were either formerly a part of the Soviet Union or the Republic of Yugoslavia, which fell apart in the early 1990s after the fall of communism. Albania, Iceland and Macedonia have formally applied to join the EU, while Montenegro, Serbia and Turkey are in negotiations with respect to EU entry. Ukraine has also sought closer economic and political relations with the EU although this has led to a fraught relationship with neighboring Russia as well as the significant Russian population in Eastern Ukraine. The Eurasian countries of Armenia and Georgia are also EU non-accession states, although their location in the *Transcaucasia* region means they are far removed from the core of Western Europe.

Physical Geography of Europe

The Greek geographer Strabo (63 BCE to 24 CE) described Europe as a very irregular peninsula. He was referring to the *interpenetration* of land and sea on three sides of the European continent. Europe is a peninsula, but it also has many smaller peninsulas and islands extending out from its coastline. Conversely, Europe has many small seas and bays that extend between these land extensions.

One way of conceptualizing the physical form of a landscape was proposed by the geographer William Morse Davis (1850–1934). Although a simplification, his approach does help in understanding Europe's diversity. Davis suggested that landform can be divided into the categories of *youth*, *maturity* and *old age*. Youthful landscapes are areas of recent and new mountain building, and they mostly consist of high mountains and deep valleys (which is seen in southern Europe). Mature landscapes usually have lower hills that long ago stopped growing and have since eroded considerably (seen in the middle of Central Europe). Old age landscapes consist of flat, low-lying plains that were once high mountain systems but have since been completely eroded (seen in northern Europe, south of the mountains of Scandinavia).

Waterways

The two largest rivers of Europe are the *Rhine* and the *Danube*. If Europe were extended to include all of the land area to the Ural Mountains, then the *Volga River* in Russia would also be one of the largest rivers of Europe. (Russia will be discussed in more detail separately in the Eastern Europe section.) Smaller rivers that are also very important include the Maine River in Germany, the major tributary of the Rhine; the Po River in Italy; the Seine River in northern France and the Rhone River in Southern France; the Elbe River in former East Germany; and the Vistula River in Poland.

Europe also has several very important canals. The *Mittelland Canal* cuts across the *North Europe Plain* from east to west. The *Rhine-Maine-Danube Canal* is an 80 mile-long (130 km) canal in southeast Germany that connects the Rhine River system to the Danube River system, allowing ships to travel from the Black Sea to the North Sea. The Emperor Charlemagne (742–814 CE) first attempted to construct this canal in 793 CE, but it did not open until July 31, 1992. The two combined river systems traverse a length of 2,170 miles (3,500 km), which is about 175 miles (280 km) longer than the Mississippi River in the United States. It travels through ten different countries and provides a significant route for tour boats of various types, from upscale canal barges to small cruise ships and day ferries.

Climates of Europe

Climate refers to the average annual pattern of weather in a place or region. There are two primary factors that influence the climate of a place. The first is its land features, including the place's location relative to oceans and other large bodies of water. The other major factor that influences climate is the atmosphere, primarily the *high and low pressure air masses* that circulate around the globe (see chapter 1).

The major locational and terrestrial features of Europe that affect its climates include the following:

1. Europe is located in a *temperate zone*. This means that Europe is located between the *Tropic of Cancer* and the *Arctic Circle*. It receives sunlight all year because it is mostly below the Arctic Circle. However, the sun is never directly overhead because it is north of the Tropic of Cancer. The very northernmost part of Europe is subarctic, while the southernmost part of Europe is subtropical.
2. Europe is located on the eastern side of the *Atlantic Ocean*, and it also has large bodies of water on its north and south sides. Because the temperature of large bodies of water changes little from one season to the next, the Atlantic Ocean, the North Sea, and the Mediterranean Sea have a moderating influence on the climate of most of Europe. In addition, the northern Atlantic Ocean contains the *Gulf Stream*, which is also known as the *North Atlantic Drift*. The Gulf Stream brings warm water from the Gulf of Mexico across the Atlantic directly into the English Channel and the North Sea and up to the island group of Svalbard, which lies north of the Norwegian mainland. This warm water keeps these water bodies ice free most of the time, which is unusual at such a northerly latitude. Without the Gulf Stream the climate of northern Europe would be more akin to that of British Columbia, Canada.

▶ *The surface physiography of Europe. This image shows the complex mountains systems of the young Alpine Orogeny in southern Europe around the Mediterranean Sea, the older mountain systems of middle and northern Europe and the European Plain from the shallow North Sea and Baltic Sea and across to the Ural Mountains in Russia.*

3. Europe is mostly influenced by winds and air masses that flow from west to east. In the temperate zones of the earth air masses and winds move from west to east. For Europe, this pattern brings moist and relatively warm (influenced by the Gulf Stream) ocean air to most of the European continent, north of the Alps. This results in higher land temperatures in the winter than one would normally find this far north.

4. Europe is located on the western edge of the world's largest land mass, the Eurasian continent. Large land masses (such as continents) gain heat in the summer and lose heat in the winter at a much faster rate than do large water bodies, such as oceans. In the winter, the north-central portion of the Eurasian landmass is the coldest place on Earth. In winter, bitterly cold and dry continental air comes out from this area and dominates the climate of Russia. It also extends into Eastern Europe and occasionally into Western Europe.

5. Europe does not have a continuous north–south mountain range with the exception of the Ural Mountains in Russia, which are often considered the natural boundary between Europe and Asia. Unlike North America and South America, air masses coming off the Atlantic Ocean do not come up against a large north–south trending mountain range that both slows them down and changes their moisture content. Instead, air masses can extend considerably far inland before their ocean influences dissipate. When the Atlantic Ocean air systems weaken, dry continental air masses (hot in summer, cold in winter) from Russia can similarly extend westward into northern Europe.

6. Europe has an almost unbroken east to west mountain chain in its south-central region. The mountain ranges, starting with the *Pyrenees* in the west and extending through the *Alps* and the *Carpathians*, form a sharp divide between the climate zones of northern and southern Europe. In particular, the mountains prevent large scale interchanges (mixing) between the warmer *tropical air masses* to the south and the cold *polar air masses* to the north.

Air Masses

Air masses from the far North Pole are either *Arctic* (from the Arctic Ocean) or *Polar* (from the far north). Air masses from areas to the south of Europe are *tropical* (originating from between the tropics of Cancer and Capricorn) and very warm. Air masses that pass over an ocean before reaching Europe are *maritime* (coming from large water bodies) and generally bring humidity and precipitation. Those that originate over land before reaching Europe are *continental* (originating over land masses) and tend to be very dry. There are five major air masses that affect the climate of Europe.

1. The principal air mass affecting most of Europe is the *polar maritime air mass* that originates over the North Atlantic Ocean and brings cold moist air to northern and central Europe through much of the year. This air mass affects central Europe more in the summer than in the winter. In the winter, it can sometimes migrate south of the Alps into the Mediterranean region.
2. In the winter, an *arctic maritime air mass* that is very cold and dry (because it is so cold) moves southward, mostly affecting the northern portions of Europe, but occasionally extending much further south.
3. Eastern Europe is mostly affected by a *polar continental air mass* from the interior Eurasian continent. This very dry air mass is also very cold in the winter, but relatively warm in the summer. It can sometimes move further westward into central Europe.
4. Mediterranean Europe is mostly affected by two air masses. The first is a *tropical maritime air mass* that originates over the Atlantic Ocean and is warm and moist, largely affecting Portugal and the Atlantic coast of France.
5. The second system affecting the Mediterranean area is a *tropical continental air mass* that originates over Africa and brings hot and dry air to the Mediterranean region in the summer, especially its eastern portions.

Seasons

In the winter, all of the air masses move southward, while in the summer they all move northward, which causes the seasonal weather patterns to change. In the summer, the Arctic and maritime polar air masses move northward and the temperature differences between northern and southern Europe are more moderate, although there are growing concerns that the regularity of jet stream movements (which move air masses from west to east) are being affected by climate change, making seasonal weather much more unpredictable.

Winds from the tropical maritime and tropical continental air masses come to dominate southern Europe and occasionally areas north of the Alps in the summer. This gives southern Europe its famous *Mediterranean Climate*, in which rainfall occurs in the winter, while summers are hot and dry. Other large areas in the world that have this type of climate include California in the United States, south west South Africa and south west Western Australia.

In the winter months, especially, the difference in temperature between northern and southern Europe becomes quite extreme. Northern Europeans seek relief from the cold weather by spending their holidays in the warmer climates of Southern Europe, similar to the so-called *Snowbird* pattern found in North America.

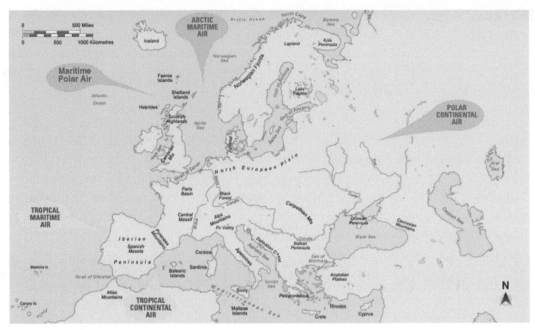

▲ *Physical Geography of Europe*

Three Climate Regions

The result of the seasonal changes in these air masses is that Europe has three major climate regions. The climate of northern Europe is a *maritime Marine West Coast climate* (which is also found in the Pacific Northwest of the United States and in British Columbia in Canada). It is mostly affected by low-pressure systems (cold fronts) bringing storms off the Atlantic Ocean. The precipitation occurs all year round, with the higher amount in winter. Northern Europe receives over 30 in. (76 cm) of rain a year, and snow is relatively rare at sea level, though it can be quite considerable at higher elevations. The region has high humidity, a lot of clouds and considerable fog in the winter.

Mediterranean Climate

Southern Europe has a subtropical and *Mediterranean climate.* In the summer, the region is dominated by a high pressure air system (from Africa), which is dry and hot. In the winter, low-pressure systems from the North Atlantic occasionally migrate south below the Alps. Summers are hot, with an average above 80°F (27°C). Winters are cool to cold, but almost always above 32°F (0°C). Because air masses tend to move from the west to the east, coastlines that face the west are much more humid than coastlines on the eastern side of peninsulas and islands. The mild and warm year round temperatures of southern Europe have made it one of the most attractive beach destinations in the world.

Continental Climate

Eastern Europe is dominated by a *Continental climate,* consisting of air masses that originate over the interior of the Eurasian continent. Summers tend to be warm and dry, though maritime air masses from the North Atlantic and moisture from the Mediterranean Sea and Black Sea can also bring rainfall during the summer months. In the winter, the region is dominated by a very cold and dry air, which often results in a permanent snow cover throughout the winter months. Eastern Europe is often colder in winter than

ISSUES AND INSIGHTS ◖ The Threat of Climate Change and GGEs

Climate change has become a major contemporary environmental concern. The Fifth Assessment Report of the Intergovernmental Panel on Climate Change (in 2013–14) projected that:

- The frequency of hurricanes and storms will increase dramatically.
- Snow will disappear from all but the highest mountains.
- Deserts will spread.
- Oceans will become acidic, leading to the destruction of coral reefs and atolls.
- Deadly heat waves will become more prevalent.
- Sea levels will rise over the century by around half a meter.

Climate change issues and their significance will be noted throughout this book. Sea level change will have a significant impact on low-lying areas—such as those of the Netherlands, northern Belgium, eastern Denmark and the German North Sea coast and the east of England (East Anglia and Kent). Just a 3.3 ft (1 m) rise would threaten one-third of the world's agricultural land, while a 6.6 ft (2 m) would make the Thames flood barrier that protects London useless and 13 ft (4 m) would drown the city of Miami, leaving it 37 miles (60 km) from the nearest continental coast. A worst-case scenario would see a return to conditions that existed about 14,000 years ago, when sea levels rose 45 ft (3.5 m) – the height of a three-story house – in the space of about 300 years. Such scenarios would have enormous implications for coastal communities.

However, recent research indicates that climate change may have other effects upon the environment and upon human beings in ways that were not previously recognized or thought related, including increases in *geologic disasters*, such as volcanic activity, earthquakes, and submarine (ocean floor) landslides that result in tsunamis. For example, following the retreat of continental glaciers around 18,000 yrs ago, which brought sea levels back up to where they are today, a 300 percent increase in volcanic activity occurred in the Mediterranean.

Even small changes in sea level can make a difference. For example, eruptions of the active Pavlof volcano in Alaska tend to occur during the winter months when, for meteorological reasons, the regional sea level is barely 12 in (30 cm) higher than during the summer. If other volcanic systems are similarly sensitive, then there would be a burst of volcanic activity as *anthropogenic* (human-caused) climate change drives sea levels upwards.

Such changes occur because the addition of the extra depth of water is often sufficient to load and bend the underlying crust. This in turn acts to squeeze the magma beneath the crust and may prove significant in triggering responses from volcanoes that are already close to erupting—and perhaps not just from volcanoes. Loading of the continental margins could also activate faults, triggering increased numbers of earthquakes, which in turn could lead to submarine or coastal landslides and the potential of greater tsunami activity.

Global Geophysical Events (GGEs) are what we call global magnitude events that, although infrequent, occur with catastrophic results. Such a situation is believed to account for the gigantic underwater Storegga Slide, when an area the size of Iceland sloughed off the Norwegian coast around 6100 BCE, sending a tsunami more than 20 m (66 ft) high in places across the Shetland Isles and onto the east coast of Scotland.

In Scotland, traces of the subsequent tsunamis have been recorded through deposited sediment discovered in the Firth of Forth, 80 km (50 miles) inland and 4 m (13 ft) above normal tide levels. The western flank of the volcano of Cumbre Vieja, on the western Canary Island of La Raima, which has already slid 4 m toward the sea, appears vulnerable to future events. If this happens, a mass of rock that may be as large as the Isle of Man in the Irish Sea may slide into the ocean and threaten the entire North Atlantic margin with a megatsunami. Is this the scenario the world faces as the ice of Greenland and the Antarctic starts to melt?

Sources

Bondevik, S., J. Mangerud, S. Dawson, A. Dawson, and Ø. Lohne. 2003. Record-breaking height for 8000-year-old tsunami in the North Atlantic. *Eos, Transactions, American Geophysical Union*, 84(31): 289–293.

Intergovernmental Panel on Climate Change (IPCC). 2013. Summary for Policymakers. In *Climate change 2013: The physical science basis. Contribution of Working Group I to the Fifth Assessment Report of the Intergovernmental Panel on Climate Change* [Stocker, T.F., D. Qin, G.-K. Plattner, M. Tignor, S.K. Allen, J. Boschung, A. Nauels, Y. Xia, V. Bex & P.M. Midgley (eds.)]. Cambridge Cambridge University Press.

Scandinavia, which is much closer to the North Pole. Continental climates are characterized by large seasonal fluctuations in temperature, and Eastern Europe is no exception to this. Low temperatures in winter are often 0°F (-17°C), while summer highs can reach over 90°F (32°C).

In between these three climate zones are *transition areas*. In these areas, weather is often unpredictable and changeable, as they can sometimes be affected by the Continental climate to the east and north, the Marine West Coast climate to the west or the Mediterranean climate to the south.

Human Geography of Europe

Europe is a major center of culture, education and history that has had global influence through colonial ties, migration, education systems and the media. Because of this, much of the tourism in Europe is culture-based. Knowledge of the cultural and historical geography of Europe is necessary to appreciate fully the experience that Europe has to offer.

Ancient Greece

The first major European civilization that historians usually recognized is that of the Greeks. Early Greek civilization was composed of independent *city-states* (cities that were also countries) scattered throughout the Mediterranean Sea, Aegean and Black Sea area but centered on Greece. They were tied together by their culture, language and history. Greek culture rose about 1000 BCE, and peaked in about 500 BCE. The Greeks were conquered by the Romans in 147 BCE.

Greek civilization laid many of the philosophical foundations for Europe. *Socrates* (469 to 399 BCE, who lived about the same time as *Confucius* in China) was a proponent of the idea that wisdom could be taught to anyone. His student, Plato, founded *the Academy* in 387 BCE, which could be considered the first university in the Western world. To the Greeks, humans were the measure of all things. Morality was based on decisions made by humans, not the gods. Greek gods were essentially mortals with superpowers — they were to be worshiped, but they did not have moral values. Socrates was put to death in 399 BCE for corrupting the young, which basically meant asking too many questions.

Greek City-States

In the Greek city-states, citizens were considered free individuals, although they were also all males, and most Greek city-states had more slaves than citizens. The Greek city-states experimented with different forms of government, including *tyranny*, which is a dictatorship by government; *aristocracy*, which is rule by the elite; *oligarchy*, which is rule by the few; and *democracy*, which is rule by the people (citizens). The Greeks also contributed major advances in *geometry*, which they used for navigating the Mediterranean Sea, and *medicine*, which produced the Hippocratic Oath to which doctors today still adhere. The Greeks also introduced the idea of the professional athlete, supported by city-state governments, in their *Olympic* competitions. Other contributions were made in the arts, architecture and education.

The Romans

Roman civilization lasted from about 500 BCE to about 500 CE. The Romans were the first to gain a true awareness of Europe as an entire continent. Through conquest they integrated much of northern Europe into the existing civilization of southern Europe, often destroying many of the existing regional cultures and religions, such as the *Druidic* traditions in Britain and France.

The Romans built a unifying road network throughout their European empire, which saw the blossoming of a major period of tourism and travel during the height of the Roman Empire. The Romans admired Greek philosophy and ideas, which they spread throughout Europe in the way they built their cities, laying out their streets in a grid pattern with a central plaza and public buildings. Greece was also a major *tourist destination* for the Romans, who purchased souvenir statues of Greek gods and models of the Acropolis to take back with them to Rome, much as tourists do today. The city of Troy, in present-day Turkey, was also a major tourist destination in Roman times (see below).

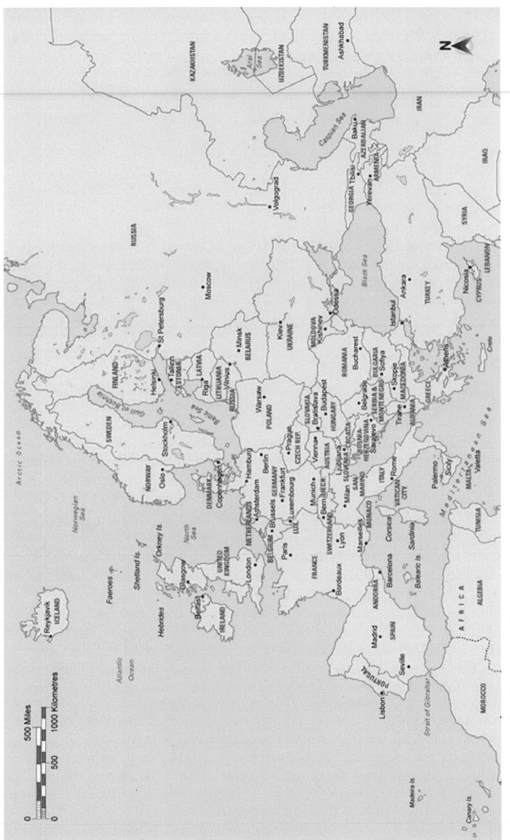

▲ *The Countries, Capitals and Major Cities of Europe*

► *Ruins of the Roman Forum in Rome, Italy, are evidence of the ancient Romans' ability to build affluent cities and influence the empire throughout Europe, the Middle East and North Africa.*

Photo: Dallen J. Timothy

Roman Contributions to Modern Europe

The Romans made major contributions to European traditions of law, public administration (including the concept of a Senate), military organization and communications. During the Roman period different parts of Europe came to specialize in different economic activities, and trade flourished along the Roman roads. *North Africa* was the major grain producing region for the Roman Empire. The Mediterranean islands were a source of iron ore. Spain was a major center for silver and lead mines, while Cornwall in southwest England was a source of tin. Further north, agriculture was the main industry. All roads from these regions led to Rome, which was the *central marketplace* for goods from throughout the empire.

Latin became the *lingua franca* throughout Europe. This was the first and only time that most of Europe spoke the same language, although the use of Latin was primarily among intellectuals and the upper classes. Interestingly, this situation is perhaps increasingly similar to that of English in present-day Western Europe. The Latin influence today is primarily seen among the *Romance languages*, including French, Italian (and Sardinian), Spanish (and Catalan), Portuguese and Romanian. All Romance languages descend from *Vulgar Latin*, which was the language of the soldiers and merchants of the Roman Empire. It was somewhat different from the Latin of the Roman elite. (In fact, the Roman legacy of language is such that in 1954 the *Latin Union* was established as an international association of nations that use a Romance language.)

Toward the end of the Empire the Roman Emperors also helped spread Christianity, often through force, throughout much of their empire. The empire's limits extended far north into Britain. Although most people associate the northern border of Roman Britain with Hadrian's Wall (near the present border of England and Scotland), there was actually a much further northern wall – the Antonine Wall – between the Firth of Forth and the Firth of Clyde. On mainland Europe the Roman control did not cross the Rhine River, on the other side of which were located the *Germanic tribes*. Trade relations, however, extended Roman influence beyond its political boundaries, as well.

Fall of the Roman Empire

The decline of the Roman Empire began about 200 CE through a combination of economic and environmental changes. A period of *hyperinflation* led to a collapse of the Roman currency and people returned to a barter society, which meant that Rome was unable to collect taxes. Necessary reforms did take place in the eastern Roman Empire, which was centered in *Constantinople* (modern-day Istanbul). This location saved it from the collapse of the Rome-based Empire during the fifth century CE.

Environmental change in North Africa, which was regarded as one of the breadbaskets of the Roman Empire, affected food supplies in other parts of the empire. Some historians also regard environmental change in northern and far eastern Europe as one of the factors that led to the migration of nomadic peoples, often referred to as the *Barbarian Invasions* or the *Great Migration*, into the Western Roman Empire. This is regarded as a significant factor that led to its eventual collapse, including the sacking of Rome in 410 CE. The latest date given for the end of the Western Roman Empire is 610 when the Emperor Heraclitus based in Constantinople readopted Greek as the language of Imperial administration. *The Eastern Roman Empire*, also referred to as the *Byzantine Empire*, survived until 1453 when Constantinople fell to the Ottoman Turks under Mehmed II.

Ancient Tourism

Although tourism is often viewed as a modern industry, dating either from the advent of the railway in the 18th century or the jumbo jet in the late 20th century, domestic and international leisure travel has in fact been taking place for at least 2000 years in Europe. In August 2007, archaeologists at a site in Turkey found a huge (4–5 m in height) carved statue of the Roman Emperor Hadrian who ruled Rome from 117–138 CE. It was about 5 m below ground, among the buried ruins of a bathhouse at Sagalassos, an ancient mountaintop town in southern Turkey. A major earthquake destroyed the bathhouse where the statue was found sometime between the late sixth and early seventh centuries CE. It is believed that the statue was likely erected to honor Hadrian as thanks for officially recognizing Sagalassos as the first city of the Roman province of Pisidia, and therefore making it the regional center for worship of the emperor. This status would have meant that the city attracted large numbers of visitors during imperial festivals, thereby boosting trade and making the town prosperous.

Troy

Sagalassos was by no means the only Roman tourism destination in present-day Turkey. Troy is famed in Greek legend for the Trojan Wars (sometime prior to about 1100 BCE) and is still today a major heritage destination, as was the city of Ilium in Roman times. Homer's epic, *The Iliad* provided the basis for contemporary cultural understandings of Troy as it is portrayed in books and film. Greek stories of also portray the role of Troy in the founding myths of a number of different European nations and their rulers. For example, in *The Aeneid*, the Roman poet Virgil traced the ancestry of the founders of Rome to the Trojan prince Aeneas. Because of this, Troy became a significant travel destination for the Roman aristocracy and wealthy middle classes, particularly from the time of Augustus (27 BCE) until about 200–300 CE. Tour guides would take them to see the locations where particular battles or scenes took place in exactly the same way that some present-day guides are doing with visitors.

The Great Migration

A period known as the Great Migration began in the fourth century CE. During this time Germanic-speaking tribes crossed the Rhine River and spread throughout Europe. The *Saxons* settled in northern France and later migrated to England along with the *Angles*. The *Franks* settled in France and Belgium. The *Lombards* settled in the Po Valley of northern Italy. The *Alemanni* settled in northern Germany. The *Burgundii* settled in the upper Rhone River Valley. The *Goths* and the *Visigoths* migrated to the Iberian Peninsula, and the *Vandals* migrated to northern Africa. In 410 CE the Goths and Vandals sacked Rome, marking the beginning of the end of the Roman Empire.

Various theories have been provided about the reason for these migrations, but the most likely explanation appears to be a combination of the decline of the military and political strength of the Roman Empire and *environmental change* in much of northern and eastern Europe. As these Germanic-speaking tribes moved north, west and south, *Slavic-speaking* people from further east moved into modern-day Eastern Europe. Slavic languages today dominate the region from the Balkan Peninsula in the south to Poland in the north, and east to Russia.

Each of the Germanic tribes established a mini-state of its own, leading to the eventual collapse of the *Western Roman Empire* Europe, and initiating the *Middle Ages*. The Middle Ages lasted from approximately 500 CE to about 1500. During this time, Europe was divided, wars were frequent and feudalism prevailed. Most of Roman high culture was lost within Europe, as travel in Europe was difficult and learning was suppressed, although Christian scholars maintained its legacy in isolated monasteries during this period. Contemporary scholars usually identify three distinct periods within this period of European history:

- *The Early Middle Ages* (500–1000 CE) is also often described as the *Dark Ages* or the *Viking Ages* because of the extent to which Roman territory was invaded by various tribes early in the period. The Scandinavian Vikings began to exert their influence from 700 CE on in a series of invasions and raids throughout Europe. Trading links broke down and famine was common as a result of the Dark Ages' cold period, which was associated with a decline in agricultural production. This period also saw a new challenge to European culture with the emergence of Islam in Arabia and the growth of the *Arab Empire* (also known as the *Arabian Empire*). Most of the Iberian Peninsula of Spain and Portugal came under Islamic control in the eighth century, leaving a legacy that lasts to this day in southern Spanish culture.

- *The High Middle Ages* (1000–1300 CE) marks a period when the last of the great migrations was undertaken, although war and invasion were still commonplace. For example, the *Norman invasion* of Anglo-Saxon England in 1066 and Ireland in 1169. This was also the time of the *Catholic Inquisitions*, which are associated with the period in the public imagination. Although inquisitors had existed in the Church since Roman times, the Episcopal (1184–1230s), Papal (1230s) and *Spanish Inquisitions* (established 1478) used particularly terrifying methods to find and destroy what the Church judged as heretical mass movements. For example, the Cathars and Waldensians in France and the Jews in Spain were often subject to the acts of the Inquisition, with many being massacred. Ironically, such events have served as tourist attractions in recent times with the Cathars being featured in tourism promotions for Carcassonne in France and profiled in a number of publications and television programs that focus on the Knights Templar and the Holy Grail.

- *The Late Middle Ages* (1300–1500 CE) began with the Great Famine of 1315–1317 and the *Black Death*, which dramatically reduced the population of Europe. Although the period was marked by great scientific achievements, it was also characterized by crises within the Catholic Church. During the Papal schism of 1378–1417 there were multiple claimants to the papacy. There were also the reform movements associated with John Wycliffe (England), John Calvin (France and Switzerland), Jun Hus (Czech) and Martin Luther (Germany). A number of other church and pilgrimage-related aspects of the High and Late Middle Ages also serve as present-day tourism attractions. Many of the great cathedrals of Europe date from this period, and a number of pilgrimage routes remain in use and are arguably more popular than ever as heritage trails. Some of the best known examples include the *Pilgrim's Way* from the cathedral city of Winchester in Hampshire, England, to the shrine of Thomas Becket at Canterbury in Kent; and the *Camino de Santiago*, also known as Route of Santiago de Compostela, or by the English name the Way of St. James. Also significant are the many walled towns of Europe, along with art and some of the first travel accounts, such as the works of Marco Polo and William of Rubruck. The initial sailing of Christopher Columbus to the New World in 1492 is often seen as a critical date marking the end of the Middle Ages.

◀ *The ruins of the Byzantine-era Basilica of Agios Achilleos on a small island in Lake Prespa, Greece, testify of the inseparability of religion and political power during the Byzantine Empire.*

Photo: Dallen J. Timothy

The Byzantine Empire

The cultural contributions of the Greeks and Romans were mostly preserved through the Middle Ages in the *Eastern Roman Empire,* also known as the Byzantine Empire. They were also preserved in the Arab Empire, which was centered in Baghdad and was possibly the world's largest and most advanced civilization of that time period. (The Arab Empire is discussed in more detail in Chapter 3.)

The Roman emperor, Flavius Valerius Aurelius Constantinus, or Constantine I (c. 274 CE to 337), made Byzantium the center of the Roman Empire in 330 CE, changing its name to *Constantinople.* The Byzantine Empire lasted until 1453 CE, when it was conquered by the Ottoman Turks, who changed the name of Constantinople to *Istanbul.* During the Dark Ages, Constantinople was the foremost city of trade and culture in Europe. The people of the Byzantine Empire called themselves Romans, and they called their territory the *Romani Empire.* They copied Roman and Greek art and literature, and Constantine made Christianity the state religion (in 324 CE). They brought orthodox Christianity to the Slavic peoples to the north and introduced the *Cyrillic alphabet* throughout their territory. They also transmitted Latin and Greek cultural traditions to the new Germanic rulers of Western Europe. While the rulers of Byzantium sought to replace Rome as the center of the old Roman Empire, they were never able to hold onto the more distant territories of Europe and Africa.

Unlike the Roman Empire, where church and state were largely separate, in the Byzantine Empire the emperor came to be considered God's representative on Earth. The defeat of the emperor would only occur by the will of God. Extremely elaborate palaces were built to impress visiting barbarians with Byzantine wealth and power. Today, Byzantine architecture can be found throughout much of southern Europe. The city of *Venice,* for example, is architecturally a Byzantine city.

European Re-emergence

The Byzantine Empire protected the eastern portions of Europe from Muslim invaders during the Dark Ages, allowing a new Europe to develop from the ruins of the Roman Empire. The fact that Europe faced a powerful Muslim empire to its south and east further contributed to a sense that Europe was a territory separate from the rest of the world. This was especially true after the Germanic tribes also adopted

Christianity. Aethelberht of Kent first became a Christian in Anglo-Saxon Britain in the early seventh century, followed by *Charlemagne*, the King of the Franks, in the eighth century.

The Nation-State

The concept of the nation-state gradually developed in Europe during the Dark Ages. A different concept than the idea of a country, a *nation* is a territory that shares a common language, history, politics and culture. The word *state* refers to a politically independent and sovereign territory. As culturally based nationalistic identities increased in Europe, the anarchy of the great migration and the subsequent divisions of feudalism gradually came to an end. *Feudal lords* gradually lost their power and independence as they came under the domination of increasingly powerful nation-state kings. Prior to this period, the king was just another feudal lord who ruled his own domain and had a slightly different title than the other feudal lords.

Merchants

The rise of the nation-state took several centuries and was the result of several different processes. One was the rise of a new merchant class. These were basically landless (and therefore, homeless) peasants who became traders and crafts-people and migrated from one village to the next. As their numbers grew, merchant towns with artisans and crafts-people began to appear, especially along the major rivers. The *mercantile class* formed a middle class between the lords and the peasants. The mercantile towns also served as centers for the sale of excess agricultural products. The sale of these products provided surplus wealth that enabled peasants to purchase crafts produced in the towns, as well as imported goods from other places. Some of these mercantile towns became university towns, signaling a revival in learning.

Kings

An increase in the ability of people to travel in Europe also signaled the decline of feudalism. This began with the *Crusades* (1095–1270), which introduced people to different parts of Europe and to places beyond Europe. New ideas and new products were brought back to Europe as a result of the Crusades. Townspeople lived outside of the feudal lord system. As such, they looked to the regional king for their security. The king gained his wealth from taxing the townspeople, so the wealthier the townspeople became, the wealthier and more powerful the king became. With time, the king's military became much stronger than that of the lords.

France and the Hundred Years' War

The Hundred Years' War between France and England (1338–1453) resulted in even stronger kings seeking to expand their territory, as well as the rise of national heroes and symbols. *Joan of Arc* was one such hero. She fought against the British in the Hundred Years War and was burned at the stake after her capture in 1430. The Hundred Years War also resulted in the gradual unification of France by the late 1400s (along with the final collapse of feudalism).

France is often cited as the best example of the nation-state. It was a political territory that shared a common language, history and culture. *Spain* became a nation-state in 1469 when the marriage of Ferdinand and Isabella united the Spanish-speaking territory of Iberia. *England* was united in 1485 after the War of the Roses was fought among rival feudal lords. Similar processes occurred in the Netherlands and Portugal. Italy and Germany, however, were not unified as nation-states until the 1860s and 1870s.

Constitutional Monarchy

Another state in the political development of Europe developed when nation-state leadership changed from a strong monarchy to a constitutional monarchy and direct democracy. The earliest constitutional monarchy form of government appeared in England in 1215 with the signing of the *Magna Carta*, which was the first document imposed upon the English monarch by a group of his subjects, the feudal barons, in an attempt to limit his powers by law and protect their rights. The Magna Carta proved to be important elsewhere in Europe, but was even more significant in the colonies of the United Kingdom, such as Australia and the United States, whose constitutions were based upon British law. The first *constitutional democracy* in Europe was the short-lived Corsican Republic of 1755. For Eastern Europe, limitations on the absolute rule of monarchy did not occur until after World War I (1914–1918). However, Iceland was the first European country to have a parliament with the *Alþing* ('Althing' in English) established in 930.

The European Renaissance and Colonialization

The year 1492 is often used to mark the end of the Dark Ages. This was the year Columbus sailed to the Americas and the *Muslim Moors* were pushed out of Spain. It was also the middle of the European Renaissance (1300s through the 1500s). The European Renaissance began in southern Europe and was marked by artists and thinkers such as Michelangelo, Raphael and Leonardo da Vinci. By the end of the 15th century, European explorers were crossing the Atlantic to the Americas and rounding the Horn of Africa to Asia. Much of this period of overseas colonization was dominated by Spain and Portugal.

By the 16th century, the center of European culture and economic dominance had shifted to northern Europe, especially France, Belgium, the Netherlands and England. In England, this was the time of William Shakespeare, Sir Francis Bacon and Thomas Moore. With the growth of an increasingly more powerful mercantile class of traders and sailors, *economic nationalism* came to be the predominant policy of the nation-states in Europe. What was good for the traders and merchants was good for the country and the king. The military power of the king was therefore used to control or dominate trade, often by directly conquering and colonizing distant lands.

The Grand Tour

The growing economic wealth of Europe resulted in rising standards of living and advances in education. It was also a period in which travel became more commonplace for the young men of the aristocracy as the *Grand Tour* through Europe came to be seen as an important part of the education process. But Europe was still largely an agricultural society, with some large cities based primarily on foreign trade. The only industrial activities were craft guilds, not factories.

The Industrial Revolution

The year 1869 was another important year of change for the European continent. That was the year in which the *Suez Canal* opened, giving European ships access to Asia without having to circle the entire continent of Africa. This was also a time when the *Industrial Revolution* reached its peak in Europe. (The Industrial Revolution began in England in the late 1700s and was soon exported to northern Europe.) The result of these changes was a huge expansion in Asian and African colonization by European countries after 1869 as Europeans sought to open both new markets and develop resources.

During the Industrial Revolution, the steam engine freed small-scale industries from river locations and enabled the building and use of increasingly larger machinery. The *mass production* of large-scale iron products and machinery, and finished goods created from raw materials shipped from colonies, made Europe the wealthiest region in the world. Europe experienced a *population explosion* at this time, as birth rates increased and death rates declined despite the high health hazards in the industrial cities of Europe. Large machinery enabled fewer people to work larger tracts of farmland, which resulted in massive migrations from the countryside to the cities. By 1800, England had the largest cities in the world. However, those cities were also badly overcrowded, with poor housing and sanitation, child labor and slavery, and dangerous working conditions.

Modern Travel

Nevertheless, this period marked the beginning of modern tourism in several ways. First, the invention of the *steam engine* allowed the development of the railway and the steamship, which revolutionized travel. Second, the industrial revolution provided for the emergence of a new middle class that had income and the desire for travel. Higher levels of education and improvements to the printing press also meant that it was much easier to promote leisure travel. Third, the transport system and time became standardized, thereby allowing for scheduling and, along with the invention of the telegraph, dramatically improved the commercialization of transport.

Finally, the *Romantic Movement* grew as a reaction to the conditions of many of the industrial cities, led by poets such as Coleridge and Wordsworth. This meant that mountainous and wild countryside became desirable to visit for the first time ever in British culture, making such places as Switzerland, Scotland and Norway attractive to visitors from industrial Europe. This artistic and intellectual development was mirrored elsewhere in Europe. Significantly, in German Romanticism as well, a notable early expression of which was the work of the writer Goethe, travel, nature, and ancient myths were all important motifs.

International Migration

England used its vast colonial territories, including Australia, North America and countries of the developing world, as sources of raw materials and as outlets for immigration out of England to relieve the population pressures there. Many Americans can trace their ancestry to the 1800s when large numbers of Europeans migrated to the United States. This migration began with immigrants from the British Isles at the start of the Industrial Revolution. Subsequent immigrants to the United States came from areas of Europe as they were influenced by the Industrial Revolution. Northern Europeans from the Netherlands and Germany migrated to the United States in large numbers in the mid-1800s, while southern and eastern Europeans migrated in the late 1800s and early 1900s. (After World War I, Europeans accounted for only a small percentage of immigrants to the United States.) The flow of migrants, as well as the mercantile connections of empires, also provided a basis for international travel and tourism, and by 1872 Thomas Cook was promoting his first World Tour.

Higher European standards of living have led to smaller families as birth rates have decreased to match death rates. Many European countries today are experiencing negative population growth because birth rates are much lower than death rates. However, as a result of their past colonial ties, most European countries today also have large immigrant populations, whose growth rates are higher than those of the native Europeans. Turks went to Germany, Algerians went to France, immigrants from Suriname and Indonesia went to the Netherlands, and England welcomed large numbers of immigrants from India, the Caribbean, South Africa and other former British colonies. The legacy of European history can clearly be seen in such changes, and a new stage of travel patterns and connections are now being formed as migrant groups seek to stay in contact with one another the way they have done for hundreds of years.

Demographic Shift

In addition to migration, the other major demographic factor that will affect tourism is the aging of the world's population. This substantial *demographic shift* has occurred because of dramatic improvements in healthcare and a decline in the birth rate in the developed world. Predictions are that this will continue well into the coming centuries so long as the world's resources are able to support the increases in population. The United Nations Population Division estimated that at the end of the 20th century, 11 percent of the world's population was aged 60 and above. By 2050, it is estimated that 20 percent will be 60 yrs or older. By 2150, approximately one-third will be 60 yrs or older. Just as significantly, the older population itself is aging. The increase in the number of very old people (aged 80 yrs and older) between 1950 and 2050 in the world is projected to increase by a factor of between 8 and 10 times the present amount. As well as a general aging of the world's population, there are also substantial regional differences in the aged population. For example, currently 20 percent of Europeans are 60 yrs or older, but only 5 percent of Africans.

Retirement Migration

Given that the majority of the world's international tourists come from developed countries, an aging population will clearly have substantial implications for the tourism industry. Not only may particular types of tourism favored by older travelers, such as cruising, continue to grow in popularity, but *second homes* and retirement homes and the provision of health facilities for retirees will likely become increasingly important in the development strategies of some destinations. For example, the following areas are already subject to substantial seasonal and permanent retirement migration, much of which is international: areas of Mediterranean Europe (Italy, Greece and Cyprus) and the Iberian Peninsula (Portugal, Spain); the southwest United States and Florida in North America; and some parts of coastal Mexico.

Population Change

A further influential factor with respect to demographic change and tourism is the estimation that among the major industrialized countries only the United States is expected to have significant population growth by 2050. The United States is expected to have reached a population of 420 million by 2050, an increase of 43 percent. But Europe is expected to have 60 million fewer people than today, and some countries could lose more than a third of their populations. Japan may shrink in size to approximately 100 million people by 2050. Over the same period Eastern Europe is also predicted to experience major population loss. Bulgaria is expected to lose 38 percent of its people, while Romania could have 27 percent fewer and Russia 25 million fewer people. Germany and Italy are expected to shrink by approximately 10 percent.

Although the world's developed countries are expected to grow in total population by about 4 percent to over 1.2 billion, the population in developing countries is predicted to grow by up to 55 percent to more than 8 billion. Under this scenario Western Asian nations are expected to gain about 186 million people by 2050 and Sub-Saharan African countries more than one billion people. By 2050, India will be the largest country in the world, having surpassed China in population growth.

2.2 | WESTERN EUROPE TODAY

Current Western Europe Issues and Resources (http://wrgeography.com/westerneurope.html)

Tourism Trends in Europe

With over 560 million international arrivals annually, Europe has more than 50 percent of the market share of worldwide tourism. However, Europe's global *market share* is slowly declining (Europe accounted for 60 % of international arrivals back in 1990, or 261 million international arrivals), mostly because of growth in tourism to Asia. Europe, however, will probably continue to be the leading, international tourist region in the foreseeable future, largely because it is so easy to cross *international borders* there and because of its historic ties to many corners of the world.

Several of the world's usual top ten tourism destinations by arrivals (France, Spain, Italy, the United Kingdom, Turkey, Russia and Germany) and by receipts (Spain, France, Italy, Germany, the United Kingdom, Turkey and Austria) are either wholly or part in Europe. Although Europe's most important long-haul markets in North and South America currently account for just over half of total overseas arrivals, numerous studies are already pointing towards a significant shift in the overseas source markets for Europe towards the emerging economies of Russia, China, and India as well as other Asian countries.

Tourism is the third largest socio-economic activity in the EU after the trade and distribution and construction sectors. In 2010, more than one in seven enterprises in the European non-financial business economy belonged to the tourism industries (the term used by the EU for statistical purposes to describe the range of businesses that supply services to tourists). These 3.4 million enterprises employed an estimated 15.2 million persons, this represents 11 % of the persons employed in the non-financial business economy and 29 % of persons employed in the services sector in the EU.

However, the tourism industries' shares in total turnover and value added at factor cost were relatively lower than their employment share, with the tourism industries accounting for 6 % of the turnover and 9 % of the value added of the non-financial business economy. The location of tourism businesses in the EU reflects the concentrations of tourists. Out of the 3.4 million enterprises in the tourism industries in the EU in 2010, 56 % were located in four member states: Italy, Spain, France and Germany.

The European Union

The European Union (EU) is comprised of 28 countries, and their combined economy is the largest in the world by nominal GDP, if the EU were treated as an individual state. The EU has a single economic market between its members, a common trade policy, a common agricultural and fisheries policy, and an extensive series of regional development policies. Although the EU has many of the trappings of a country, such as the *European Parliament*, the European Court of Justice and the European Central Bank, it is important to recognize that the EU is a supranational body with control only over a number of policy areas that its member states have acceded to. For example, although the *euro* is a common European currency, it has only been adopted by 18 member states. Accordingly, a number of member states, including the United Kingdom, Denmark and Sweden, have retained their own central bank. In addition, Monaco, San Marino, the Vatican City and Andorra have formal agreements with the EU to use the euro as their official currency and issue their own coins, while Kosovo and Montenegro have adopted the Euro unilaterally but are not formally part of the Eurozone.

Joining the EU

To join the EU, a country needs to fulfill economic and political conditions called the "Copenhagen criteria" (after the summit at which they were confirmed in June 1993). "Membership requires that a candidate country has achieved stability of institutions guaranteeing democracy, the rule of law, human rights, respect for and protection of minorities, the existence of a functioning market economy as well as the capacity to cope with competitive pressure and market forces within the Union. Membership presupposes the candidate's ability to take on the obligations of membership including adherence to the aims of political, economic, and monetary union."

Under the *Maastricht Treaty*, each member state and the European Parliament must agree to any enlargement. Croatia was the most recent country to join the EU in 2013. As of early 2014, Iceland, Macedonia, Montenegro and Turkey are all official candidates, while Albania and Serbia have applied for membership. Bosnia and Herzegovina has concluded an association agreement, and Kosovo is beginning negotiations for an agreement.

Creating the European Union

The formation of the EU has taken several decades and reflects the desire of European countries to work together toward economic, social and political goals. Established in 1957, the founding members were Belgium, France, Germany, Italy, Luxembourg and the Netherlands. In 1973 it was expanded to include

Photo: Dallen J. Timothy

◀ *Although there is a fairly strong movement in Iceland to join the European Union, there is also a strong voice against it, as this billboard outside of Reykjavik illustrates. It says 'No thank you, to EU membership.*

Denmark, Ireland and the United Kingdom. Greece joined in 1981 and Portugal and Spain in 1986. Other Western European nations to join included Austria, Finland and Sweden in 1995. After the fall of the Soviet Union in 1991, most Eastern European countries sought entry into the EU.

The EU offered a large, affluent market and development subsidies to poorer regions. The prospect of membership has provided a stimulus to democratic reforms and market economic policies in *Eastern Europe*, which must be achieved prior to admission to the EU. European Union expansion made a major leap on May 1, 2004, when ten eastern and southern European countries joined: Latvia, Lithuania, Estonia, Poland, Hungary, the Czech Republic, Slovakia, Slovenia, Malta and Cyprus. Together, they increased the EU's population by 20 percent to 450 million. Bulgaria and Romania were admitted in 2007, although their admission was delayed because of the greater challenges that they faced in tackling post-communist economic, political and social issues. Croatia joined in 2013. The number of member countries, the EU's economic and political significance and its role as a market mean that its history and development are important to understanding European tourism.

The European Coal and Steel Community

Shortly after the end of World War II, Belgium, Luxembourg and the Netherlands formed an alliance (BENELUX) that was meant to help each of these small countries recover from the socio-economic devastation of the war. The primary goal of the BENELUX agreement was to bring down the trade barriers that existed between the three countries and to allow a freer flow of goods and people across their common borders. In this way, each of the member states could focus more on what it produced best and could more freely import what its neighbors specialized in. At the same time, the European Coal and Steel Community was founded with France, West Germany, Italy and BENELUX, in order to pool their steel and coal resources and create a common market for those products as part of post-war reconstruction.

The Three Pillars of the EU

As part of this pan-European movement of the 1950s, two further organizations were created in 1957 by what is known as the Treaties of Rome. These were the European Economic Community (EEC), which established a Customs Union to eliminate or lessen trade barriers and tariffs between member countries

▶ *This derelict Portuguese customs and immigration office at the Spain-Portugal border is a commonplace scene throughout the European Union where Schengen countries have abolished their border controls. Many border stations have been demolished, although a few have been turned into tourist information offices, museums, cafes, or souvenir shops.*

Photo: Dallen J. Timothy

in Western Europe, and the European Atomic Energy Community (EAEC or Euratom), which was established for cooperation in the area of nuclear energy. The EAEC was later merged into the EEC in 1967 in what was known as the *Merger Treaty* that developed a single membership with a combined set of treaties and institutions. This new collectivity was originally referred to as the European Communities, although now it is known just as *the European Community*. It is one of the three pillars of the EU. The others are *Common Foreign and Security Policy*, and *Police and Judicial Co-operation in Criminal Matters*. However, the term European Community is often used in many diplomatic areas instead of the EU.

European Supranationalism

Supranationalism is strongest in the first pillar: a combined set of treaties and institutions (as defined through the European Community). In the other two areas, the powers of the European Parliament (elected every 5 yrs), the Commission and the European Court of Justice are more limited, as many countries are holding onto their national authorities and institutions. The relative powers of the EU and its member states are a hotly debated issue.

The still unratified European Constitution, signed by representatives of the member states in 2004 but not yet approved by all member countries as of 2014, proposed the merger of the European Community with the other two pillars of the EU, making the EU the legal successor of both the European Community and the EU in its current form. However, its members did not approve the proposed Constitution because French and Dutch voters disagreed with several of its tenets and rejected it. Substantial opposition to the Constitution also exists in the United Kingdom as a result of further concerns about the perceived loss of national sovereignty over a number of issues. A Draft Treaty Amending the Treaty on EU and the Treaty Establishing the European Community, better known as the Reform Treaty or the Lisbon Treaty because of the proposal to sign it in Lisbon in October 2007, replaced the proposed European Constitution but retained a number of its elements.

Schengen Agreement

The debate over the balance between EU and national authority has been integral to the development and history of the EU, and the areas over which the EU system has authority have slowly grown since the 1950s. Since its foundation, the union has expanded its goals to include citizenship issues, foreign policy and security arrangements. In an effort to create a completely border restriction-free zone in Western Europe, the Netherlands, Belgium, Luxembourg, Germany and France signed the *Schengen Agreement* in 1990, followed later by Spain, Portugal, Italy and several other countries.

While the goals of the treaty were many, its primary purpose was to abolish *immigration and customs* controls at border-crossing points inside "Schengenland" and to initiate negotiations on establishing a common visa. Even though the Schengen Agreement was meant to create a borderless region of contiguous states, it took several years for the complete removal of border controls between member nations. The 1992 Maastricht Treaty reinforced the Schengen goals and changed the alliance's name to the EU on January 1, 1993. The birth of this union officially eliminated immigration and customs and trade barriers between member states and created a single market, which in theory at least guaranteed the free movement of products, people, services and capital within the EU.

Reducing Regional Disparities

The *European Commission* has worked to promote social and economic solidarity and to reduce regional disparities. Many of their efforts in recent years have included tourism. The industry was solidly positioned in EU policy with its recognition as a separate economic entity in the 1992 Maastricht Treaty. Perhaps owing to its status as the most truly integrated of all supranational alliances, the EU has demonstrated the greatest emphasis on developing pan-European tourism policies. Many studies and development projects have been commissioned and policies formulated by the EU Commission, and billions of euros have been devoted to community development through tourism in the peripheral regions of Europe through structural funding programs, such as *Interreg* (Interregional) and Leader.

In the EU, the elimination of border formalities and the creation of a common EU/*Schengen visa* have created a freer flow of Europeans and non-Europeans alike, which some observers believe will have the long-term effect of increasing intra-European tourism and tourist arrivals to the continent from the rest of the world. The importance of agreements such as Schengen for mobility is indicated in that a neighboring or nearby country is the preferred foreign destination for trips of nearly all European tourists.

Exceptions to this for tourism are Hungary (Germany the most popular), Romania (Italy) and the U.K. (Spain). Domestic tourism is also of importance. In 2012 just over 3 out of 4 trips (76 %) by Europeans were domestic trips, although domestic trips accounted for 60 percent of nights spent away from home and only 47 percent of tourist expenditure. International travel was responsible for the majority of trips for Belgium, Luxembourg, Malta and Slovenia. Interestingly, Belgian and Luxembourgish tourists made more trips to France, their main foreign destination, than in their own country.

Trips, nights spent, and expenditure of EU residents by destination, EU-28 (2012)

TABLE 2.1			
	Domestic	Outbound: EU28	Outbound Non-EU28
Trips	76.0%	18.0%	6.0%
Nights Spent	60.2%	25.2%	14.6%
Expenditure	46.8%	30.7%	22.5%

Source: Eurostat

Intra-Regional Travel in Europe

In 2012, EU-28 residents of the European Union made 1.196 million trips (for personal or professional purpose) and spent 6.165 million nights during those trips. Tourist expenditure amounted to EUR 400 593 million. On average for the EU, 24.0 percent of all trips were international: 18.0 percent in another EU Member State and 6.0 percent outside the EU. Spain was the EU residents, number one foreign destination, in terms of number of trips as well as number of nights spent and expenditure. 3.1 percent of all trips made by Europeans in 2012 were spent in Spain (equivalent to 12.9 % of all trips abroad), followed

TABLE 2.2 Top 3 destinations for outbound trips from EU28 countries			
	1	**2**	**3**
EU28	Spain	France	Italy
Belgium	France	Spain	Netherlands
Bulgaria	Greece	Turkey	Germany
Czech Republic	Slovakia	Croatia	Italy
Denmark	Germany	Sweden	Spain
Germany	Austria	Italy	Spain
Estonia	Finland	Sweden	Russia
Ireland	UK	Spain	France
Greece	Albania	UK	Germany
Spain	France	Portugal	Italy
France	Spain	Italy	Germany
Croatia	Bosnia/Herzogovena	Germany	Italy
Italy	France	Spain	UK
Cyprus	Greece	UK	Russia
Latvia	Lithuania	Estonia	Germany
Lithuania	Latvia	UK	Belarus
Luxembourg	France	Germany	Belgium
Hungary	Germany	Austria	Romania
Malta	Italy	UK	Spain
Netherlands	Germany	France	Spain
Austria	Germany	Italy	Croatia
Poland	n/a	n/a	n/a
Portugal	Spain	France	UK
Romania	Italy	Bulgaria	Greece
Slovenia	Croatia	Italy	Austria
Slovakia	Czech Republic	Croatia	Hungary
Finland	Estonia	Sweden	Spain
Sweden	n/a	n/a	n/a
United Kingdom	Spain	France	US

Source: Eurostat

by France and Italy. Spain was the main destination for more than one out of five outbound trips of Irish and U.K. residents. The United States is the second ranked EU-28 destination in terms of expenditure and fourth with respect to number of visitor nights.

Tourism and Economic Development

A major goal of the EU is to reduce regional socio-economic disparities among member states and to improve the standards of living in the less affluent areas of Europe. Financial assistance is provided in various forms throughout the EU, as well as to non-member neighboring countries, through various structural funds. These are the European Social Fund (ESF), the European Regional Development Fund (ERDF), the Financial Instrument for Fisheries Guidance (FIFG) and the European Agricultural Guidance and Guarantee Fund Section (EAGGF).

Perhaps the most influential of these for tourism, the ERDF, was established in 1975 to create economic balance in the European Community (EC-the precursor of the EU) by helping to develop areas that fell behind the normal standard. This is especially the case in declining industrial regions and peripheral rural regions where tourism has come to be regarded as a potential mechanism for economic development and employment generation.

While tourism itself has been a fairly minor focus of the EU's overall economic development goals until quite recently, the industry has received considerably more attention in recent years, and many projects that include tourism have been funded through the ERDF. In addition, the EU Commission has established ERDF-based policies for tourism that aim to utilize the industry as a course for economic development and to assist tourism-dependent regions in diversifying their economies to offset the effects of seasonality and outside forces that cause other shifts in demand. The EU has initiated several programs through the ERDF and other structural funds targeted at strengthening social, economic and ecological balance, including tourism. Programs such as *Interreg*, *Leader* and *Envireg* have been especially important in the context of tourism.

The main focus of the Interreg program was the creation of jobs through cooperation between EU states in anticipating a decline in employment in border-related activities with the finalization of the single market. Tourism was seen as a key industry for new jobs. Cooperation in tourism development between EU states and non-member states along the EU periphery was another important element of Interreg.

These efforts were especially important along the borders of eastern and western Europe and between Switzerland and its EU neighbors. Leader began in 1991 as a way of funding integrated rural development projects conceived and initiated by community action groups. Tourism was treated as a fundamental component of this program as it provided a wider range of linkages to rural and countryside tourism and recreation and assisted with improvements in technology and equipment provision, service quality and information. Communities and tourism benefited from Envireg through the project's efforts to reduce the ecological impacts of tourism and other industries in coastal regions and other sensitive areas.

Although the European Parliament has been relatively slow in establishing policies specifically for tourism relative to other economic, social and environmental areas of interest, partly as a result of not being included in considerations of the first European Treaties, the extent of EU involvement in tourism is considerable. The official recognition of tourism's importance by the European Commission in the 1992 Maastricht Treaty elevated the industry in the Union's administrative structure and paved the way for new endeavors in the areas of education, culture, transportation, environmental protection and job training.

The responsibility for tourism policy in the European Commission lies with the Tourism Unit of the Directorate General Enterprise. The European Parliament has a Committee for Tourism and Transport, and there is also a European Parliament inter-group for tourism, which brings together members of the European Parliament that share an interest in tourism issues. The attention of EU tourism policy since 2010 has focused on the development of a new political framework for tourism in Europe. In May 2010,

EU Ministers for Tourism supported the "Madrid Declaration," which established a series of recommendations concerning the implementation of a consolidated European tourism policy. It also stressed the need to strengthen sustainable competitiveness in the sector and recognized the added value of action by the EU on tourism, including a more integrated approach to tourism. To achieve these objectives, actions promoting tourism are grouped under four priorities:

1. Stimulate competitiveness in the European tourism sector.
2. Promote the development of sustainable, responsible, and high-quality tourism.
3. Consolidate the image and profile of Europe as a collection of sustainable and high quality destinations.
4. Maximize the potential of EU financial policies and instruments for developing tourism.

Although the administrative and political structure that supports tourism is not as well developed in the EU as in some countries, the relative lack of official recognition of tourism has not meant that the EU has been without significant influence. The representative body of the hotel, restaurants and cafe sector in Europe, HOTREC, identified over 250 EU measures that directly impact the hospitality and tourism industry. These measures have been developed in a number of Directorate-Generals of the European Commission, including Agriculture (farm and rural tourism), Environment (impact assessment, climate), Transport (Single European Sky) and Enterprise (entrepreneurship and innovation policy).

A Common Currency

One of the highest priorities of the EU fairly early on was to establish a common currency among its member countries. Finally, in 1990, the Economic and Monetary Union (EMU) was established within

ISSUES AND INSIGHTS   Supranationalism

International trade unions, economic blocs or economic communities, as they are variously known, such as the European Union (EU), are manifestations of supranationalism. This refers to national governments or organizations working together to achieve common goals at an extra-national level, usually in areas of trade and economic development. In addition to the EU, other examples of such organizations include the OAS (Organization of American States), ASEAN (Association of South East Asian Nations) and SADC (South African Development Community). The notion of a supranational organization is often used interchangeably with international organizations, but there are differences. A supranational organization is an international multilateral organization consisting of three or more members.

Although international in scope, they are often not global organizations in the sense that countries from around the globe are members, such as the United Nations. Many supranationalist alliances exist throughout all parts of the world and will be discussed in more depth in other sections in this book. Usually, a major goal of most supranational alliances is to facilitate the flow of products and services, capital and investment, and people. In the context of tourism, the people most affected by a decrease in the barrier effects of borders are tourists who reside in the trading bloc countries, tourists from outside the alliance and tourism employees from member countries.

The term "supranational" was first used in an international treaty in the founding documents of the EU the Treaty of Paris, 18 April 1951, which created the European Coal and Steel Community and was enshrined in the "European Declaration" made the same day. The 1985 Schengen Agreement among European states allows for the abolition of systematic border controls between participating countries; it includes provisions on common a common policy on the temporary entry of persons (including visas), the harmonization of external border controls and cross-border police cooperation. Citizens of non-EU (or non-European Economic Area) countries who wish to visit Europe as tourists can obtain a common Schengen visa on arrival at a border and are then not required any further visas within the Schengen area.

Because supranational organizations and agreements tend to increase the flow of tourists, they are seen very favorably by the tourism industry. However, passports and IDs are still usually required for tourism movement from one country to another and the movement of a person as a tourist may also be much easier than moving as a worker or a migrant. Heightened concerns over national security since 2001 along with illegal immigration often have made it more difficult for nationals of some countries to cross national boundaries within their supranational regions. A major challenge for members of such regions is to balance the rights of tourists, the economic interests of the tourism industry and their broad security needs.

the EU, with all but Sweden, Denmark and the United Kingdom signing the accord. These three countries chose instead to keep their own currencies, at least for an extended time. In 1995, the euro was established as the EMU's new currency among its twelve members, and the European Central Bank was established in 1998 to oversee the union's finances and the creation of the euro.

While the euro has been operating as a recognized currency on world markets since 1999, it was only introduced physically in January 2002, when euro notes and coins replaced the currencies of the EMU's twelve member states. The adoption of the euro by most of the EU's member states has potentially had a positive result on tourism for consumers by making the industry more competitive and standardizing prices better throughout the region. Perhaps, however, the most notable benefit has been for tour operators, hotel companies and other service providers that operate in a trans-European market and who no longer need to deal with foreign exchange regulations, or exchange currencies each time a border is crossed or when more than one country is involved.

Western Europe: The Nordic Countries

Physical Landscape

Northern Europe is an old age landscape in its southern seas and a mature landscape in its northern mountains. The old age landscape is the region between the highlands of central Europe and the Western Uplands of Scandinavia known as the North European Plain. Here the European landscape flattens out into a broad coastal plain. Elevations seldom reach over 500 ft (152 m), and the continental shelf extends under the relatively shallow Baltic Sea and North Sea. In fact, these seas become exposed land whenever ocean levels are lower than they are today, and most of the North European Plain is covered with water when sea levels are higher than they are today. The British Isles are continental shelf islands.

The oldest exposed rocks in Europe (up to three billion years old), which were formed in the Precambrian period, are in Scotland and in the Baltic. Known as the Lewisian complex or the Lewisian Gneiss, they are metamorphic rocks located in the northwest of Scotland on the mainland and on the offshore

◄ *Part of the cultural landscape, this traditional farmhouse in Norway is an important element of that country's historical identity and heritage tourism milieu.*

Photo: C. Michael Hall

islands, most notably the Outer Hebrides. The significance of the area has been recognized by UNESCO, which awarded the area Geopark status in 2004.

The Baltic Shield

The oldest rocks in mainland Europe are on the Baltic shield, which extends under the Baltic Sea. Some of these are also three billion years old and over. Some of the oldest rocks are located in the Kola Peninsula, and in Russian and Finnish Karelia. The Baltic shield is related to the Canadian Shield and the Siberian shield. All of these contain exposed rocks that are over one billion years old and were part of Laurasia, the supercontinent that formed three hundred million years ago out of North America and Eurasia. These massive rock shields also contain large areas of fossil fuels in the form of oil and natural gas. These fossil fuels were formed by dense vegetation that grew on the edge of the shallow seas that once covered large areas of these rock shields. In Europe, this can be seen in the large oilfields found in the North Sea that have mostly benefited Norway and the United Kingdom.

The Nordic Cultural Landscape

The Nordic countries of northern Europe comprise five states: Denmark, Finland, Iceland, Norway and Sweden, and three autonomous territories: the Åland Islands, the Faroe Islands and Greenland. The term Nordic refers to the northern lands and characterizes the countries not only in terms of location but also in terms of other aspects of their geography, culture and history that lead to substantial present-day political and societal commonalities. During the early and High Middle Ages Denmark, Iceland, Norway and Sweden shared a common language (Norse), cultural behaviors and shared belief systems often described as Nordic, or by some, as Viking, although the latter term is sometimes interpreted in a negative fashion because of misinterpretation by outsiders.

From the Late Middle Ages onward, Finland also became increasingly drawn into Norse culture as a result of Swedish settlement and becoming a part of the Kingdom of Sweden. At the same time, Greenland, the Shetland and Orkney Islands (now part of Scotland) and much of mainland Scotland and Ireland were also parts of the Danish Kingdom. For much of the following 600 years, Denmark and Sweden contested for political supremacy in the region with substantial shifts in borders. However, the structure of the present-day Nordic countries started to emerge when Sweden ceded Finland to Russia in 1809 under the Treaty of Fredrikshamn. Finland emerged as a fully independent nation in 1917, although Finnish nationalism, particularly through the promotion of Finnish language and culture, had been expressed well before this. In 1905, the union of Norway and Sweden was dissolved, while Iceland declared its independence from Denmark in 1944.

The Nordic Council

The Nordic countries are also a supranational political entity with a cooperation forum in the form of the Nordic Council and the Nordic Council of Ministers. The Council was formed in 1953, with Finland joining in 1956. The formation of the Council has had a number of significant implications for tourism, including a common labor market and free movement across borders for the countries' citizens. In addition, the Nordic Council has supported a number of initiatives to support tourism between the countries, to promote the region and to encourage tourism as a form of regional development.

Although the region includes countries and regions both inside the EU (Denmark, Finland and Sweden) and outside (Iceland, Norway, the Åland Islands, the Faroe Islands and Greenland), the region can still be considered a distinct geographical and political unit, particularly as a result of the ease of travel without passports between the countries.

A significant feature of the Nordic region is the presence of indigenous peoples. On the European mainland, the north of the Scandinavian Peninsula and northern Finland and Russia are often described as Lapland. This is in reference to the presence of the Lapps, or Sami, who traditionally herded reindeer, but for whom cultural tourism is also now a major activity. The population of Greenland is primarily Inuit (sometimes referred to as Eskimo).

Denmark

Denmark is the southernmost of the Nordic countries and is located north of Germany on the Jutland Peninsula. In addition to the mainland, there are also well over 400 islands that are part of the country, many of which, such as Bornholm, are significant tourism areas. The capital, Copenhagen, is on the island of Sjaelland in the western part of the country. It is also the smallest Nordic country (just over 16,600 mi^2; 43,000 km^2), if the autonomous regions of Greenland and the Faroe Islands are excluded, although it is the largest if they are included! The population of Denmark is approximately 5.5 million people.

Denmark is a constitutional monarchy and became a member of the EEC, a forerunner to the EU, in 1973. However, a 2000 referendum rejected monetary union with the EU. Therefore, the currency remains the Danish krone. Greenland and the Faroe Islands are outside of the EU, including the customs zone. Like many of the Nordic countries, Denmark has a strong *welfare system* and is highly *unionized*, although a period of conservative government has changed a number of aspects of state involvement in society. However, it also has a strong developed economy with high-technology industries as well as a very strong agricultural sector, with Danish ham and butter being premium food products.

Copenhagen

Copenhagen is a major transport hub for Scandinavia, as well as a significant tourist destination in its own right. Copenhagen is one of the major hubs for SAS airlines (Scandinavian Airline Systems), which is jointly owned by the governments of Denmark, Norway and Sweden. Copenhagen is also a major rail hub with a direct rail link to Malmö, Sweden, across the Øresund Bridge, and from there to throughout southern Sweden. There is an integrated Øresund rail system between the two countries going as far as Kalmar in southeast Sweden, some three hours travel time to Copenhagen. However, possibly the most obvious form of transport is the enormous number of bicycles in the city with people riding along the extensive series of bicycle paths. The city also provides bicycles for hire. The focus on bicycling and public transport is also an important part of Copenhagen's positioning as a leading sustainable city.

The city has long been an urban tourism destination of international renown. The Tivoli Gardens amusement park and pleasure garden was opened in 1843 and was one of the inspirations for Disneyland. It has one of the oldest wooden roller coasters in the world as well as the world's tallest carousel. Although open year-round, the Gardens attract the greatest number of people in the summer months when the park also hosts concerts and other events.

The Danish capital, Copenhagen, also hosts a number of museums and galleries, as well as royal places and centers of Danish design. Some of the most popular attractions include the Carlsberg Brewery and the iconic Little Mermaid, which is on a rock in Copenhagen harbor. Erected in 1913, the statue commemorates the fairytale of The Little Mermaid written by Hans Christian Andersen, who is probably Denmark's most well-known writer and an important factor in the international image of the country and of the song, Wonderful Copenhagen.

In addition to specific attractions, Copenhagen also has a number of distinct shopping areas as well as a vibrant restaurant and cafe scene, including Noma, which has been rated as the world's number one restaurant for several years by *Restaurant magazine*. One of the significant features of Copenhagen is

▶ *The Little Mermaid statue commemorates the famous story by the Danish author, Hans Christian Anderson (1805–1875). Despite its relatively small size, it has become an iconic symbol of the country of Denmark and is one of the most important tourist attractions in Copenhagen.*

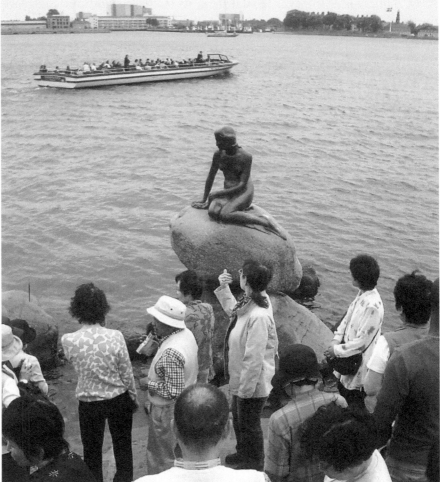

Photo: C. Michael Hall

the extent to which the old waterfront areas have been redeveloped for tourism/entertainment/heritage purposes and housing. The former naval dockyards area is growing in importance with respect to heritage tourism while the Copenhagen Opera House that opened on the island of Holmen is also part of the dockland redevelopment.

Although Copenhagen is a focal point for international tourism, there are a number of other significant tourism sites in the country, especially in relation to its historic military and naval rivalry with its neighbors Sweden, Germany and Great Britain. There are numerous historic castles in Denmark, but possibly the most famous is Kronborg Castle, known by many as Elsinore, the setting of Shakespeare's *Hamlet*, The castle is located in the town of Helsingör on the northeast coast of Zealand and at the narrowest point of the Øresund, the narrow sound of 2.5 mi (4 km) that separates Denmark and Sweden. There are ferry services approximately every 20 min across the sound to Helsingborg on the Swedish side. The castle is regarded as one of the most significant Renaissance castles in northern Europe, although the site also includes a series of interesting battlements and military barracks. The site was made a World Heritage Site in 2000 and is frequently used for cultural events.

One of the other significant tourism features, and one common throughout the Nordic countries, is the importance of second homes or summer cottages for domestic tourism. Many of these are located in coastal areas and on some of the islands, such as Bornholm, the easternmost island of Denmark located in the Baltic Sea between Sweden and Poland; summer visitation is a very important part of the economy.

As with many of Denmark's coastal areas, its strategic significance has meant that it has a number of important historical sites, best known of which is probably Hammershus Castle—northern Europe's largest medieval fortification.

Finland

The Republic of Finland is a modern country in more ways than one, having only achieved full independence in 1917. Until then it had been a part of Russia and Sweden. Finland is the most sparsely populated country in the EU, which it joined in 1995. It has a population of 5.4 million people and an area of over 130,500 mi^2 (338,000 km^2). Finnish is one of the few official languages of the EU that is not Indo-European in origin; instead it is a member of the Finno-Ugric group of languages (that includes Estonian and Hungarian). Finnish is also an official minority language of Norway and Sweden, while Swedish is an official language of Finland, although only about 6 percent of the population speak it as their native tongue.

Like all of the Nordic countries, Finland is noted for its low corruption levels, its substantial gender equality (it was the second country in the world to have universal suffrage), its high quality education system, its freedom of the press and civil liberties, and a strong state welfare system. Finland is probably best known internationally these days for its design, shipbuilding (many of the world's largest cruise ships are Finnish), forest products (wood and paper) and technology companies.

Finland even sometimes used to be jokingly referred to as *Nokialand* prior to the sale of part of the company to Microsoft in 2013. The northern position of Finland and the effects of ice age glaciation mean that agricultural production is limited by climatic factors and poor quality arable land. Notwithstanding, rye is grown far to the north, while forest foods such as berries and mushrooms are also important traditional foods. Fish, such as herring and salmon, have also been important culturally and economically, but the pollution of the Baltic Sea and over-fishing have severely damaged some of the commercial fisheries.

Helsinki

Helsinki is the capital and largest city of Finland with a population of 620,000, although the capital region, which also includes Vantaa, Espoo and Kauniainen, has well over 1.4 million people. Helsinki-Vantaa Airport is the international gateway to Finland, along with ferry services that connect Helsinki with Rostock (Germany), Tallinn (Estonia) and Stockholm. The city has an impressive architectural heritage ranging from neoclassical buildings of the 19th century, such as the Helsinki Cathedral, through to the Art Nouveau buildings of the early 20th century and the modernist and functionalist architecture of the latter parts of the 20th century, perhaps best represented in the architecture of Alvar Aalto. As with many capital cities, there is also an assemblage of museums and art galleries that give full expression to Finland's rich cultural heritage.

Saunas

Perhaps unusual for many international visitors, yet most representative of Finnish culture is the Museum of the Sauna. Saunas are integral to Finnish lifestyle and are integrated into domestic bathrooms. Yet the tradition of having a sauna available at lakeside summer cottages also remains strong. The second home is an extremely important part of domestic tourism and is often waterside, simply by virtue of the immense numbers of lakes in the country; some 9.4 percent of the country's area is actually water. According to Statistics Finland, there are 187,888 lakes larger than 500 m^2 and 179,584 islands. Therefore, it is perhaps not surprising that lakes are an important element of the promotion of Finland internationally.

Taiga

Another important element of the landscape is the forests that cover about 75 percent of the country. Comparable to the northern forests of Norway, Russia and Canada, these are categorized as taiga and the common tree species include conifers such as larch, spruce, fir and pine, and deciduous species such as birch and willow. In addition to being harvested, the forests are also important in cultural and ecological terms. An extensive system of national parks is now established to conserve significant species, such as the brown bear. In the southeast section of the country, the Saimaa Lake system is home to the endangered Saimaa Ringed Seal, the emblem of the Finnish Association for Nature Conservation, and one of only three lake seal species in the world.

ISSUES AND INSIGHTS ⟩ Planning for Tourism in National Parks: The Example of Finland

Nature is an extremely important part of Finland's international tourism promotion, as well as being extremely important for domestic tourism. As of 2014, there were 38 parks in Finland. The total number of visits in Finnish national parks was 1.7 million people in 2007, the most recent date for which accurate figures are available. The average number of visits in Finnish National Parks has doubled since the 1990s, and the national park-related tourism development in remote areas such as Northern Finland has been even stronger. This has been the recognized tourism significance of national parks that regional and local stakeholders and policy actors usually support plans for opening parks, for they believe the national park status increases the attractiveness of the area and promotes nature-based tourism. Consequently, national parks have become a significant tool for regional economic development. In addition to national parks, there are twelve wilderness areas, all of which are located in Lapland, as well as 19 strict nature reserves.

However, while tourism is obviously important for national parks in Finland, it is not the sole rationale for their establishment, for biodiversity and landscape conservation are obviously also significant. As in other national park jurisdictions, the original Finnish legislation included provision for both conservation and visitation, a balance that has become increasingly difficult given the growth in tourism numbers in recent decades.

The 1923 Nature Conservation Act aimed to preserve untouched nature although it also stated that national parks were meant for the pleasure and enjoyment of all citizens, as they are in the United States, and that they should have value as an attraction and be easily reached by people. Metsähallitus, the Finnish Forest and Park Service, reports that the Finnish network of protected areas forms a varied network intended to preserve for present and future generations a suitable number of representative and ecologically viable areas of all the ecosystems and natural habitat types occurring in Finland; they should also take into account geographical variations and the various stages of natural succession.

The following should be preserved as part of the aim of protected conservation:

- Natural gene pools and ecosystem diversity.
- Species, geological and geomorphological features, especially species and features that are either naturally rare or threatened or declining as a consequence of human activity.

- Landscapes and habitats shaped by previous generations, including the cultural heritage associated with the Finnish countryside, along with endangered domesticated plant and animal breeds.
- The natural succession of ecosystems and other natural processes at various stages.
- Areas of outstanding natural beauty.
- Wild areas.

The growth of ecotourism and an increase in the number of visitors to protected areas is regarded as indicating a more favorable attitude toward nature conservation. Yet tourism is regarded as only one out often different uses of the Finnish protected area system that require a policy statement. (The others are every man's right to access, fishing and hunting, photography, local residents, traffic, forestry, mineral prospecting and mining and leasing land.) These different use demands are usually managed via a master plan for a given protected area. The Finnish experience mirrors the approaches of the Nordic countries and other national park agencies in the developed world with respect to the relationship between tourism and national park and protected area planning. The difficult task of balancing conservation aims while still encouraging visitation is one that is almost universal at destinations that are either national parks or have a significant protected area set aside.

The Finnish approach of providing land use zoning with particular zones for specific activities and visitor experiences is an extremely common strategy. However, the Finns also promote the perspectives of local communities, the tourism industry and other government bodies, as well as an understanding of the market in the planning process.

Sources

Gossling, S., and J. Hultman (eds.) 2006. *Ecotourism in Scandinavia*. Wallingford: CAB International.

Hall, C. M., D. Muller, and J. Saarinen. 2008. *Nordic Tourism*. Bristol: Channelview.

Metsähallitus (Forest and Park Service). 2000. *The Principles of Protected Area Management in Finland: Guidelines on the Aims, Function and Management of State-owned Protected Areas*. Metsähalliuksen luonnonsuojelulkaisuja Sarja B No.54. Vantaa: Metsähallitus-Forest and Park Service, Natural Heritage Services.

Metsähallitus (Forest and Park Service). 2014. http://www.outdoors.fi/destinations/nationalparks

Iceland

The Republic of Iceland is home to the world's northernmost capital city, Reykjavik, and it has been very much influenced by its landscape. Iceland has an area of a little over 38,600 mi^2 (100,000 km^2) and a population of approximately 310,000 people. Although the country's government strongly encourages decentralization, the population is concentrated in the capital. Iceland is built upon a strong tradition of self-sufficiency, but it is also extremely international in outlook. The economy is strongly connected to natural resources in various ways either through thermal and hydroelectric energy or via its fishing industry. The country is not a member of the EU, primarily because it wishes to maintain control of its fisheries, and the currency is the Icelandic krona. However, the effects of the 2008–2010 global financial crisis have reopened debate about EU membership.

Rural areas with grasslands do still produce niche products such as lamb, mutton, and milk products. But the rural economy is increasingly connected to tourism, particularly for fishing, wilderness hiking (approximately 11 % of the country is glaciated while the country has an extensive set of volcanic features) and horse riding on Icelandic ponies–an activity geared to American and European tourists. The volcanism for which the country is famous has also generated another set of tourist attractions in the form of hot springs. It is possible to take a leisurely drive around Iceland on the country's ring road (Route 1), which is just under 870 miles (1,400 km) in length. This journey has long been popular with domestic tourists but is also increasingly attractive to international tourists.

Iceland's international airport, Keflavik, which is near the capital, is the international gateway. Its trans-Atlantic position has also made it a stopover point and a weekend getaway destination from both New York and London. *Stopover tourism* is popular in the area of Keflavik, where the world-famous Blue Lagoon thermal spa is located. Icelandair is a popular airline for trans-Atlantic travel. All Icelandair flights to or from Europe stopover in Iceland, often with a layover of several hours for passengers. Many people take advantage of this opportunity to visit the Blue Lagoon or to shop in Reykjavik. This form of

◀ *Gollfoss (Golden Falls) is one of Iceland's most significant tourist attractions. It is surrounded by the tundra and volcanic landscapes that have made the island known throughout the world.*

Photo: Dallen J. Timothy

day-tripping, stopover tourism, or *transit tourism*, is an important element of Iceland's tourism economy. Reykjavik has the reputation of being one of the world's best clubbing destinations, but it also has a number of cultural institutions and provides easy access to Iceland's natural landscape.

Volcanism

Iceland has received considerable international prominence in recent years, first, because of the collapse of several of its banks as a result of the global financial crisis in 2009. However, this actually made tourism even more important to the Icelandic economy, especially as its currency was devalued. The second reason was because of the April and May 2010 eruptions of Eyjafjallajökull, which, although relatively small for volcanic eruptions, caused enormous disruption to air travel across western and northern Europe because of the mountain's location underneath the jet stream. Some estimates put the number of cancelled international tourist arrivals in Europe at over 2 million, and the direct cost to tour operators at approximately 1 billion euros.

Although the imposition of "no fly" areas by the EU was criticized by a number of members of the tourism and aviation industry as being overcautious, subsequent studies of the volcanic ash from the eruption indicated that it included a high proportion of abrasive particles with the potential to melt and cause severe damage to a jet aircraft engine.

Norway

Norway is the westernmost country of the Scandinavian Peninsula. It has an elongated shape with the easternmost part on the border with Russia and Finland being as far east as Istanbul. However, the vast majority of the land border is shared with Sweden and runs through various mountain ranges. Physical geography has played an enormous role in shaping the history of the country. Dominated by heavily glaciated mountainous terrain with substantial valleys and deep Fjords, transport linkages between settlements have been either via the sea or along the valleys, with many valleys historically quite isolated in winter.

As with Iceland though, much of the country is made habitable by the presence of the Gulf Stream, which warms the land. The coastline is extensive and stretches over 15,500 mi (25,000 km). However, the coast is also an important marine tourism resource and the location of second homes, particularly in the south. Another important influence of the Gulf Stream is its contribution to coastal precipitation and snowfall further inland. Bergen, the country's second city, has the nickname of Regnbyen, which means city of rain.

Oil Wealth

The country is extremely well off economically as a result of its substantial oil and gas wealth and a relatively small population of approximately 5.1 million. However, the relative current wealth belies the fact that for most of its history Norway has been one of the poorer countries in Europe. Fishing and shipping have been some of the most important industries, a tradition that continues to the present day, and Norwegian cruise ships are an important part of the cruise industry throughout the world.

The strength of the Norwegian currency, the krone, and a high rate of taxation mean that Norway is regarded as an expensive tourist destination compared to other Nordic countries, particularly to its neighbor Sweden. However, international tourism is still a significant part of the economy because of its nature-based tourism through coastal cruising and visits to the Fjords, as well as travel to the northernmost point of mainland Europe, the North Cape (which because of the Gulf Stream is often shrouded in mist). Ecotourism promotion in Norway faces difficult promotional obstacles. For example, while activities such as whale watching or observing reindeer and moose are attractive to many visitors, eating them may not be, although they are a traditional food.

Furthermore, the independent spirit of the Norwegian people is such that external pressure from conservation groups in other countries may well be counterproductive, for these foodstuffs are a long tradition, together with wild berries and mushrooms from the forests. This is typical throughout the Nordic countries.

Second Homes in Norway

Also important in Norway for domestic tourism, the summerhouse or cottage has become a feature of leisure and holiday activity. Originally, such second homes were often very primitive and associated with summer work, such as grazing cattle or sheep in the mountains or fishing along the coast, to prepare for winter and earn extra money. However, over time the summerhouses came to be regarded more as a cheap form of holiday and weekend leisure to allow activities such as fishing or skiing. Even though many Norwegians could afford to buy holiday homes in the Mediterranean (and some do), the traditional cottage in the mountains or along the coast continues to play an important part in Norwegian life and travel.

Sweden

At almost 174,000 mi^2 (450,000 km^2), Sweden is the third largest country in Europe and, excluding Greenland, the largest Nordic country. It is, therefore, perhaps not surprising that the country has a very low population density outside of the major urban centers of Stockholm (the capital), Gothenburg and Malmö. The Laponia region of northern Sweden, for example, is one of the largest areas of relatively unmodified natural environment in the world and is primarily populated by Sami. The majority of people live in the south of the country where the climate is more benign and allows for better agriculture and better access to markets.

Although an empire in the 17th century, Sweden, like many of the Nordic countries, has historically been relatively poor for much of its history. However from the second half of the 17th century on, the industrial development of Sweden, particularly due to its large iron ore reserves and engineering expertise, helped provide a substantial boost to the economy; this has given it one of the best standards of living in the world. It should be noted that, as in the case of Switzerland, criticism is often made of the extent to which Sweden's neutrality during World War II helped provide the framework for subsequent developments.

Stockholm

The imperial and mercantile heritage of Sweden is a focus for much of its urban tourism. Stockholm, sometimes described as the Venice of the North because of its location on a number of islands, has a range of old districts, royal buildings and palaces (the changing of the guards at the royal palace is a significant attraction) and historic waterfronts. The Vasa Museum on the island of Djurgarden is probably one of the best-known heritage attractions in Scandinavia. It displays the *Vasa*, an almost fully intact 17th century, 64-gun warship that sank on her maiden voyage in 1628, as well as other features of Sweden's maritime history. Gothenburg, on the west coast of Sweden, also has a strong maritime influence, as it has been a major trading port and industry center for hundreds of years. This has also made it a major center for business travel.

ISSUES AND INSIGHTS   **Transterritorial Governance: The Case of Øresund, Denmark and Sweden**

Trans-territorial organizations are organizations with area boundaries that include the territories of constituent members. A number of trans-border regional agencies have been established within Europe with the support of the EU and with national and regional governments. One of the most recognized is that of Øresund.

The Øresund region links the region of Skåne in southwestern Sweden with Zealand in Denmark. The most concrete, as well as symbolic, example of the Øresund region was the construction of the cross-border Øresund bridge across the Øresund Strait between Denmark and Sweden, which connects the cities of Malmö (Sweden) and Copenhagen (Denmark) by road and train.

Although a physical connection between Denmark and Sweden had been suggested since the 19th century, a new cross-border regional governance structure did not emerge until the 1950s. Various economic and political factors delayed both the physical and political connectivity of the cross-border region until the convergence of a number of interests in the late 1980s. These interests were expressed in a number of institutions that promoted the Øresund concept.

Foremost among these is the region's governing body, the Øresund Committee (Øresundskomiteen), which is a forum for local and regional politicians and authorities from both sides of the Øresund Strait. Also of great importance are a number of other organizations that contributed to the development of the Øresund Region concept. As well as various business, scientific and innpvation networks, these include Øresund University which is a network composed of universities in the region, four Swedish and eight Danish. The Lund Helsingborg campus is a significant center of tourism research in the region.

Although the region has a number of historical, cultural and agricultural attractions, as well as a world-class restaurant scene that makes it a major center for gastronomic tourism, the development of new businesses and education and research initiatives has created further benefits for tourism. The development of new organizational networks and structures has practical implications for tourism as they encourage the greater movement of people within the region both directly through requirements for business travel and meetings and also as a justification for further transport provision. In addition, running parallel to the organizational and political development of the Øresund regional concept are the development of place marketing and promotional strategies that aim to reinforce an Øresund brand and identity both within the region and externally to potential visitors and investors.

Malmö and the province of Skåne, which includes such cities as Lund and Helsingborg, also had a substantial industrial legacy that continues to the present day. However, tourism has become increasingly important for the area, particularly following the completion of the Øresund Bridge linking Sweden with Denmark, which allows even greater mobility between the two countries. For example, it is estimated that 10 percent of Malmö's population works in the Copenhagen area.

Second Homes in Sweden

Like Norway, one of the most notable features of tourism in Sweden is the importance of summerhouses or cottages (second homes), which are integral to Swedish domestic holiday-making and are found in all areas of the country. This practice is in common with other Nordic countries. Since Sweden joined the EU in 1995 there have been substantial second home purchases by Germans in the south of the country. Other major non-Swedish second home owners include Norwegians, particularly in some of the mountain areas near the Norwegian border.

Second homes are particularly important for tourism and the economy in the north of Sweden, which is otherwise marked by its dependence on forest products and mining, though in recent years new forms of tourism are developing. For example, the city of Umeå, the gateway to the north, is a prominent university town, business center and a European city of culture, which also has a regional cheese (Vasterbotten- named after the county) that is a significant gastronomic tourist attraction. Perhaps more famously, Kiruna in Norrbotten County in the far north is home to the ice hotel every winter and the proposed site of a European space tourism center to be developed in conjunction with Virgin Galactic.

Western Europe: The Northern Countries

The countries of northern Europe have significant cultural and historical commonalities that have been brought back together in recent years following the expansion of the EU. Although Rome did influence the region through its occupation of some areas and by trade, the cultural heritage was primarily Celtic and Gothic in origin and marked by the Druidic faith until Christianized. Though Norse influence was extremely significant in the early Middle Ages, and Roman language has been influential, the Germanic language forms dominate, providing a linguistic as well as a geographical divide between northern and southern Europe.

The Benelux Countries

The term Benelux refers to the customs and economic union between Belgium, the Netherlands (Holland) and Luxembourg that entered into force in 1947, although Belgium and Luxembourg had established an economic union as early as 1921. The Benelux Union was an important precursor to the creation of the EU, as all three countries were founding members, along with West Germany, France and Italy. The three countries are also sometimes referred to as the Low Countries, which is an historical reference to the lands around the delta of the Rhine, Scheldt and Meuse rivers. A number of other cultural, linguistic and political similarities also contributed to the region's identification.

The three countries have been united several times, although the most recent period was after the Napoleonic Wars. Belgium and the Netherlands became separate kingdoms in 1830, while the Grand Duchy of Luxembourg left its union with the Netherlands in 1890. Luxembourg is the world's only sovereign Grand Duchy. As befits the term Low Countries, the outstanding geographical feature of much of the region (and particularly the Netherlands) is how flat and low-lying the land is. Much of it has been reclaimed with the lowest point being almost 7 m below sea level. However, given the threats of climate change some of the lowest lying areas of reclaimed land are now being returned to wetlands.

The romantic image of the Netherlands with a windmill beside a dike (a canal or channel used for draining land) was based very much on the reality of wind power being used to drain low-lying areas of farmland. Such images still remain these days for one can see many modern windmills in the Dutch countryside; they are as important for power generation as they are for drainage.

The rich Rhine delta soil has provided the foundations of a strong agricultural economy with the Netherlands being world famous for cheese production (Edam and Gouda) and wooden clogs worn by agricultural workers in former times. But the extensive network of waterways also means that sailing, riverboats and other forms of water-based tourism are extremely popular in the countryside. In addition, the flat country means that the bicycle is a very important form of transport for shorter journeys, while an extensive railway network is also a very significant part of tourist transport.

Amsterdam

Amsterdam, the Dutch business capital, is a major center for international tourists and is based on a curious mixture of heritage attractions and contemporary culture. Amsterdam's Schipol Airport is also a major aviation hub and one of the busiest airports in the world, as it is one of the hubs of Air France-KLM. Historic cultural attractions are found in both Amsterdam and the seat of government in The Hague (Den Haag). (Somewhat bizarrely, the Dutch constitution states that the official capital is Amsterdam even though the government has been based at The Hague since 1584.)

Amsterdam is extremely popular as a destination because of the large number of major art galleries, including the Van Gogh and Rembrandt museums, and the attraction of the old town area and the canals.

► *Bruges, Belgium. Leisure tour boats are a popular way of touring the historic cities of Europe. This reflects the importance of waterways and port facilities in the historical development of Europe. The rivers facilitated internal transportation and access to the continent's natural resources, while the coastal ports connected the workshops of Europe to the world.*

Photo: Alan A. Lew

More contemporary cultural attractions include the Anne Frank House de Wallen, the city's red-light district; the city's excellent coffee shops; and the sex museum.

Brussels

Brussels, the capital of Belgium, and Luxembourg, the capital of Luxembourg, are also major business and diplomatic centers, but primarily with respect to their roles within the EU system. Luxembourg is the seat of several key EU institutions, including the European Court of Justice, the European Court of Auditors and the European Investment Bank. Brussels is home to the European Parliament (along with Strasbourg), the European Commission and the Council of the EU, the North Atlantic Treaty Organization (NATO) and the European Organization for the Safety of Air Navigation (EUROCONTROL).

A significant cultural feature of Belgium is that it straddles the divide between the Germanic languages of northern Europe and the Romantic languages of the south. The two largest regions are Dutch-speaking Flanders in the north and the French-speaking region of Wallonia in the south. The Brussels-Capital Region, with a population of over a million people, is an officially bilingual enclave. There is also a small German-speaking community in eastern Wallonia near the German border. Belgium's linguistic diversity has contributed to the development of a complex system of federal government based along linguistic lines, while the linguistic divisions also contribute to differences in culture. These divisions have also led to significant political differences at times between the French and Dutch speaking regions.

Belgium is a heavily industrialized country, which has contributed to poor water and environmental quality in some regions. However, many of the old industrial regions are now being restructured and industrial heritage tourism is being developed as a response. Belgium is home to several interesting ancient cities (e.g., Bruges) and the famous hot springs town of Spa, known as one of the most desirable holiday destinations in Western Europe during the medieval period. Belgium is also known for the quality of its chocolate, a product that developed from its colonial empire. Another famous product is beer, of which there are over 500 varieties, many of them fruit beers. The brewing traditions in Belgium have meant that in the same way countries with vineyards have wines associated with regions, so in Belgium regions and towns are associated with particular types of beer.

Germany

Although Germany is a modern economic superpower–it is the world's fourth largest economy and the third largest exporter of goods–it comes as a surprise to many people that it is a country that only achieved the status of a nation-state in 1871. However, it has already been divided and reunified since that time. In fact, the history of the area occupied by present-day Germany and much of Austria has been one of shifting political boundaries made up of different amalgamations of political sub-units of independent duchies or kingdoms. It should therefore be of no surprise that, befitting its political heritage, Germany is currently a federal republic consisting of 16 different states.

Länder

In German political tradition, states are called Länder (countries), and the concept of Germany refers to the alliance or confederation of the autonomous Länder. The relative independence of the states holds to the present day, with the Länder retaining the right to act independently at an international level under Articles 23, 24 and 32 of the German Federal Constitution (The Basic Law of the Federal Republic of Germany). Such political traditions also underlie Germany's historical enthusiasm for the EU, of which it was a founding member.

History of Germany

The notion of a German identity exists in a broader sense from 900 CE and was often associated with the Holy Roman Empire. In its heyday, it was an amalgamation of hundreds of political entities, such as duchies, kingdoms, principalities and cities, and it covered most of Europe. The emperor could be elected from throughout the realm of the empire, but by the beginning of the 15th century, emperors were elected nearly exclusively from the Austrian Habsburg Dynasty. A critical point in German history was the effect of the Reformation, which by 1530 resulted in a separate Protestant Lutheran church. It came to be acknowledged as the official religion in many German states, although significant religious conflict between Protestants and Catholics lasted for many years and culminated in what was known as the Thirty Years' War (1618–1648). The War is regarded as one of the bloodiest periods of German and European history, with Germany's population reduced by almost a third through military actions and associated famines and diseases.

The Peace of Westphalia (1648) ended religious warfare among the German states. However, it had the long-term effect of dividing the various principalities and territories of the empire along political and religious lines, with the Austrian Habsburg monarchy dominating the southeastern German states and what became modern-day Austria and Hungary; the Kingdom of Prussia dominated much of what is now modern Germany.

German Confederation and Empire

The German Confederation, a loose alliance of 39 states, was created in 1814 following the Congress of Vienna, which convened to redraw the map of Europe after the defeat of Napoleon Bonaparte. A tariff union developed at the time also helped encourage German integration. However, the most critical feature of the period was the growing power of the state of Prussia (much of which is now in present-day Poland and the Kaliningrad exclave of Russia on the Baltic Sea). Prussian victory in the Austro-Prussian War of 1866 enabled the creation of the North German Federation (Norddeutscher Bund) of 22 states and excluded Austria, formerly the leading German state, from the affairs of the remaining German states. It cemented the political and, to an extent, cultural divide between the two countries that exists to the present day.

The nation-state of Germany can be identified with the creation of the German Empire in 1871 and the height of German power with the acquisition of a number of colonies around the world. Yet despite close family ties between the royal families of Europe, German military and colonial ambitions contributed substantially to World War I. Defeat meant the loss of large portions of German territory, especially its overseas imperial possessions. The loss of territory combined with the conditions imposed on Germany following the Great War helped provide fertile ground for Adolf Hitler and the Nazi movement, but defeat in World War II led to a further loss of territory and the forced migration of many Germans from territories that had come under Soviet control.

West and East Germany

The division of Germany along the lines of the military occupation of different sectors led to the sectors controlled by France, the United Kingdom and the United States merging in May 1949 to form the Federal Republic of Germany (commonly referred to as West Germany). The Soviet Zone became the German Democratic Republic (East Germany) later that year. Berlin was also divided into West Berlin and East Berlin. The East-West division ended in 1989 with the collapse of the Berlin Wall and mass migration to the west from East Germany. Germany was officially reunified in 1990, although some of the social, economic and political divisions from that time are still visible.

Germany is better known for outbound travel than for inbound tourism, with German tourists being extremely important for the coastal areas of the Mediterranean and destinations throughout the world. Nevertheless, there is a significant inbound tourism market for which business travel is extremely important. There is also a substantial cultural tourism market that visits many of the palaces and castles of the various states, art galleries and museums that contain works from those periods. Many of the most prominent castles are found in the mountains of southern Germany and along the banks of the Rhine; the latter also attract visitors to the vineyards located along the river valley.

World War II internment, concentration and extermination camps are also a significant feature, with continued interest in the Holocaust as well as in other aspects of the war. Other sites of significant World War II events have become tourist attractions. Berchtesgaden in the Bavarian Alps near the border with Austria is famous as Hitler's mountain residence, the Berghof, which was situated there.

The Black Forest

Like much of alpine southern Germany, the mountains and areas such as the Black Forest also serve to attract visitors to present-day national parks, skiing and walking opportunities. However, increasing competition means that many locations are seeking to diversity their product. For example, Kirchzarten is seeking to expand its tourism base by developing conference tourism and romantic walking getaways as an alternative to its traditional domestic base.

Although one of the main "gateways" to the Black Forest region Freiburg has developed a different tourism base that mixes heritage and contemporary interest in the environment. Freiburg im Breisgau is situated in a major wine growing region and has a medieval cathedral dating from around 1200 and a university that was founded in 1457. This heritage is significant for tourism but the city is also known for its sustainable urbanism and promotes itself as an "eco-city" as a result of substantial investment in public transport, renewable energy and urban design practices. This has meant that many of the local hotels also have a strong focus on sustainability. For example, Hotel Grüner Baum in Merzhausen has a particular emphasis on local foods, with a special type of poached egg being a breakfast specialty.

The Berlin Wall

Another form of heritage tourism is the growth of interest provided by former East Germany. The sense of nostalgia created by the elapse of time since the fall of the Berlin Wall has allowed the development of hotels that recreate the accommodation experience of East Germany, while some of the former workers' hotels and camps on the Baltic Sea have been converted into backpackers (youth hostels), second homes, or hotels.

Austria

Austria shares language and cultural ties with Germany, but, apart from the time Austria was incorporated into Nazi Germany, it has been politically separate from Germany since the beginning of the 19th century. Like Germany, Austria is a former imperial monarchy, but it is now a parliamentary representative democracy consisting of nine federal states. Along with Finland, Ireland, Sweden and neighboring Switzerland and Liechtenstein, Austria has declared permanent neutrality.

However, since joining the EU in 1995, the interpretation of neutrality has come to be reassessed as a result of EU security commitments. Nevertheless, such a position, along with Vienna's status as a meeting place between Western and Eastern Europe, has allowed Austria to host a number of major international organizations. These include the secretariat of the Organization for Security and Cooperation in Europe and the headquarters of the International Atomic Energy Agency, the United Nations Industrial Development ment Organization, the United Nations Drug Control Program, the Organization of Petroleum Exporting Countries and the Comprehensive Test Ban Treaty Organization.

Austria's imperial past provides a substantial cultural heritage to explore, particularly in the capital of Vienna, which has an impressive array of cathedrals, galleries, museums and many heritage buildings from a range of historical periods. Vienna is a noted cultural and educational center, which means that the city is blessed with an impressive range of art galleries and performance centers for both classical and contemporary music. The Danube River flows through the city and is a significant focal point for visitor and local recreational activities, and it provides opportunities for international travel via cruise ships that travel up and down the river.

Health Spas

Austria has a long tourist tradition dating from the 16th century, when some of the first health and spa resorts were developed for the aristocracy. More recently, it has been developing summer and winter alpine tourism. The country hosts a range of ski runs to suit a variety of different markets, and it offers opportunities for summer hiking in the cool mountains, which many Austrians and central Europeans enjoy to avoid the heat of summers on the plains.

Vineyards

Like in Germany, another growing aspect of rural tourism in Austria is the development of food and wine tourism, particularly near such major centers such as Vienna where vineyards are accessible as day trips for visitors. These are primarily found along the Danube Valley and in the southeast of the country near Lake Neusidl, close to the Austria-Hungary border, which is an area famous for sweet wines. The lake, which is about 22 mi (36 km) long, and between 3.5 and 7.5 miles (6 km and 12 km) wide, is extremely shallow with a depth of no more than 6 ft (1.8 m); some of its surrounds have been declared a World

ISSUES AND INSIGHTS   Dark Tourism

Dark tourism is travel to sites of human tragedy, such as the sites of the Holocaust or the killing fields of Cambodia, or even taking a walking tour in the London of Jack the Ripper. Dark tourism is not a new phenomenon. During the Middle Ages, Christians used to go on pilgrimages to Rome and see the catacombs where the early Christians were tortured and imprisoned. More recently, Ground Zero in New York has become an essential part of the tourism itinerary for many visitors. However, media and transport now put the places shown on the History Channel or in the history books onto the tourist itinerary. To some, visiting such locations may be an act of reverence or even atonement, while to others it is probably voyeurism.

Over half a million people visit the Auschwitz-Birkenau German Nazi Concentration and Extermination Camp (1940–1945) in Poland each year. The camp commandant, Rudolf Hoss, testified at the Nuremberg Trials that up to 2.5 million people had died at Auschwitz. This figure was revised in 1990 by the Auschwitz-Birkenau State Museum, which put the figure at between 1.1 and 1.6 million. It is estimated that 90 percent of them were Jews. Large numbers of Poles, Gypsies and homosexuals were also killed. A July 2, 1947, Act of the Polish Parliament established the Auschwitz-Birkenau state

Museum on the grounds of the two extant parts of the camp, Auschwitz I and Auschwitz II-Birkenau. The site was given World Heritage status in 1979. In 1996, Germany made January 27, the day of the liberation of Auschwitz-Birkenau, the official day for the commemoration of the victims of National Socialism.

The standard guided tour of Auschwitz-Birkenau lasts for two-and-a-half-hours. However, given the history of the place and the significance of the events that took place there, the question can be asked whether this is sufficient or whether tourism in such a place is appropriate at all. Alternatively, it could be argued that even this time may be sufficient to convey significant interpretive messages.

Of course, another issue is that such places are interpreted through present-day eyes. In fact the passage of time may be important in even enabling dark tourism locations to be promoted to visitors; otherwise, their transformation to a tourism attraction may conflict with the values of those affected by them. Regardless of the arguments, it is apparent that dark tourism sites will continue to be a significant part of the tourism experience. The challenge for the attraction managers is to be able to handle the conflicting stakeholder values and interests that surround such sites.

Heritage Site, and it is a significant tourist attraction for fishing, sailing and windsurfing. It is also the site of an annual crossing of the lake from Morbisch, Hungary, to Illmitz, Austria.

The British Isles: Ireland and the United Kingdom of Great Britain and Northern Ireland

The British Isles comprise more than 6000 islands off the northwest coast of continental Europe, which include the countries of the United Kingdom of Great Britain (England, Scotland and Wales) and Northern Ireland, and the Republic of Ireland. The group also includes the United Kingdom crown dependencies of the Isle of Man, and, by tradition, the Channel Islands (the Bailiwicks of Guernsey and Jersey), even though these islands are strictly speaking an archipelago immediately off the coast of Normandy (France) rather than part of the British Isles.

It should be noted that the term British Isles is unpopular among many people in Ireland as a result of that country's succession from the United Kingdom in 1922 and the continued differences over control of Northern Ireland. No branch of the government of Ireland officially uses the term in a way that includes the Republic. However, the term is widely used in the international setting, often without realizing the political offense it may cause the Irish.

Common Travel Area

Despite Ireland's independence, there are still several areas of political cooperation between the countries. Most notably for tourism, this includes what is known as the Common Travel Area between all the countries and dependencies in the British Isles. This allows passport-free travel between them with only minimal or no border controls. This relationship is important, because neither the United Kingdom nor Ireland is party to the Schengen Agreement (passport-free travel).

◀ *St. Andrews, Scotland, the Home of Golf. St. Andrews is known as the home of golf because of its world-famous golf courses, and its royal and ancient golf club, which establishes golfing rules worldwide (Except for the United States and Mexico).*

Photo: C. Michael Hall

Nevertheless, the various nations that have occupied the Isles since Roman times have had a close-knit set of relations that remain the basis for the development of the British Empire and the consequent spread of English as a language around the world. Their history too is a major factor for tourism in the region.

Kingdoms in the Middle Ages

Following the end of Roman rule in Britain, the land became divided into numerous kingdoms. These added to the existing kingdoms of Scotland and Ireland that had never been under full Roman control. The seven major kingdoms of southeast and central Britain—Northumbria, Mercia, East Anglia, Essex, Kent, Sussex and Wessex–along with minor kingdoms, eventually came to form the nation of England, although this process was not completed until the tenth century when King Alfred of Wessex united the various kingdoms against a common Danish (Viking) enemy.

By the end of the 13th century, the four major kingdoms of England, Wales, Scotland and Ireland dominated the British Isles. Norse areas, such as the Kingdom of Mann, the Isles of Orkney and the Shetland Islands, remained independent until 1266 and 1472, respectively. The Norman invasion of England in 1066 not only served to bring England under one clear government authority but also provided impetus to bring Wales, Scotland and Ireland under the English Crown.

Cathedrals and Castles

Many of the cathedrals built in the high to late Middle Ages have become major focal points for tourism, such as Canterbury Cathedral in southeast England, which is the seat of the head of the Anglican Church. In addition, the conflicts that occurred during this period led to the development of another type of infrastructure that is now a significant attraction for tourists, that is, the numerous castles in major towns and throughout the countryside. Although they became obsolete with the development of artillery, many castles were converted into country grand houses and have become an important part of the tourist promotion of Ireland and the United Kingdom.

Colonization of Ireland

The religious conflicts that affected the European mainland and Germany with respect to the Reformation also had tremendous impact in the British Isles. Apart from conflict over the throne itself, James I encouraged the settlement of Scottish and English Protestants in Ireland on land confiscated from Irish Catholic landowners in what was known as the Plantation of Ulster. This was done in an attempt to prevent further Irish rebellion against the English colonists.

This colonialism has had a substantial historical legacy, as the present-day partition of Ireland into Ireland and Northern Ireland is primarily a result of the settlement patterns of the plantations of the early and late 17th century with the Protestant descendants of settlers in the north favoring linkages with Britain. In Northern Ireland, the Catholic-Protestant divide was a major contribution to "the troubles," the term used to describe the communal violence involving Irish Republican (loyal to Ireland) and Loyalist paramilitary organizations (loyal to the United Kingdom), and the British Army for which the peace process is still underway, although a ceasefire was enacted officially in 1998.

Religiously inspired conflicts and wars often tied to competition for the throne were also significant in England and Scotland for much of the 17th century and did not finally subside until the late 18th century. England did have a republican government for a brief period of time (1649–1660) following the English Civil War, but the monarchy has long had a central place in English political life. There is, however, a growing republican movement. Today the monarchy is also a very important tourist attraction, with palaces and royal events, such as royal weddings, being a major drawing card. The Queen's residence in London, Buckingham Palace, is invariably one of the must see attractions, particularly if it coincides with the changing of the Royal Guard.

▶ *Souvenir shop in London, England. London is an important gateway not only for the British Isles, but also for all of Europe.*

Photo: C. Michael Hall

ISSUES AND INSIGHTS   Margate: The first modern coastal tourism resort?

Margate is an ancient coastal town on the Isle of Thanet in Kent in southeast England. The name is thought to refer to a gap in a cliff where pools of water can be found. Margate was an established fishing and trading port by 1300. However, its significance in the history of tourism derives from being an exemplar of temperate coastal resort change and possibly being the first modern coastal tourism resort. Margate is recorded as being Britain's first commercial sea bathing resort in 1736 when a channel was cut from the sea to a waterfront tavern where bathing and drinking salt water was advocated for medical conditions.

From 1750 onwards, a bathing machine, invented by Benjamin Beale and first used at Margate, revolutionized sea access and immediately popularized sea bathing. It allowed the occupant to be driven into the sea and enter the water in privacy and safety. The Royal Sea Bathing Hospital founded in 1791 was Britain's first specialized hospital and advocated sea bathing and sea air in cases of consumption (tuberculosis), which was of epidemic proportions until well into the early 20th century. Before the advent of the railways in the 1850s visitors would arrive by sea. However, once access was improved, Margate rapidly became a mass tourism resort for London's working class, along with Southend, Eastbourne and Brighton.

Margate was also the first resort to have donkey rides (1890) and the first to introduce deck chairs for hire (1898).

From the 1930s on, sunbathing grew in popularity as a result of Margate's sandy beach, the Golden Sands, which is unlike many United Kingdom coastal resorts that have shingle beaches. Margate also receives a relatively high number of sunshine hours. Growth continued after World War II, with Margate becoming the site of several Butlin's hotels, as well as Dreamland, a theme park that was once the most visited charging attraction in the United Kingdom.

Like many British seaside resorts, its attractiveness as a beach destination declined rapidly once access to Mediterranean coastal resorts increased in the early 1970s due to cheap airfares, holidays and direct flights. Consequently, in the 1970s Margate experienced a large downturn in tourism and a substantial loss of people and capital through emigration.

Since 2004, regeneration has been focused on the arts through the development of the Turner Contemporary Art Gallery and Museum, named after the Romantic landscape painter J. M. W. Turner who lived in Margate for many years, and a heritage precinct supported by noted British artist Tracey Emin, who grew up in Margate. More than 45, 000 people visited the Turner Contemporary Gallery in the first ten days of its opening in April 2011. The success of the Turner Gallery has led to a rejuvenation of the old Market Square and Town Hall area and the development of new restaurants and cafes, boutique accommodation establishments, and arts and crafts stores.

Regional Nationalism in the United Kingdom

In recent years, there has also been a revitalization of Welsh and Scottish nationalism, which has seen the devolution of power from Westminster (the Parliament of the United Kingdom). The National Assemblies of both Scotland and Wales were created in 1999 following a referendum. A referendum on independence in Scotland in 2014 saw the Scottish voting only marginally to reject it on the promise of even greater devolution of power. The Welsh Assembly is based in Cardiff, the largest city in Wales. Although the largest city of Scotland is Glasgow, the Scottish capital is Edinburgh. Other major centers are Aberdeen, the third largest city in Scotland, also known as Europe's oil capital because of its role in North Sea oil development, and St. Andrews, historically one of the major religious and political centers of Scotland, but now better known for its university and golf.

London

The end of civil wars and fighting between the various kingdoms also saw the development of British colonialism and the growth of the empire. Even though America was lost in the 18th century, the rise of the industrial revolution, mercantile trade and the power of the British navy all contributed to the expansion of the empire and reinforced the importance of London, the capital, as a political and economic center. The imperial legacy remains important to London even today.

London has long been a cosmopolitan city with a rich heritage. Many of the government buildings and monuments to the empire remain despite bombings during World War II. It is home to many museums and key historic sites. Some of the most visited attractions include the reconstruction of The Globe

▶ *The beach at Budleigh Salterton in Devon, England. Southwest England is the warmest part of the country due to the relatively warm waters of the Gulf Stream. This has made possible the development of several noted coastal resorts.*

Photo: C. Michael Hall

Theatre of William Shakespeare, the Tower of London (that served as a royal prison and is the home of the British Crown Jewels), the Tate Gallery, the National Portrait Gallery and the Victoria and Albert Museum. More recently developed attractions include the home grounds of English soccer clubs, such as Chelsea's Stamford Bridge that supporters around the world follow, and shopping. For example, Harrods in Knightsbridge is possibly the most famous department store in the world and is commonly visited by tourists, although locals often regard Fortnum and Mason in Piccadilly, often shortened to just Fortnum's, as higher quality. It is famed for teas, high-quality foodstuffs and meals.

Like the days of the British Empire, London is today a major European transport hub, although now it is aviation rather than passenger ships. Heathrow Airport is one of the world's busiest airports and getting through United Kingdom customs or through security and gate control on departure is a memorable experience for many visitors in its own right. Even with the completion of a new terminal (Terminal 5), it is likely that the airport will remain congested. Although a new runway proposal was vetoed by the incoming government in 2010 as a result of public opposition, in part based on concerns over climate change, the issue has not disappeared and there are also proposals to build more runways and airports in the south of England along with the development of new train lines.

Rural Britain

Although cities are the focal point for tourism, the countryside is often a key factor in the attractiveness of the British Isles. The rural landscape developed from a combination of physical and human factors. Key physical factors include the Gulf Stream that passes the west coast of Ireland and Scotland and helps keep the climate warmer than equivalent latitudes in North America by several degrees. The warm ocean current also ensures that there is plenty of moisture in the air, thereby providing plenty of rain, which has had the consequence, for example, of ensuring good grass growth and contributing to Ireland's green fields image. Another important item is *islandness*. There is actually a relatively low number of indigenous species because the British Isles have been islands since the Ice Age, and most species were introduced from Europe and elsewhere and became acclimatized.

Physical factors notwithstanding, the human influence on the rural landscape has been remarkable. Much of the countryside with its fields of hedgerows and farm holdings has only existed in its present form since the enclosures at the time of the Industrial Revolution. Landholdings were amalgamated, and tenants, who had often been on the land for generations, were forcibly removed to provide more efficient farming. This is a process referred to in Scotland as the clearances, and it still conjures up substantial emotions in many of the Highland areas. In much of northern Scotland the present-day population remains far below what it was prior to the clearances. Many went to the industrializing cities while others became part of the great migrations to North America, Australia and New Zealand to find a better life and a land they could call their own.

There also remains a significant movement to reclaim some of the lands for small holders rather than have them held by large landowners. However, the landscape had been changing significantly before this time. For example, many of the areas of grassland or moorland that have become national parks or scenic areas were formerly forests. However, the combination of grazing and tree cutting, often for mining or for charcoal, as in Dartmoor in southwest England, has meant that the uplands became denuded of trees, with constant grazing keeping it that way. Although there is a *wilding movement* in the United Kingdom to return such sites to forestland, many people oppose such changes because of their present landscape preferences.

Little remains of the extensive original forests of the Isles, although there are attempts to conserve the remaining Caledonian forests in Scotland, for example. Today only 1 percent of the original forest survives, covering 180 square kilometers (44,000 acres) in 84 locations. Even the famous New Forest of Hampshire was a combination of extant and planted woodlands that were protected for royal hunting since the 13th century. Nevertheless, despite being highly developed, the countryside remains a very important part of the attractiveness of the United Kingdom and Ireland.

Western Europe: The Southern Countries

Physical Landscape

Southern Europe has a youthful landscape. It consists of steep mountains with peaks reaching over 10,000 ft (3,000 m) and separated by deep river valleys. Coastal plains are narrow, if they exist at all, because the mountains extend vertically from the oceans and seas. The mountain systems of southern Europe were created during the most recent geologic time period, the Alpine Orogeny, which started 136 million years before the present (m.y.B.P.). *Orogeny* refers to a geologic time period of mountain uplifting. Not only were the Alps formed during the Alpine Orogeny, so were the Himalayas in Asia, the Rocky Mountains in North America and the Andes in South America. The major mountain ranges of southern Europe include the Pyrenees in Spain and France; the Alps; the Carpathian Mountains in Eastern Europe; the Apennines in Italy; the Dinaric Alps in Croatia and Bosnia-Herzegovina; and the Anatolian Plateau of Turkey.

About 136 million years ago, the Alps started to rise, and the Mediterranean Sea started to sink. The Alps pushed up land that was formerly the bottom of the Tethys Sea between the ancient continents of Gondwana and Laurasia. Today, the Ionian Sea, part of the Mediterranean Sea, is 15,000 ft (4,600 m) deep. Other portions of the Mediterranean are over 10,000 ft (3,000 m) deep. The Apennine Mountains, which form the spine of the Italian Peninsula, used to be part of a mountain chain that connected the Alps to the Atlas Mountains in northwest Africa. The island of Crete is a remnant of the land bridge that once connected Turkey and Greece, but which has since been pushed down below sea level.

Human Geography

The human geography of southern Europe is saliently affected by the Mediterranean, which has for many years provided trade routes between the countries of the region, and North Africa and Asia Minor, and allowed the development of ancient civilizations to spread their influence throughout all parts of the region. These included the Phoenicians, the Greeks and the Romans. The Roman legacy is still felt over most of the western Mediterranean because of the dominance of Romance languages that are rooted in Latin.

The Mediterranean Sea is also a focal point for human settlement. Over a third of Europe's total population lives within 30 miles (50 km) of the coasts, and that figure is growing. By 2025, the percentage of the population of Spain, France, Greece, Italy and Croatia living in coastal cities is projected to be more than 85 percent on average, and as high as 96 percent in Spain. Coastal populations are also growing not only as a result of internal migration, but also because of international retirement migration and second home purchases within the EU. All countries in southern Europe are major tourism destinations and rank among the largest international tourist destinations in the world. The Mediterranean region, therefore, accounts for about 30 percent of all international tourist arrivals, and estimates suggest this figure is growing in absolute terms. This has contributed to significant urbanization in the most popular coastal resort areas on the Mediterranean Sea.

Food and Wine

Another common characteristic of the southern countries is the role of food and wine in the region's cultural geography and tourism. Such synergies should not be surprising, for wine, food and tourism are all products differentiated on the basis of regional identity. Wine is often identified by its geographical origin (e.g., Burgundy, Champagne, Porto, Rioja), which in many cases has been formalized through a series of appellation controls, in turn founded on certain geographical characteristics of place. Foods (e.g., cheese such as Parmesan) are also identified by their place of origin. Within the region and other parts of the EU, such geographical place designations have been given special legal protection. This is also significant for tourism because the cuisine of a country can be a source of attraction in its own right with respect to consuming the authentic article, while place name protection is also useful for marketing and branding.

France

Prior to the emergence of China as a major international tourism destination France was usually identified as the world's top tourist destination with respect to numbers of international visits. In many ways metropolitan France provides a geographical link between northern and southern Europe, although culturally it is very much a part of southern Europe. France also has a number of overseas departments in the Americas, the Indian Ocean, and the South Pacific.

France has been a separate political entity since 843 and the division of Charlemagne's Carolingian Empire. The name France originates from the Franks (Francs), a Germanic tribe that occupied the northern border of the Rhine in a region called Francia. After the fall of the Roman Empire, the Western Franks moved westward, gradually becoming the Kingdom of France, while the Eastern Frankish Kingdom evolved into what it is now Germany.

The language of the Franks was Germanic. Contemporary French has primarily evolved from Vulgar Latin, although some Frank words were incorporated into the language as it developed. In addition, the influence of the Vikings (Norse) on French has also been substantial, although more recently French has become notable for its lack of capacity to absorb new words from other languages, which is perhaps a reason for its replacement by English as the language of international trade and business.

The French Revolution

Since the *French Revolution* in 1789, when King Louis XVI and his wife, Marie Antoinette, were executed, France has been politically organized as a series of republics. The exceptions were the First

French Empire under Napoleon Bonaparte (1804–1814), the July Monarch (1830–1848) of Louis-Philippe and the Second French Empire of Napoleon III (1852–1870). The core of the various French republics has been one of the fundamental documents of the French Revolution, The Declaration of the Rights of Man and of the Citizen (La Declaration des droits de l'Homme et du citoyen), which defined a set of individual rights and the collective rights of all of the three estates (clergy, aristocracy and the general populace) as one. Men are born and remain free and equal in rights. Social distinctions can be founded only on the common utility (Article 1). This is often simplified to the words Liberty, Equality, Fraternity.

However, while these are laudable sentiments and extremely important to the broader growth of the concept of political rights, the extent to which they have been achieved in French society has been highly variable. In fact, France, like many EU countries, is grappling with the issues that come with the growth of large immigrant populations and their relative rights, particularly with respect to religion and culture. Some sets of beliefs, such as Scientology, are not recognized as religions and are instead regarded as cults.

The Largest Country in the EU

France has an estimated population of over 66.5 million people and is the largest country in the EU. Unlike many other European countries, it is experiencing considerable natural population growth, which is occurring in such urban areas as Paris, Marseille, Lyon, Nice, Nantes and Toulouse. Rural areas are suffering from substantial depopulation though. Partly in response to this, but also in light of the place that rural and peasant culture has in French society, the French agricultural sector remains highly protected from foreign competition—an issue that causes international trade negotiation difficulties not only in the EU but also with other agricultural nations.

Nevertheless, some French agricultural products are appreciated around the world, such as wine, cheese, truffles, escargot (snails), frog's legs and *foie gras*, as well as French cuisine which is recognized under the World Heritage Convention. The culinary diversity of France is also reflected in its cultural diversity, which, in turn, is reflected in the differences between the provinces and the regions; for many visitors though Paris remains the central expression of French culture and lifestyle. In part, this is because of its function as a gateway and the center of government over many centuries.

Paris

Because Paris has been the capital for many years and the center of various French empires and kingdoms, it has an incredible concentration of built heritage, ranging from religious edifices, such as Notre Dame Cathedral, to secular sites, as in world-class art galleries (e.g., the Louvre). In addition, its long tradition of political and cultural tolerance means that Paris is exceptionally cosmopolitan. The city is also recognized, together with London, Milan, New York and Sydney, as an international fashion center. Nevertheless, in recent years the development of new attractions, such as Euro Disney, has led to a significant change in the profile of visitors to the city, including day-trippers from southern England who travel to Euro Disney and Paris itself on trains through the Channel Tunnel.

Other significant areas of France for international tourism include the wine regions of Alsace, Champagne, Bordeaux and Burgundy, as well as Provence, which has become a heavily romanticized region owing to various films and novels, and the growth of holiday homes. Southern France is also featured strongly in tourism promotions, largely because of its climate and association with glamour and the film industry, most famously the international Cannes Film Festival held there each year.

Greece

Greece is often regarded as the home of European civilization, and the sites of ancient Greece and Greek mythology have historically provided a major reason to visit. However, in recent years Greece has become a mass destination particularly for sun and climate, although there are some mountainous areas that offer winter skiing. The landscape of Greece is often considered in terms of the Aegean islands, coastal peninsulas and rugged inland terrain. The mountains and uplands of Greece were once well wooded, but deforestation, resulting from such agricultural practices as sheep and goat farming, has led to soil erosion. Agriculture is significant in the fertile plains and valleys, with olives, olive oil and grapes being significant crops and exports.

Greek identity is substantially shaped by the legacy of ancient Greece and by many years of occupation by the Ottoman Empire. Greece only gained its independence in 1831. In the modern historical period, Greece has been one of Europe's poorer countries and has often experienced periods of authoritarian leadership, the most recent being a military coup d'état against the elected government and King Constantine II on April 21, 1967. Between 1967 and 1974, a United States-supported military junta, known as The Junta or the Regime of the Colonels, was established.

Authoritarian Past

As in Spain and Portugal at the same time, tourism in Greece was encouraged by an authoritarian government to attract foreign exchange, to create employment and to help provide a positive image to the rest of the world. The role played by tourism continues to the present day, with the service sector being the largest sector in the Greek economy, especially following the period of austerity in Greece following the 2008 financial crisis from which the country is only gradually recovering,

On many of Greece's islands, tourism is the primary economic sector, and tourism in the overall Greek economy is extremely important, more than in many other European countries. However, tourism in Greece has been substantially affected by the 2008–2010 global financial crisis and the wider public response. Initially, this was partly because visitor arrivals have been down from other impacted European countries, but also because the government's *austerity measures* severely hit the domestic economy and caused substantial domestic unrest that received European-wide coverage. Since 2012 as other parts of northern Europe, especially Germany, have enjoyed a degree of economic growth and started travelling again so the relatively cheap cost of vacations in Greece has served to encourage significant growth in tourism again.

Italy

Although Italy is usually associated with the Roman Empire and its unifying forces, for most of its history since Roman times Italy has been a series of different kingdoms and states fighting among themselves for territory and power while subjected to foreign intervention. Unification of Italy began in 1861, with the Venetian region added in 1866 from the Austrians, and Rome in 1870 from the French. Following the period of fascist dictatorship under Mussolini and the end of World War II, Italy became a republic after a popular referendum held on June 2, 1946, a day celebrated since as Republic Day.

The period from 1946–1992, known as *The First Republic*, was characterized by its highly unstable national governments, although economic growth was still being achieved in part because of Italian membership in the EU. The period from 1992 to the present, known as The Second Republic, has had a slightly more stable government, although there was considerable voter disenchantment with politics in the 1990s as a result of corruption. This accusation has continued to plague several time Prime Minister Silvio Berlusconi, who is Italy's richest person and who was the owner of three of the seven national television channels during his time in office.

Photo: Dallen J. Timothy

◀ *The scenic canal village of Burano, Italy, is a popular side attraction for visitors to Venice. The unique Mediterranean cultural landscape and architectural heritage lend significant appeal to many destinations in Italy, Spain, Portugal, Greece, Malta, France and Cyprus.*

The North–South Divide

Because of their relatively recent unification, Italians think of themselves as much in provincial and regional terms as they do with respect to being Italian. There are also substantial historical north–south divides in economics and politics. Industry and most of the population are concentrated in the wealthy north, where industrialization and modernization began in the 19th century; the south is poorer and more agriculturally based. As in France, this divide is also partly reflected in what people eat. Dairy products such as butter are associated more with the mountainous north and olive oil with the south.

Although Italy is the world's sixth largest exporter of manufactured goods, the country is often perceived in rural terms. Indeed, one interpretation of the origin of the term Italy derives from the Greek word meaning land of cattle. In great part this is because of the international success of Italian cuisine and various books and films that focus on the region of Tuscany. Undoubtedly, the rural areas of Italy are important because of their agricultural production, especially olive oil, cheeses and wine.

However, in recent years many rural areas of Italy have started to focus on tourism as a means of economic development, and like other parts of the Mediterranean, such as Spain and Portugal, there has been substantial retirement and second-home development. Such processes have also been encouraged as Italy has one of the lowest birth rates in Europe, and many rural communities have therefore had difficulty maintaining their population bases. Nevertheless, considerable concerns exist in Italy with respect to illegal migration, especially across the Mediterranean from North Africa.

Tourism in Italy

Italy is regularly ranked as the fifth major tourist destination in the world with over 40 million international visitors each year. Many of these, along with domestic tourists, concentrate in areas with high amenity values, particularly the coast. However, there are increasing concerns over the environmental sustainability of coastal tourism development. In Italy over 43 percent of the coastline is completely urbanized, 28 percent is partly urbanized, and less than 29 percent is still free of construction. There are only six stretches of coast over 12 miles (20 km) long that are free of construction and only 33 stretches between 6 and 12 miles (10 and 20 km) long without construction.

In addition to its coastal and mountainous regions, Italian cities also host many heritage attractions. In addition to Roman sites, the art and architecture of the Renaissance is extremely important, with sites in Milan, Rome, Venice and Florence being major drawing cards. Rome is also significant as the seat of the Vatican (see below), which is therefore a site of significant religious importance for many Catholics and culturally significant in its own right. The role of heritage in Italy is evidenced by the fact that it has more World Heritage Sites than any other nation.

The Iberian Peninsula: Portugal and Spain

The Iberian Peninsula, or Iberia, includes Portugal, Spain, Andorra and the British Overseas Territory of Gibraltar. Iberia has been occupied by humans for thousands of years as evidenced by prehistoric remains. In historic times Iberia was successively settled by waves of colonists and invaders, including the Phoenicians, Greeks, Carthaginians, Romans, Germanic tribes, and the Moors. The Muslim Moors ruled parts of the peninsula for 700 yrs, with Seville and Granada being their last strongholds. The Moors left an architectural and culinary presence that lasts to the present day. Following the defeat of the Moors in 1492, the various small states gradually amalgamated, although the only time the entire peninsula was united as a single political unit was from 1580–1640 in a period referred to as the Iberian Union. Since then Portugal and Spain have remained independent states.

Gibraltar

Spain has called for the return of Gibraltar, a small but strategic British territory near the Strait of Gibraltar, known colloquially as *The Rock* or *Gib*, which was ceded by Spain to Britain in perpetuity in 1713 under the Treaty of Utrecht to help end the War of Spanish Succession. However, in referendums held about the future of the territory, the majority of Gibraltarians have rejected unification with Spain or even co-sovereignty. Tourism is important in Gibraltar, although it was hurt by the unilateral closure of its border with Spain (by Spain) between 1969 and 1985, when it was reopened. Since 1985 traffic has flowed consistently across the border in both directions with occasional interruptions by Spanish border police.

Catalonia

The Autonomous Community of Catalonia is one of Spain's most economically developed regions. Owing to various economic, political and cultural forces within the Spanish state, Catalonia has since 2010 (although it was brewing earlier than that) developed a strong desire to secede from Spain and become an independent country. In much the same way Scotland did in September 2014, a referendum on Catalonian independence occurred in November 2014 but was met with serious opposition by the Spanish national government. Should Catalonia become an independent country, it will have major implications for tourism and its potential membership in the European Union.

Overseas Empires

Portugal and Spain had significant overseas empires as a result of maritime expansion and exploration in the 15th and 16th centuries. Both countries controlled large territories of the Americas, while Portugal also had possessions in Africa and Asia. Spain had minor holdings in Africa and more in Asia. The colonial territories were important sources of wealth for the metropolitan powers and provided income vis-a-vis gold, silver, spices and slaves. The loss of the American territories was a major financial blow to both countries, but their empires had been in decline long before then, partly as a result of European

This wine route in Portugal is an example of the growing importance of wine tourism and themed routes throughout Europe, particularly in Portugal and Spain. Many such wine routes and heritage trails have developed all over Europe during the past quarter century and are becoming a more important part of regional tourism products.

Photo: Dallen J. Timothy

wars, but also because of the rise of the French and British Empires. However, civil wars and disastrous international wars led both countries into relative decline.

Modern Portugal and Spain have also been marked by periods of autocratic dictatorship. In Spain, General Miguel Primo de Rivera ruled between 1923 and 1931. Following the bitter Spanish Civil War (1936–1939), which ended the Second Spanish Republic, General Francisco Franco ruled until 1975. Franco won the civil war with the support of Italy and Germany, but he re-entered the international arena from the mid-1950s on as a result of the United States' desire to base forces in Spain during the Cold War.

In Portugal a revolution deposed the monarchy in 1910, but following considerable economic problems a military coup d'état occurred in 1926, which led eventually to the establishment of a right-wing dictatorship by Antonio de Oliveira Salazar. The Salazar regime was based on Catholic social doctrine and corporativism, in which power is given to civic assemblies that represent particular economic, industrial, agrarian, social, cultural and professional groups. The Portuguese authoritarian government was overthrown in 1974, and Franco died in 1975, which revived the deposed Spanish. Both Spain and Portugal made extensive use of secret police and repression to remove opposition. However, the political stability provided by dictatorship, along with the need for foreign currency and economic development, provided the impetus for tourism development in both countries in the late 1960s and early 1970s.

EU Membership

Spain and Portugal joined the EU in 1986, which proved to be immensely beneficial to their economies both in terms of EU assistance to marginal regions and to the encouragement of second home ownership in the countries' coastal areas. Such developments led to substantial capital investment and employment generation, particularly in the Spanish construction industry. However, the environmental impacts of these coastal developments are increasingly being questioned, particularly as there are increased concerns over competition for scarce water supplies between agricultural users and tourism and leisure industries, especially golf courses. The construction industries in both countries were also severely affected by the 2008–2010 global financial crisis, which led to high levels of unemployment with which both countries are still affected by in 2014.

Viticulture

Winegrowing (*viticulture*) is a major agricultural sector in Portugal and Spain. Spain is well-known for the production of sherry and a variety of white and red wines. Rioja in northern Spain is probably the most famous wine region internationally, but other areas are becoming better known, especially as a result of large numbers of tourists traveling to Spain and drinking local wines. Portugal, by virtue of its long trade links with Britain, has long used wine as a means to link wine regions with tourism.

The city of Oporto in northern Portugal uses the port wine connection extensively in its promotions and in hosting events and festivals. From the perspective of local politicians and government officials, such relationships create opportunities for long-term loyalty from visitors to the region in terms of their wine purchasing behavior. However, while port has a high profile in the British market, it has diminished in popularity as consumers have begun to focus on lighter, fruitier styles of wine for immediate drinking.

Nevertheless, market awareness of port has provided an opportunity for Portugal to promote its other red and white wines. The Alto Minho region in northern Portugal, which is best known in wine terms as producers of vinho verde, has also been attempting to develop linkages between wine and tourism. The Vinho Verde DOC region is Portugal's largest demarcated wine region. The area has received substantial EU funding through various regional and rural development programs to improve tourism product development and promotion, including wine tourism. Interestingly, the region's wine tourism has mainly targeted international and domestic tourists who have already arrived in the region to see culture and heritage, rather than specifically using the region's wine as a branding tool to attract tourists. Wine tourism-related development has included the founding of wine routes, home-stay accommodations and guides to the region's gastronomic opportunities.

World Heritage

The imperial legacies of both countries, the role of the church and the presence of many different colonists have provided a wide range of heritage attractions, including very important Catholic pilgrimage sites (e.g., Santiago de Compostella, Spain and Fatima, Portugal). Spain has the third highest number of *UNESCO World Heritage Sites* in the world (as of 2014 Spain had 44).

Portugal has 16 World Heritage Sites, with three of them (Torre de Belem, Sintra, Mosteiro dos Jeronimos) located near to the capital, Lisbon. Both Spain and Portugal have also sought to promote tourism and urban redevelopment via the hosting of mega events such as World Expos and Olympic Games. Barcelona in particular focuses on events as a means of promoting its culture. However, the legacy of these events is being revisited given the recent years of austerity that followed the period of unsustainable economic development based on construction and real estate speculation.

The Island States of the Mediterranean: Malta and Cyprus

Malta and Cyprus are the southernmost parts of the EU, becoming member states in 2004. Even though extremely densely populated (almost 3,300 people per mi^2 (1,300 people per km^2)), Malta is the smallest country in the EU in population and area. An archipelago of seven islands situated in the Mediterranean Sea between Italy in the north, Libya in the south and Tunisia to the west, Malta's strategic location has long been fodder for fighting between other countries to control the Maltese islands. The United Kingdom was the last colonial power until independence was declared in 1964.

Malta became an independent republic in 1974. For much of the 19th and 20th centuries, Malta was primarily economically dependent on the presence of the British navy and other military postings. The economy is presently based on tourism, freight (as a transshipment point from the Suez Canal) and manufacturing. The country is also seeking to develop its economy by investing heavily in education and

TIPOTA ΔEN KEP ΕΤΑΙ ΧΩΡΙΣ ΘΥΣΙΕΣ ΚΑΙ ΕΛΕΥΘΕΡΙΑ ΔΙΧΩΣ ΑΙΜΑ
NOTHING IS GAINED THOUT SACRIFICES AND FREEDOM WITHOUT BLOOD

Photo: Dallen J. Timothy

◀ *A viewing platform in Nicosia, Cyprus. From this platform in Greek Nicosia, tourists can look across the buffer zone (the Green Line) into the Turkish Republic of Northern Cyprus. This is an example of a border of conflict that has become a significant tourist attraction for both sides of the Cypriot divide.*

developing a knowledge-based economy. However, a significant constraint to the island's development, including tourism and vacation home development, is the lack of freshwater supplies in summer and the need to import nearly all of its energy.

Divided Cyprus

The divided island of Cyprus in the eastern Mediterranean Sea provides an interesting case study of political unrest and its effects on tourism in a popular seaside destination. In response to a coup d'état against the government of Cyprus by revolutionaries who desired unification with Greece, Turkish armed forces arrived on the island in July 1974. In an effort to protect Cyprus' Turkish population from annexation by Greece, they occupied the northern third of the island. This was an unwelcome advance in the eyes of Greece and among the island's Greek population. Fierce fighting ensued between the Greek and Turkish populations of the island, resulting in thousands of lives lost on both fronts.

A ceasefire was eventually declared in August 1974, and a de facto boundary was drawn across the island, separating the northern third of the territory from the southern two-thirds. By the end of 1975, some 185,000 Greek Cypriots had migrated from the northern portion of the island to the Greek-controlled south (Republic of Cyprus), and approximately 45,000 Turkish Cypriots moved northward into the Turkish zone. The new political landscape of division, dominated by the scar (ceasefire line) known as the Green Line, was a stark reminder of the conflict and its effects on everyday life, including tourism.

Tourism in Cyprus: North and South

This conflict and the new border became major barriers to tourism, which had already begun to flourish and provided a major source of foreign exchange earnings. In 1970, the number of international tourist arrivals to Cyprus was approximately 140,000. Three years later, in 1973, the number had grown to nearly 300,000. However, this figure fell drastically with the July 1974 Turkish invasion/intervention on the island. Visitor arrivals dropped to 150,000 in 1974, most of which arrived before July; and in 1975, only 47,100 visitors arrived on the island.

This sliding trend lasted only a few years, however. Since 1975, the Republic of Cyprus (the Greek south) has seen considerable tourism growth in both arrival numbers and infrastructure development. The industry has in fact recovered fully in the south, while the north's (Turkish Republic of Northern Cyprus–TRNC) tourism industry has been beleaguered by slow growth in arrivals, a sluggish tourism economy and minimal investment in tourism infrastructure and services.

The sluggish growth of tourism in the north is a result of several factors: (1) the Turks were seen as the aggressors in the conflict, which has affected the international image of the Turkish-controlled north; (2) international economic sanctions against the TRNC has severely curtailed visitor arrivals and foreign investments; and (3) in response to the world community's embargo, the government of the north has opted to focus on economic development through agriculture, food processing and textiles, rather than on tourism, although tourism is continuing to grow in the north.

Following the 1974 incident, non-Cypriots visiting the island were permitted to cross from the south to the north during daylight hours, but they were not permitted to purchase any products north of the border, and the government of the south strictly prohibited entry to the island in the north. For the island's residents, the Green Line became a very significant barrier. Greek Cypriots were forbidden from crossing into the north, and the Turkish population was not allowed south of the border. Thus, travel between the north and south and vice versa for Cypriots was almost impossible, until 2003.

On April 23, 2003, the situation changed as the border opened up in both directions, thereby allowing Greek Cypriots to visit the north and Turkish Cypriots to visit the south. At the time of writing, the situation had begun to normalize, and hundreds of Cypriots from both sides were crossing the border in both directions each day for shopping, work, gambling (in the north) and various other activities. However, outsiders are still not permitted to enter the island through the north.

Western Europe: The Non-EU Countries

The non-EU countries of Western Europe (excluding the Nordic countries of Iceland and Norway) are Switzerland, San Marino, Vatican City, Andorra, Monaco and Liechtenstein. With the exception of Switzerland, the non-EU countries of Western Europe are all *microstates*. The term microstate generally refers to countries and islands with populations of less than one million people. However, in the context of Europe, the term has additional meaning and often refers to the smallest of the microstates, in that they are both physically and demographically very small, albeit independent, countries that function legally and in practical terms as sovereign nations. The microstates are discussed in more detail below.

Switzerland

Switzerland is a landlocked country bordered by Germany to the north and Italy to the south, France to the west and Austria and Liechtenstein to the east. Switzerland has a population of 8 million people and occupies an area of 15,940 mi^2 (41,285 km^2). The history of Switzerland has been strongly influenced by its geography, which is primarily alpine with many glacial valleys. Even the northern part of the country, where most of the population resides, remains quite hilly and is also characterized by numerous lakes. The largest cities, such as Geneva, Bern, Basel and Zurich, are located in this area.

The country's official name is the Swiss Confederation. Politically it is a federal republic with a direct democracy based on 26 cantons that are the equivalent of states or provinces. The cantons have their origins in geographically determining features such as urban areas, forests and valley systems. Switzerland is a multilingual country with different language groups concentrated in different regions largely because of

◀ *This picturesque village in the Swiss Alps illustrates the important role of rural landscapes in mountainous areas. Such areas are especially popular for skiing in winter, and for hiking and alpine walking in the summer months.*

Photo: Dallen J. Timothy

its position in Europe and the relative isolation of the various cantons from each other for much of Swiss history. Historically, there have been few transport connections, and travel in winter has been difficult. The official languages are French, German, Italian and Romansh (a Romance language that developed from Vulgar Latin). The range of languages spoken in the country has become an important factor in Switzerland's international orientation, as has its long history of neutrality.

Switzerland has not been involved in a foreign war since 1815, although a civil war in 1845 primarily between Catholics and Protestants was a pivotal point in national history. It provided the impetus for a revision of federal relations, as well as for the development of *direct democracy*, a prominent feature of the Swiss political system. As a result of Switzerland's position of neutrality and its democratic and political orientation, it has long been the headquarters of international institutions.

UN and Red Cross

The International Committee of the Red Cross was founded in Switzerland in 1863, and the International Federation of Red Cross and Red Crescent Societies (IFRC), founded in 1919, is based in Geneva. The United Nations also has several institutions in Geneva, although Switzerland did not join the United Nations itself until 2002. The location of numerous international bodies in the country along with Switzerland's substantial business community, including internationally branded companies, such as Nestle, Credit Suisse, Novartis and Swatch, means that business travel is an important element of tourism.

Competitive Economy

Switzerland's economy is regarded as one of the most competitive in the world. Key industries include banking and finance, pharmaceuticals and biomedical research, and chemicals and engineering, with Swiss watch-making being world famous. There is also significant agriculture and food production. Swiss wine is developing an international presence, although cheese and chocolate probably remain its most famous international food items. Tourism is an important industry particularly in rural areas that have otherwise been affected by global agricultural restructuring.

International Tourism in Switzerland

International tourism in Switzerland dates from the Romantic period of the late 18th and early 19th centuries. Prior to this time the country was generally passed through rather than sought after as a destination in its own right. However, the massive shift in attitudes toward natural and rural landscapes helped create a favorable perception of mountain landscapes for the first time (see the previous discussion with respect to the Industrial Revolution in Europe). This became extremely important for Switzerland as the development of positive attitudes towards the mountains led to the country being regarded as a sports destination for mountain climbing, a health destination because of the clear air and a location for sightseeing. The development and promotion of skiing saw a new stage in Swiss tourism development with the establishment of many ski resorts, which in summer are also used for alpine walking.

One of the features of Switzerland and the Swiss tourism industry has been its environmental emphasis. Considerable attention is being given to recycling and anti-littering, and significant effort is being given to conserve landscapes and moderate the effects of ski field development. However, in recent years considerable concerns have been raised about the potential impacts of climate change in the country's alpine areas because of the loss of guaranteed snow.

This has affected the financial viability of ski resorts, especially with low altitude runs. Yet climate change, pressure from visitors and overgrazing are also impacting alpine ecologies, with tree lines, alpine grasslands and herb fields shifting in response. In addition, there is increased concern over the long-term future of alpine glaciers, which have an important tourism role as landscape features and for recreation as well as providing a significant source of water supply.

The European Alps and Climate Change

The European Alps are regarded as being particularly sensitive to climate change, with recent warming being approximately three times the global average. The years 1994, 2000, 2002, and 2003 were the warmest on record in the Alps in the last 500 yrs, but climate model projections show even greater changes in the coming decades. There are predictions of less snow at low altitudes and receding glaciers and melting *permafrost* higher up. From 1850–1980, glaciers in the region lost 30–40 percent their area. Since 1980 a further 20 percent of the ice has been lost. The summer of 2003 led to the loss of a further 10 percent. By 2050 about 75 percent of the glaciers in the Swiss Alps are likely to have disappeared, rising to close to 100 percent in 2100.

Andermatt, a popular ski resort in central Switzerland, has decided to act on the threat to winter sports tourism that climate change represents and has started to experiment with a high-technology protective blanket to stop the Gurschen glacier from melting away. Yet the amount of area that can be covered remains limited. Changing the terrain by grading the slopes and rerouting natural streams also carries risks to the natural environment and increases the chances of natural hazards such as flash floods and rockslides.

The European Microstates

There are five mini microstates in continental Europe, namely San Marino, Vatican City, Andorra, Monaco and Liechtenstein. These range in size from 180 mi^2 (468 km^2) for Andorra to 0.17 mi^2 (0.44 km^2) for the Vatican City. Tourism is a very important part of the economies of all five of the European microstates. In fact their smallness and novelty value is something that makes them attractive to many visitors.

Located in the Pyrenees Mountains between Spain and France, *Andorra* is popular as a major duty-free shopping, skiing and hiking destination, and it has several major ski resorts, such as Soldeu/El Tarter, Pas de la Casa, and Pal/Arinsal. Tourism is an important part of the economy, although its tax-haven status means that the banking sector is also significant. Andorra has a population of just over 70,000 people, and the official language is Catalan.

Surrounded completely by Italy, *San Marino* is famous for its archery festivals, historic buildings and smallness. Founded in 301 CE, it is regarded as the oldest constitutional republic in the world. The country has an area of 23 mi^2 (61 km^2) and a population of just below 30,000 people. Although not a member of the EU, it has been allowed to use the euro as its official currency. The language of San Marino is Italian.

Monaco is a major seaside destination known for yachting, gambling (it has a major casino and a few secondary casinos) and hosting a Formula 1 Grand Prix on its street circuit. One of the most expensive places in the world, it is often regarded as a playground of the rich, as there is no income tax on individuals. This has proven attractive to many sports stars and business people. However, the tax-haven status has raised concerns over the control of money laundering and other criminal activities.

Monaco's currency is the euro, although it is not a member of the EU, because it has a close relationship with France, including a customs union, with which it shares its land border. Monaco has a population of approximately 32,000 people and an area of .78 mi^2 (2.02 km^2), although this is expanding as a result of reclamation activities that will see a new ward, Le Portier, constructed by 2015. The language of Monaco is French along with Italian. The traditional national language is Monégasque which is a dialect of Ligurian, although it is only spoken by a minority of residents.

Liechtenstein is a landlocked country of alpine Europe bordered by Switzerland to the west and Austria to the East. It is most commonly visited for skiing, mountain scenery, shopping, museums and hiking. Liechtenstein is a constitutional monarchy with a population of approximately 35,000 people and an area of 61 mi^2 (160 km^2). However, a notable feature of its history is that even though the lands that make up present-day Liechtenstein were purchased by the Liechtenstein dynasty in 1699 and 1712, no member of the family visited the principality until 1818. As with many other microstates, the country is a tax haven with advantageous business tax rates, and it is unusual in that it has more registered companies than citizens. The official language is German and the currency is the Swiss Franc because of a customs union with Switzerland.

Photo: Dallen J. Timothy

◀ *The tiny Principality of Monaco, the second smallest country in the world, is entirely urbanized. It is considered the most densely populated country on the planet, with 42,000 people per sq mile / 16,135 people per sq kilometre.*

The Vatican City, the smallest sovereign country in the world, is best known as the center of the Roman Catholic Church and as a major pilgrimage center for Catholics from around the world. An argument could be made that the Vatican is the most visited country in the world as well, based on the assumption that nearly all tourists in Rome also visit the Vatican, and that the thousands of Italians who visit the country each day for religious purposes are, technically, international day visitors. The Vatican is the last widely recognized state without full UN membership.

While most consider the Vatican City (located in Rome) to be the smallest country, there is one that might be considered to be even smaller, depending on the definition of "country." The Sovereign Military Hospitaller Order of Saint John of Jerusalem, of Rhodes and of Malta (often referred to as the *Order of Knights of Malta* or SMOM, for short), which consists of several buildings located in the Palazzo Malta in Rome. The Order issues passports and has diplomatic relations with over 104 other countries and other international bodies, such as the EU and the International Committee of the Red Cross. The Order has observer status at the United Nations and is a Catholic charitable and social service organization, membership of which includes many of the wealthy and elite from around the world.

The Value of Being Different

The microstates of Europe have been able to use their non-membership of the EU as a source of differentiation in the international marketplace. Because they are not members of the EU and therefore subject to its laws, they can use differences in taxation and regulation to attract investors or to provide products to visitors. For example, Andorra is able to sell a wide range of duty-free products much cheaper than France or Spain. Nevertheless, the microstates in one sense can have the best of both worlds by still being included in various agreements with the EU, particularly in terms of trade and customs. Tax-haven status, in the form of low-cost company registration and minimal business tax rates, is also important for tourism as it often encourages significant flows of business travel.

2.3 | EASTERN EUROPE TODAY

Current Eastern Europe Issues and Resources (http://wrgeography.com/easterneurope.html)

Eastern Europe's Transformation

Physical Landscape

In contrast to the mountains of the south, the middle part of Europe, extending from central France well into Eastern Europe, Europe is a mature landscape. It has low-lying hills and plateaus that generally range from 500–2,000 ft (150–600 m) in height. A few of the peaks in this region reach 4,000 ft (1,200 m). The mountain ranges are interspersed with major basins where considerable settlement has taken place. These mountains are older than those of southern Europe, having been created between 570 and 136 million years ago, during the Caledonia and Hercynian Orogenies. At that time, these mountains were as high as the Alps. Subsequently, they eroded and became mostly submerged below sea level, along with most of Europe and Russia. As Europe began to rise from the sea during the *Alpine Orogeny*, these older mountains were again exposed. Organic deposits in the ancient basins in this area became sites of major coal deposits that contributed to Europe's Industrial Revolution.

The major mountain systems of Central Europe are heavily forested and include the *Central Massive* in France; the *Central German Uplands*; the *Jura Mountains*, which include the *Black Forest* in Germany; and the *Bohemian Massive* in the Czech Republic. The Western Uplands of Scandinavia, Ireland and

Scotland, as well as the *Ural Mountains* in Russia, were also created during this geologic orogeny. Outside of Europe, the Appalachian Mountains in the United States and the mountains of eastern Australia are of the same geologic time period.

The Fall of Communism

Perhaps the most remarkable socio-political event to have occurred in Europe during the past 20 yrs was the collapse of communism (or state socialism) in Eastern Europe in the early 1990s. It dominated the Eastern Bloc countries since the 1940s and functioned as one of the most repressive forms of government in the world. Under communist control, food and fuel were scarce, health care was rudimentary and few citizens were permitted to travel outside their own countries. At best they were allowed periodic holidays in other socialist states of Eastern Europe, Cuba and the Union of Soviet Socialist Republics (USSR).

The exceptions to this rule were Hungarians and Yugoslavs, who enjoyed the most freedom to travel among all the citizens of the Eastern Bloc countries; occasionally they were even permitted to travel to Western Europe on their annual vacations. Nonetheless, in most cases, considerable efforts were made by the various eastern European regimes to curtail contact between the people in the west and those in the east. In many cases, eastward travel by citizens of Western Europe, North America and other parts of the capitalist world was made difficult by strict government regulations, including difficult visa requirements and mandatory minimum per diem currency exchange rules.

Glasnost and Perestroika

Among the earliest events to bring about change in the communist east was the establishment of workers' solidarity unions, which began working for more rights and freedoms for blue-collar employees. Many people were imprisoned for their efforts in these movements, but the efforts had the intended effect of getting citizens in Poland and other strict communist states to think for themselves and to begin asserting their rights. In the late 1980s, changes began to occur throughout the entire Eastern Bloc as Soviet President Mikhail Gorbachev initiated his policies of *glasnost* (openness) and *perestroika* (restructuring), which aimed to create an air of openness in public discussions about past and present problems facing the USSR. Perestroika was meant to encourage a new way of thinking about the social, political and economic structure of the communist state and to begin establishing a free-market economy with some degree of private ownership and profit making.

Once the seeds of a market economy had been planted, however, the floodgates opened and people began demanding more freedoms and more accountability from their governments. Unintentionally, Gorbachev's policies became the medium for bringing down the totalitarian system that had governed the Eastern Bloc countries for decades. The resultant new economic and social liberties related to religious worship, freedom of speech, freedom to strike and freedom to assemble destabilized the autocratic governments of Eastern Europe.

The fall of communism began in Poland and the German Democratic Republic (East Germany) in 1989 was underscored by the fall of the Berlin Wall, and was followed closely by the USSR's Baltic republics (Lithuania, Latvia and Estonia) and other states lying east of the *Iron Curtain* declaring their independence. While these upheavals resulted in newfound freedoms, they also destabilized a long-established system of despotic control and predictability, resulting in increased crime rates, ethnic unrest, economic downturns, increased scarcity of food and supplies and growing poverty levels. Finally, during 1991 and 1992, the Soviet Union was dissolved, forming 15 new independent countries, some of which are discussed below. Likewise, in a ripple effect, Czechoslovakia split into two new countries on January 1, 1993 (the Czech Republic and Slovakia).

Yugoslavia's Disintegration

Yugoslavia, whose diverse religious and ethnic tensions had been kept in check by the power of communism and the personality of General Tito, dissolved, erupting into a series of bitter civil wars during the 1990s and into a series of new states (Bosnia and Herzegovina, Croatia, Slovenia, Serbia, Montenegro and the Republic of Macedonia). The Republic of Macedonia, which declared independence in 1991, was the only former republic to gain sovereignty without resistance from the Belgrade-based Yugoslav authorities.

Kosovo

The long-term future of the province of Kosovo is still to be decided, although it declared independence from Serbia in 2008. Some parties within Kosovo wanted independence, others desired a merger with a greater Albania, and the Serb minority still wishes to have a political relationship with Serbia. As of 2014, 108 UN member countries and Taiwan officially recognized Kosovo's independence. Most of these countries have established diplomatic relations with Kosovo. Many of the countries not acknowledging the province's independence are ones who themselves face separatist movements within their own territories, such as Spain, who themselves face separatist movements within their own territories.

The Republic of Kosovo has joined several multinational organizations, including the World Bank and the International Monetary Fund. The European Union and the United Nations in general remain neutral toward the status of Kosovo, although Serbia and its ally, Russia, believe the separation to be in violation of international law. In 2010, however, the International Court of Justice concluded that Kosovo's declaration of independence was legal.

Travel in Post-Communist Eastern Europe

These changes were extremely profound for tourism for several reasons. First, many of the post-communist countries of the east (e.g., Poland, the Czech Republic and Hungary) skyrocketed onto the World Tourism Organization's top 20 list of tourist destination countries in the mid-1990s. This is primarily because people of Eastern European descent, who had earlier migrated to the west, were now free to return to their homelands to visit relatives and places they had not seen for nearly half a century. However, Hungary has always been the most westernized country and has a potentially large tourism market because of its cultural ties with other European countries and its accessibility to the markets of Austria, Germany and Italy.

Second, the citizens of Eastern Europe became free to travel to the west to explore the lands they had heard of but had never before had opportunity to visit. Third, traveling to the east from the west became easier with fewer visa and currency regulations. As a result, the countries of Eastern Europe opened up to floods of tourists from the west and became especially popular owing to their status as former communist states, as well as to their relatively inexpensive prices in terms of transportation, food and lodging. Today, most countries of the former communist bloc have embraced tourism and have begun to thrive as important destinations. Others, however, such as Belarus and Moldova, have resisted the urge to develop tourism as a major economic growth catalyst. Other countries struggle to develop tourism against a background of negative ongoing civil unrest (e.g., Serbia, Albania, Georgia, Armenia and Azerbaijan).

Communist Heritage

Soon after the collapse of communism, many countries of Eastern Europe and the former USSR desired to abolish signs and symbols of their oppressive heritage (e.g., statues, communist factories). Since the late 1990s, however, the heritage of communism has become an important tourism product in many of parts of the former communist states of Europe. In cities throughout the region, tourists can purchase "communist nostalgia tours" that will take them through communist-era neighborhoods, view Soviet-inspired architecture and visit preserved parks, war memorials, socialist-era homes or apartments, factories, eateries and other public spaces that typified life under communism. The tours are sometimes led in Soviet or East German vehicles, provide old-fashioned food and drink, and allow tourists to dress in period clothing for photo opportunities.

Eastern Europe: EU Countries

EU Enlargement in Eastern Europe

In 2004 a number of the former state communist countries of Eastern Europe joined the EU in a process known as accession. In addition to Malta and Cyprus, the Czech Republic, Estonia, Hungary, Latvia, Lithuania, Poland, Slovakia and Slovenia joined the EU. This was the largest number of countries admitted into the EU at one time. On 1 January 2007, the former communist countries of Bulgaria and Romania joined. Croatia joined in 2013. To join the EU, countries must fulfill the Copenhagen criteria (discussed above). Other candidate countries in Eastern Europe include the Republic of Macedonia, which applied in 2004. The other successor states of the Socialist Federal Republic of Yugoslavia (Bosnia and Herzegovina, Montenegro and Serbia) have all adopted EU integration as an aim of foreign policy. Albania, Moldova and Ukraine have done likewise.

However, given that the EU has had difficulties with its own constitution and member country economies, there are concerns from existing members that the accession and enlargement process has already been overstretched, particularly as the admission of Albania and the rest of the Western Balkans into the EU has already been set as a priority by the alliance. This situation may make it extremely difficult, or at least delay for many years, the potential future accession of the former Soviet Republics and the south Caucasus: Armenia, Belarus, Georgia, Moldova and Ukraine. Including these countries in the EU would likely to receive opposition from Russia, with the 2014 conflict in Ukraine and the return of the Crimea to Russia, illustrating the tensions between the EU and Russia over international relations.

Hungary

Hungary is a land-locked country in central Europe. It was part of the Austro-Hungarian Empire in the mid-19th century and became an independent country after the First World War. However, the period when it was part of the Habsburg Empire and, prior to that, part of the Ottoman Empire has left the country with a rich cultural and architectural heritage that is a significant drawing card for tourists.

Budapest, Hungary's capital city, is one of Europe's gem cities and a very popular tourist destination for western Europeans. Its historic architecture and location on the Danube River provides considerable appeal. Hungary was also recognized as being one of the more liberal states of the former Eastern bloc and played a critical role in the downfall of communism in Eastern Europe. In 1989 Hungary opened its border with Austria, allowing thousands of East Germans to escape to the West. With such a hole in the so-called "Iron Curtain," it was only a short matter of time before the Berlin Wall would also fall.

Hungary was initially the focal point for much of the foreign investment into Eastern Europe after the collapse of state communism. However, in 2008, Hungary, like a number of other European countries, experienced considerable economic difficulties. These forced the government to appeal to international financial institutions, such as the International Monetary Fund and the World Bank, for massive loans in order to stave off economic collapse.

However, despite such difficulties, considerable attention has remained on tourism as a significant means of economic development. These measures have also encouraged other Europeans to purchase

▶ *A tourist boat makes its way on the Danube River through Budapest, the capital of Hungary. The Danube, which flows from the Alps to the Black Sea, is the most important river connection between Western Europe and Eastern Europe.*

Photo: Jarkko Saarinen

holiday homes in the country for they are considerably cheaper than many of the existing second home areas in Spain, France and Italy. Hungary is home to Lake Balaton, the largest lake in central Europe and a major spa tourism destination. It is also increasingly recognized as a food and wine destination, particularly in the Tokaj-Hegyalja region, which is world famous for its Tokaji wine.

The Czech and Slovak Republics

As noted above, on January 1, 1993, Czechoslovakia split into Slovakia and the Czech Republic. It created overnight a new international border, complete with customs and immigration stations and border-crossing formalities. This is often referred to as the "Velvet Divorce" in reference to the 1989 "Velvet Revolution" when Czechoslovakia's state communist government fell. With the collapse of communism in the early 1990s, the Czech Republic took off as a tourist destination, partially because of the desire of people of the Czech diaspora to visit the homeland that had been essentially inaccessible for several decades. National parks and woodlands provide much of the tourist appeal of these two new countries, although their capital cities, Prague (Czech Republic) and Bratislava (Slovakia), are well known for their unique urban architecture, picturesque city squares, castles and fortresses, although Bratislava also contains a number of areas of Soviet brutalist architecture, especially for housing.

Food and Beer

Prague is well known for its food and beer and is a popular destination for short-break holidays from Western Europe. Farmers' markets made a return to the city in 2010 and have proven extremely popular with locals and tourists alike. Prague is famous for its historic coffeehouses. The Czech Republic is also home to another beverage–beer. Czech beer has been brewed since at least 993 CE for written documentation indicates that Benedictine monks brewed beer in the Bfevnov Monastery. The first historical evidence of beer making on the territory of the Czech Republic is King Vratislav II's foundation deed of

ISSUES AND INSIGHTS Spas and Health Tourism in Hungary

Traveling for health reasons is an important motivation for tourism throughout many parts of Europe. The supply of health tourism products depends on the availability of health-care facilities and natural resources, including appropriate landscapes, climates, mineral springs and the development and promotion of resorts. Hungary currently considers itself one of the key players in global health tourism. This positioning is actively supported by both the Hungarian National Tourist Office and the Ministry of Local Government (the governing body responsible for tourism and tourism development). Hungary is actively promoting health tourism and rebranding the country as an international center for spa and wellness tourism. It should be noted that the term "spa" in the Hungarian context refers more to medical/thermal baths than to beauty or leisure facilities, which is how the term might be interpreted in North America.

Hungary has a long history of spa and health tourism due to its large number of thermal springs. People traveled to Hungary for its healing waters for thousands of years. Today, almost 300 out of 1300 thermal springs are used as public baths. The thermal and medicinal waters are known to have a healing effect on rheumatic problems, cardiovascular disorders, arthritis, skin problems and asthma. About 130 mineral and medicinal springs are located in the capital, Budapest and there are 13 Hungarian settlements formally recognized by the government as "health resorts": Balf, Biik, Kekesteto, Lillafured, Balatonfiired, Debrecen, Eger, Gyula, Hajdiiszoboszlo, Harkany, Heviz, Parad and Zalakaros.

Germans and Austrians have also been traveling to Hungary for dental treatment combined with visits to traditional spas since the fall of the Iron Curtain. Because of lower wages and utility costs, dental treatment in Hungary is far less expensive for German patients, and "tooth tourism" has become a major contributor to the Hungarian economy. Currently, Hungary draws large numbers of patients from Western Europe and the United States for high quality cosmetic and dental procedures that cost half of what they would in the tourists' home countries.

Spas and wellness have been identified by the Hungary Tourism Organization, the national tourism agency, as one of the country's four most important international tourism products, alongside convention and incentive tourism (MICE), and recreation in the Budapest and Lake Balaton regions. To promote the segment, 2003 was designated and promoted as the Year of Tourism of Health. From an international perspective,

the HTO focused its efforts on large-scale press campaigns in Germany, France, Italy, Sweden, Denmark, and Finland.

In 2002, the Hungarian government developed the so-called Szecheni Plan to support health tourism-related projects. The Plan aims to achieve the following objectives for health tourism:

- Modernize existing health resorts and hot springs, and improve their services.
- Promote the use of known thermal water resources not yet used for tourism purposes.
- Develop the infrastructure of health resorts and hot springs through a coordinated and tourism-oriented development of health and thermal water-related services and tourism-related infrastructure.
- Develop an institutional and regulatory framework. For example, the Hungarian government has introduced a rating system to provide an enhanced level of health facilities throughout the country. It assesses facilities and services such as baths, saunas and steam baths, fitness equipment, skin care services, therapy services and vegetarian food services.
- Ensure appropriate marketing for health and hot spring tourism.
- Develop human resources.

In the National Development Plan II (2007–2013), subsidies were available from the government to develop health tourism in Hungary. Forty percent of the amount was for developing attractions and services, and 60 percent for developing lodging facilities (wellness, medical hotels). The six-year (2007–2013) National Health Tourism Development Plan defined several priority areas:

1. The conscious product development of baths with international significance.
2. Strengthening the contact of traditional medicine and tourism.
3. Product development at wellness services.
4. Capacity development of medical and wellness hotels.
5. Development of marketing activity in the area.
6. Development of an educational system for wellness and medical service providers.
7. Development of a national and regional operating system of health tourism (like destination management).

the Vysehrad Cathedral. Issued in 1088 CE, the deed granted the canons of the Vysehrad Cathedral a tithe of hops for brewing beer.

Given beer's long history and the fact that beer is cheaper than bottled water in Prague, it is not surprising that the Czech Republic consumes more beer per capita than any other country in the world, averaging about 161 liters of beer per person each year. Budweiser Biirgerbrau, or Budweiser Bier, was

▶ *The Charles Bridge in Prague, the capital of the Czech Republic, was built in the 14th century CE and crosses the Vltava River, connecting the Prague Castle with the city's commercial downtown. It is lined with 30 baroque-style statues and protected by gothic-style guard towers, as seen in this photo.*

the original Bud, founded in the city of Budweis in 1785. In 1876, the U.S. company, Anheuser-Busch, borrowed the name for its Budweiser beer. The Budweiser Budvar Brewery sold Czech Budweiser from 1895 on, although in North America and elsewhere in Europe it is sold under the name of Czechvar. Another well-known beer variety that is Czech in origin is Pilsner, named after the city of Plzen (Pilsen in German) some 90 km west of Prague.

Other significant attractions in the Czech Republic include the Kokofinsko Nature Reserve, an area of primeval forest and sandstone, 60 km northeast of Prague. It has a number of walking tracks, but surprisingly few international tourists visit it.

Dark Tourism

Like a number of countries that were under German occupation in the Second World War, there are also sites of "dark tourism." The Nazis used the town of Terezin (Theresienstadt in German), an 18th century garrison town 60 km north of Prague, as a concentration camp. About 140,000 Jews from all over Europe were interned here, and the entire town was turned into one enormous ghetto. Although it was not an extermination camp, because most of its prisoners were sent on to Auschwitz or other death camps, some 34,000 people still died there. In 1947, the National Suffering Memorial, later renamed the Terezin Memorial, was opened there. The key mission of the Terezin Memorial, the only institution of its kind in the Czech Republic, is to commemorate the victims of Nazi political and racial persecution during the occupation of the Czech lands in World War II, to promote museum research and educational activities and to look after the memorial sites connected with the suffering and death of thousands of victims of violence.

Slovakia

The Slovak Republic, or Slovakia, first achieved independence during the World War II, as a dependency of Nazi Germany. The present Slovak Republic became independent on 1 January 1993. Its capital Bratislava on the banks of the Danube borders both Austria and Hungary. Although initially it had relatively poor political relationships with Western Europe following independence, Slovakia is now the only former communist country to be part of the European Union, Eurozone, the Schengen Area and NATO simultaneously. Tourism has not developed into a major industry in Slovakia, although there is substantial cross-border travel, particularly to and from the capital and from the Czech Republic. Like Hungary, Slovakia is developing its health and spa tourism sector while extensive forests and mountain areas in the north offer outstanding alpine and nature-based tourism opportunities.

The Baltic States: Estonia, Latvia and Lithuania

Prior to their forced absorption into the Soviet Union, Estonia, Latvia and Lithuania were independent states from 1917–18. Prior to that time, they had been part of the Russian Empire, which had, in turn, defeated Sweden in the Great Northern War of 1700–1721 to gain the northern territories that comprised most of present-day Latvia and Estonia. Lithuania, however, has a stronger history of independence. In the 14th century, the Grand Duchy of Lithuania was the largest country in Europe. It included present-day Belarus, Ukraine and parts of Poland and Russia.

In 1569, Poland and Lithuania formed the Polish-Lithuanian Commonwealth. The Commonwealth lasted until 1795 when the Russian Empire annexed most of Lithuania's territory. Despite the historical claims of Russia to the Baltic States, the United States and other countries never recognized their Soviet integration at the end of the World War II. They kept their embassies open in Washington, DC, anticipating the time when they would once again become self-governing and return ambassadors. These three republics were among the first to declare independence from the USSR, which led to violent forceful tactics by the Soviet military. Soon after, all three were recognized by the world community as independent states, and the USSR had little choice but to let them go.

◄ Tallinn is the capital and largest city of Estonia. It is situated on the Gulf of Finland, which is part of the Baltic Sea, and it is an important tourist-historic city. Tallinn is especially popular as a weekend or short-trip destination for Finns, with several ferries travelling the Gulf of Finland each day, providing easy access between Helsinki and Tallinn.

Photo: Jarkko Saarinen

Tourism in the Baltics

Today, tourism is a high priority for all three countries. Estonia particularly has close cultural and historical ties to Finland, and therefore is a significant destination for Finns who often travel by ferry boat across the Gulf of Finland for day trips to Tallinn or weekend getaways. The old city of Tallinn has been listed as a "World Heritage Site" and is a popular tourist destination and major gateway to the Baltic region. Estonia's coastal islands are popular destinations and are growing in popularity as holiday second-home locations. Visitors from Sweden are common, for the region has long had close ties with that country.

Soviet Era Tourist Attractions

Latvia and Lithuania are also burgeoning tourist destinations. Both countries' capitals, Riga and Vilnius, are vibrant cities with interesting architecture and history. There is also a growing interest in the heritage of the Soviet era for tourists. For example, "1984: Survival Drama in a Soviet Bunker" (http://www.soviet-bunker.com/en/), is a three-hour long, quasi-theatrical experience available to visitors in a genuine Soviet bunker in the forest some 40 kms from Vilnius.

One of the most interesting examples of Soviet era heritage is the Karosta Prison, which was a detention facility until 1997 but is now a tourist attraction. Originally built as an infirmary in 1900, the Soviets, Nazis and the Latvians themselves used the place as a military prison. The prison is supposedly haunted and the authenticity of the experience is enhanced by the graffiti left on the walls by inmates. It is also possible to pay to spend the night in jail as a prisoner. Such "guests" are subjected to regular bed checks and verbal abuse by guards in period costume. The facilities, such as latrines, are of a standard prisoners would have expected. Regular tours are available for the more fainthearted. Although not so staged, the KGB past is also part of tourism attractions in Lithuania.

The Museum of Genocide Victims in Vilnuis is housed in a former KGB headquarters and prison where dissidents were tortured, often before being executed. Grutas Park near Druskininkai (sometimes referred to as "Stalin's World") is an outdoor collection of the country's Soviet era statues as well as buildings housing a range of other exhibits from the communist era. The statue collection is undoubtedly of growing heritage significance, especially as the only Soviet statues left standing in the capital of Vilnius are the socialist-realist figures that adorn the four corners of the famous Green Bridge. These are frequently doused in green paint by nationalist protesters.

Kaliningrad (Russia)

Given their strategic location on the Baltic coast, the Baltic States have long acted as a cultural bridge between Eastern and Western Europe. Castles, music festivals and scenic rural areas contribute substantially to their international tourist appeal, but Estonia is the state with the strongest international tourism sector. Although not a nation-state as such, an interesting and potentially significant territory to the south of Lithuania is the Russian exclave of Kaliningrad. Originally named Königsberg in German, the town was founded in 1255 and was part of Prussia and then Germany until 1945. Following the transfer of the territory as part of the peace settlement at the Potsdam Conference, it became part of the Soviet Union, which renamed it Kaliningrad after Mikhail Kalinin, one of the original Bolsheviks.

Following the collapse of the Soviet Union, Kaliningrad became geographically separated from the rest of Russia, and this distinctness became even more pronounced when Poland and Lithuania became part of the EU. Economically, however, these developments have probably become one of the best things that could happen to the territory. In 1996, Kaliningrad was designated a *Special Economic Zone* with tax and duty benefits. It also benefits from a number of projects funded by the European

Commission. The growing economy and the city's rich cultural heritage, which includes many museums and the reconstruction of some of its Prussian heritage, provide a strong basis for tourism growth in the future.

Poland

Poland is one of Europe's larger countries and home to many cities with global appeal, as well as scenic rural areas that keep numerous traditions alive. Krakow's old city possesses architectural treasures from a variety of eras and churches. (Poland was one of the few eastern countries where religion, i.e., Catholicism, was able to flourish under communism.) Warsaw, the capital, which was nearly destroyed in World War II, was rebuilt as a drab communist capital, but today it is a cosmopolitan city with shops, hotels and a variety of tourist attractions. The historic centers of Krakow and Warsaw have been designated UNESCO World Heritage Sites. Gdansk, one of Poland's largest cities and a major seaport, is very historic, with unique architecture and religious edifices. A German-ethnic city before World War II, Gdansk was a semi-independent city-state between 1919 and 1939, when it was annexed by Germany. The city's German heritage is still part of its appeal for German tourists and others.

Concentration Camps

Auschwitz-Birkenau is one of Poland's most important heritage destinations. It was the largest concentration camp in Hitler's German Reich (Germany occupied Poland at that time), and some 1.5 million Jews and other minorities were tortured and killed there. National parks near the borders with the Czech Republic and Slovakia have been designated International Parks, owing to their status as cross-border

Photo: Dallen J. Timothy

▲ *The historic center of Warsaw is an important heritage place in Poland that attracts both local residents and tourists. Much of the historic zone has pedestrian-only access, and the entire Historic Centre was designated a UNESCO World Heritage Site in 1980.*

nature preserves. This has led to a great deal of bi-national cooperation between Poland and its neighbors in the areas of environmental improvements, conservation and nature-based tourism development.

In common with many other countries of Europe, the eastern bloc countries, including Poland, have begun to promote themselves as significant diaspora destinations. This means that they are attempting to cater to people of Polish descent who live in Australia, Canada, Israel, the United States, the United Kingdom and other diverse places who return to the land of their ancestors. Their reasons include discovering their own roots or personal heritage, meeting distant relatives, studying Polish, conducting family history research and visiting villages and churches where their ancestors might have lived.

Slovenia

Unlike its former Yugoslavian partner republics (e.g., Croatia and Bosnia-Herzegovina), Slovenia was spared much of the violence and bloodshed meted out by the Yugoslavian military when the republics began to break away in the early 1990s. Today Slovenia is known for beautiful mountain scenery, lakes, wine and food, and its colorful capital city, Ljubljana, which has many baroque and Viennese Secession buildings. Per capita, Slovenia is the richest Slavic nation-state. Slovenia has a substantial manufacturing base, but tourism is becoming increasingly important for the country of two million people. In great part this is because there is a wide variety of landscapes available to visitors in a small space, ranging from the coastal Mediterranean to alpine areas.

As with many central European countries, there is a rich spa tradition. Cave tourism is also significant. (The Karst Plateau in Slovenia provided the name for *karst topography*, which describes a landscape shaped by water that dissolves the carbonate bedrock-usually limestone or dolomite-to form caves.) At 5.3 km (3.3 mi), Postojna Cave is the longest publicly accessible depth of any cave system in the world that is open to the public, while the Škocjan Caves, which host one of the largest known underground canyons in the world, are inscribed on the UNESCO World Heritage List. The other significant form of tourism in the country that contributes substantially to government revenue, is gaming. Slovenia has one of the highest numbers of casinos per capita in Europe.

Romania

Romania is the ninth largest country of the European Union by area, and it has the seventh largest population of the European Union, with 22 million people. The capital is Bucharest, a city of two million people on the banks of the Dâmboviţa River in the south of the country. Bucharest has never been a port city although there is an unfinished canal project. When finished, the 73-km (45 miles) long Danube-Bucharest Canal will link Bucharest, Romania, to the Danube via the Argeş River. Begun in 1986, the project was halted in 1990 following the ending of communist rule. It would have regulated river flow, prevented flooding and generated hydroelectricity. Completion of the project is now seen as important for the transport and economic development of Bucharest, and it is likely it would be a significant link to the Black Sea.

Transylvania

Romania's image was terribly tainted under the dictatorship of Ceausescu, but it has recovered in the nearly two decades since the fall of communism. Since then it has become an important tourist destination. Romania is a popular destination on Danube River cruises that reach the Black Sea. The Transylvania region is home to the vampire myth of Count Dracula (based on an actual person, *Vlad the Impaler*, but not a vampire). Tourists arrive by the busload to see the famed castle and region made increasingly popular by dozens of movies and novels set in the beautiful, yet feared, forests of Transylvania. Romania's rural lifestyle and peasant traditions are important for tourism, as are its cities, which are major gateways.

Although Romania has a significant cultural heritage, with Bucharest once referred to as the Paris of the East, its image has been affected by the architectural brutalism of the Ceausescu era.

Most notable is the Palace of the Parliament (Palatul Parlamentului in Romanian) in Bucharest, which is a multi-purpose building that presently contains both houses of the Romanian Parliament as well as the National Museum of Contemporary Art and the Museum and Park of Totalitarianism and Socialist Realism. Construction began in 1983 and resulted in the demolition of much of Bucharest's historic district. According to the Guinness Book of World Records, the Palace is the world's largest civilian administrative building, the most expensive administrative building and the heaviest building.

Bulgaria

Since its communist rule collapsed, tourism has become an important priority for Bulgaria as well. Like Romania and other neighbors, Bulgaria is a favored stop on Black Sea-Danube cruises, and its rural appeal evokes a similar *mental landscape* of horse-drawn carriages, farmers and small villages. However, recently both countries have been promoting the development of coastal tourism based on the Black Sea. Kavarna on the northeast coast of Bulgaria is one of the most entrepreneurial of the seaside resorts. It plays host to an annual heavy metal music festival as well as new golf courses. Although these resorts have historically been geared more to domestic and Eastern European tourists, there is growing interest from Western Europe, especially for second homes because of their relative affordability in comparison with the Mediterranean.

Eastern Europe: Non-EU Countries

Albania

Before the 1990s, Albania was often considered the strictest communist country in Europe. U.S. citizens were forbidden entry, although Europeans could visit if they managed to get a visa, and all visitors were required to look a certain way. For example, in addition to the need for special permission and the difficulty in obtaining visas, men were required to have short hair and clean-cut appearances. Likewise, all visitors were required to be accompanied continually by a guide and were virtually prohibited from speaking to Albanian citizens. This built Albania's reputation as the most isolated country in Europe and, in the company of North Korea, in the world. Following the 1992 free elections, a new, more democratic government was put into power and the nation opened up a capitalist market economy.

Although Albania suffered some political and economic instability in the 1990s, and some regions were marked as dangerous to visit, tourism is growing rapidly and has become a priority for the present government. Most of Albania's tourism resource base is founded on ancient Greek and Roman sites, intriguing village life, and sun and sea along the Ionian Coast. Nature-based tourism is an important emerging activity given the country's *biodiversity* and geography.

However, the openness of the economy and the relative cheapness of the country compared with other Mediterranean destinations have provided a major boost to the tourism industry as well as to second-home development. Albania's increasing profile in tourism is perhaps best evidenced by the fact that it was chosen the top country in Lonely Planet's list of the ten top countries for 2011 and continues to receive positive reviews and media coverage.

The Former Yugoslavia Today

The *dissolution of Yugoslavia* in the 1990s was a major turning point for the Balkan region of southeastern Europe in geopolitical and tourism terms. While, as noted above, *Slovenia* was spared significant devastation, *Croatia* and *Bosnia-Herzegovina* were not. The Croatian War of Independence (1991–1995) was a brutal conflict where Croatia fought with the Yugoslav army for separation from Yugoslavia. During the latter part of this period, the battle was between Croatia's nascent military and the new country's Serb minority, which desired to remain connected to Serbia and greater Yugoslavia.

Croatia

Croatia won the war and is now a thriving, cosmopolitan nation populated primarily by ethnic Croatians. The Bosnian War (1992-1995) involved many players in Bosnia and Herzegovina and was known for its brutality and neighbor-against-neighbor fighting tactics. In this conflict, Croatians, Serbian and Bosnian Muslims fought against each other and destroyed each other's villages and religious buildings and icons. In 1996, a *peace treaty* (the Dayton Accord) was signed, ending the Bosnian War and forming a federal state that includes large populations of Serbs, Croatians and Muslims. While conditions have much improved, there are still some underlying ethnic tensions in Bosnia and Herzegovina, and keeping peace is a delicate balancing act.

Tourism is important for Croatia and Bosnia-Herzegovina. Croatia's coastal areas, particularly its famed cities of Dubrovnik and Split, are important urban heritage destinations, some replete with notable remains of the Roman Empire. The Yugoslav military targeted and destroyed much of Dubrovnik during the Croatian War because of its nationalistic importance to Croatian identity. The city has healed, however, and many of the damaged buildings have been repaired. The old town is a UNESCO World Heritage Site and one of Croatia's best-known destinations. Croatia's membership in the EU in 2013 and the Schengen area in 2015 is also an important boost for tourism development. The dramatic scenery associated with the Dalmatian Alps is a particularly significant tourist draw, and the country's Adriatic coastal and island resorts are booming, particularly popularized by German and other west European tourists.

Bosnia and Herzegovina

Bosnia and Herzegovina is a country of significant contrasts. Its beautiful mountain and rural scenery is often a backdrop to the dramatic ruins of the Bosnian War, many of which are still visible in the landscape. Cities such as Mostar and Sarajevo (the capital) were devastated as warring ethno-religious groups

▶ *The historic city of Mostar, Bosnia-Herzegovina, was heavily damaged in the Yugoslavian wars of the mid-1990s. Its historic bridge was destroyed but later rebuilt and is visible in this photograph. Today Mostar is becoming a popular tourist destination in the former Yugoslavia.*

Photo: Dallen J. Timothy

and military factions targeted and destroyed many ancient structures and historical sites. Sarajevo was popularized when it hosted the 1984 Winter Olympics, but later it became plastered in the world's news headlines as a site of aggressive wartime fighting. The country is recovering economically and socially from the war though and focusing more on tourism as a tool for development. The war provides much of the backdrop for tourism. Mostar and Sarajevo both use it as a tourist commodity, promoting destroyed buildings and souvenirs made from artillery shells and other wartime devices.

Macedonia

Following its 1991 independence from Yugoslavia, the Republic of Macedonia was called the Former Yugoslav Republic of Macedonia (FYROM) in an effort to appease the Greeks. They objected to the name Macedonia because of their widely held contention that Macedonia was a historically Hellenic (Greek) region and name. While the official name of the country now is the Republic of Macedonia, Greece still officially recognizes it as FYROM, and many Greeks informally refer to it as "so-called Macedonia."

Macedonia did not experience the conflicts of the Yugoslavian wars in the 1990s, although it did experience some minor skirmishes with ethnic Albanian rebels in 2001. It was also shaken slightly by the 1999 NATO fighting with the Yugoslav military in Kosovo, which led thousands of Kosovo Albanians to flee to Macedonia. The country today is waiting to become a member of the EU. Macedonia is gaining importance as a European tourist destination and is especially notable for its Roman and Byzantine ruins, churches, natural environment, excellent cuisine and wine, and massive construction projects in the capital, Skopje, to erect new "historic" monuments and statues throughout the city.

Eastern Europe: Russia and the CIS

Russia, the world's largest country, has a vast diversity of natural and cultural landscapes that create widespread appeal for tourists. The expanse of Siberia (see below), with its mountains and volcanoes, deserts, forests, coastlines and navigable rivers, contains important natural elements. Numerous cultural regions, languages, ethnicities and religions contribute to a wide array of folk practices (e.g., music, art, cuisine, dress, village layout and agricultural landscapes) that provide considerable charm for tourists. Its cities are

The Winter Palace in St. Petersburg, Russia, was the royal residence of the Russian monarchs until the overthrow of the royal family by the Bolshevik Rebellion of 1917. It now serves as part of the State Hermitage Museum, housing priceless artworks and ancient artifacts.

perhaps its most visited districts, the most important being Moscow (the capital) and St. Petersburg (known as Leningrad during the state communist period). St. Petersburg possesses some of Russia's most valuable architectural and artistic gems, many associated with the ruling czars and its national heritage. Moscow houses the government and most central organizations for the arts, sciences and sports in the country.

For potential visitors, Russia still holds on to many of its old communist-era policies and practices. For example, strict rules are still in place in applying for a visa. Visitors often must make and pay for accommodation reservations in advance, and letters of invitation frequently must accompany visa applications. This old-style approach to tourism differs significantly from several of the former USSR states that have opened up to international arrivals and eliminated most visa requirements. In spite of these tight regulations, the political instability in several regions (e.g., Chechnya), accusations of corruption at all levels of government, economic crises and a globally recognized organized crime establishment, there is considerable world demand for travel to Russia.

Commonwealth of Independent States

CIS stands for Commonwealth of Independent States. The CIS was formed in December 1991 after the breakup of the USSR, and it includes most of the countries that were formerly Soviet Socialist Republics, the largest of which, by far, is the Russian Federation (Russia). The only former Soviet Republics not part of the CIS are the Baltic countries of Latvia, Lithuania and Estonia—all now members of the EU, and Georgia. Ukraine is a founding state of the CIS but along with Turkmenistan has not ratified the charter but has participated in meetings. Following Russian occupation of parts of Moldova, Georgia and Ukraine, Moldova is also reviewing its involvement.

The CIS is primarily an economic union (similar to the North American Free Trade Agreement), which fosters tariff-free trade among member states and facilitates the movement of goods, services, capital and labor. The CIS can basically be divided into two major sub-regions: the Slavic language region, which lies west of the Ural Mountains and is therefore part of the European continent, and the Turkic languages region that lies mostly to the east of the Urals.

This easternmost region is sometimes referred to as Eurasia, partly because in October 2000 the heads of Belarus, Kazakhstan, Kyrgyzstan, Russia and Tajikistan signed an agreement on the creation of a Eurasian Economic Community. Uzbekistan stated its intent to join in the future, while Armenia, Moldova and Ukraine presently have observer status. The term CIS has not gained popularity worldwide, and is generally much more recognized in Europe than it is in North America.

The western portion of the CIS is closely related historically to Europe, although it also has had a history of closer relations with the countries and peoples of Central Asia than have other European countries. The countries that comprise the European CIS today include Belarus, Ukraine, Russia, the Romanian-speaking country of Moldova and the Transcaucasian countries of Georgia, Azerbaijan and Armenia, which have their own distinct ancient languages.

There are many Central Asian minority enclaves throughout the region of the CIS. The people in these ethnic enclaves settled many hundreds of years ago as part of migrations and invasions that crossed the Russian lowlands from Central Asia, including Mongolia and Southwest Asia, and from Iran (Persia) and Turkey. Further to the west, European culture and history were closely tied to the seas and oceans through their many peninsulas, bays and rivers because of the opportunities provided for trade. European influences here included ancient Roman communities and, more significantly, substantial numbers of Viking communities from the Baltic to the Black Sea. The CIS part of Europe, on the other hand, is far from large bodies of water and is instead closely tied to the large Eurasian landmass, and the unique opportunities and challenges this environment presents.

◀ The surface physiography of the Asian land mass. This is the largest contiguous land area on the planet, and it includes every major land, water and climate variation. Europe is one of the peninsular extensions of this land mass, along with the Arabian Peninsula, South Asia and Southeast Asia. Russia dominates the northern third of the Eurasia land mass.

© AridOcean, 2011. Used under license from Shutterstock, Inc.

European CIS

The European portion of the CIS covers half the land area of the continent of Europe, and a quarter of the land area of the former USSR. The country of Russia spans this European section and then extends far to the east beyond the Ural Mountains and all the way to the Pacific Ocean. Russia is the largest country in the world, and it covers nine times zones (reduced from 11 in 2010), which is more than a third of the 24 time zones that extend around the globe.

Central Asia CIS

To the east are the Asian countries of the CIS. These include the Central Asian republics (or countries) of Kazakhstan, whose population is half Russian and half Kazakh, Uzbekistan, Turkmenistan, Kyrgyzstan and Tajikistan. (Mongolia and China's far western Xinjiang Province are sometimes considered part of Central Asia, but these will be discussed in the East Asia reading.) The culture, history and physical geography of Central Asia have much in common with those of the Middle East. The major exceptions include a much colder winter climate and a long period of domination by Russia and the USSR.

Because the land area of the CIS is so massive, it is well endowed with a great diversity of natural resources. Its size, however, made integration difficult for the Soviet Union because transportation and communication networks were very expensive to build across such a vast territory. Another challenge for this region is that 80 percent of it lies further north than the Great Lakes of North America. St. Petersburg, formerly known as Leningrad, is at about the same latitude as Anchorage, Alaska. Moscow is at the same latitude as Ketchikan, Alaska. Ukraine is at the same latitude as the United States-Canada border, while the Caucasus Mountains are at about the same latitude as Chicago and New York. Russia also has a longer

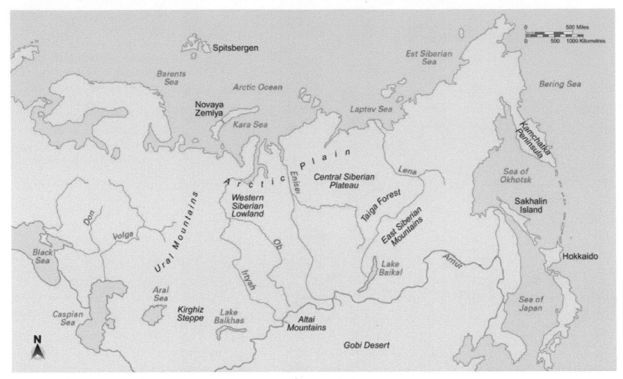

▲ *Major Physical Geography Features of Russia.*

border than any other country in the world, and the third longest coastline, much of which is frozen and inaccessible in the winter months although increasingly accessible in summer, leading to the growth of Arctic summer cruise tourism.

Physical Landforms

The dominant landform in the CIS region is the Great Russian Lowland, which is an extension of the North European Plain. It extends across the eastern half of the former USSR and into the Western Siberian Lowland, which is a vast marsh and taiga forest area situated east of the Ural Mountains. The lowlands end at the Siberian Plateau is an area of rolling hills and permafrost soils. Continental glaciers covered the northern half of the Russian lowland, as well as most of northern Europe and North America, 10,000 years ago. These glaciers have left behind marshlands and irregular river networks, along with a variety of moraines and other glacial deposition features across the northern lowlands of Russia.

Like Central Europe, the older mountains of the Urals and other smaller hills often are rich in mineral deposits. Just north of Ukraine is the world's largest mass of iron ore. Coal and peat are found throughout the Russian lowlands, which together with the iron ore formed the basis of the region's industrial core that centered on Ukraine and the Black Sea.

Waterways

Moscow is connected to this region through the Moscow Canal, which connects the Baltic Sea at St. Petersburg to Moscow and the Volga River that flows south to the Caspian Sea. Other canals connect the Volga River to the Black Sea, which leads to the Mediterranean Sea. This region, extending from St. Petersburg, through Moscow, to Ukraine and the Volga River Valley, to the Caspian and Black Seas, is also the former USSR's breadbasket. It is a rich agricultural land, with mixed dairy, vegetables, grains and livestock.

Ural Mountains

The Ural Mountains are the only significant mountain range across the Great Russian Lowland and are regarded as the border between the European and Asian continents. Although there are some smaller hill areas, such as the Ukrainian Uplands, the Central Russian Uplands and the Volga Uplands, none of them is over 600 ft (180 m) in elevation. The highest peak in the Urals, by comparison, reaches 6,184 ft (2,080 m), though most other peaks are below 3,000 ft (915 m). The Urals are similar in age to the mountain ranges of central Europe (north of the Alps), as well as to those of northern Scandinavia and the Appalachian Mountains in the United States. The Urals are also rich in iron ore and bauxite. Bauxite is the most common metal ore on the planet and is the core ingredient used in making aluminum. Large oil and coalfields lie to the east and west of the Urals, making this mountain system the most important industrial region in Russia.

Russian Siberia

Siberia starts on the eastern side of the Ural Mountains with the West Siberian Plain. This plain is an extension of the Great Russian Lowland/North European Plain. It is flat, vast, and covered with taiga forest. Russia's taiga forest is the largest continuous forest in the world, containing 20 percent of the world's timber. Much of the West Siberian Plain is completely uninhabited, and large parts are virtually unexplored. North of the taiga forest, Siberia's landscape turns to one of desolate and barren tundra and permafrost. Although there is some conjecture that much of this area could return to grassland if it were more densely grazed. Indeed, the whole future of northern tundra is being reevaluated in the light of climate change and the northern movement of species as temperatures increase.

About 1,000 miles (1,600 km) or more to the east of the Ural Mountains, the West Siberian Plain gives way to the Central Siberian Plateau. The Central Siberian Plateau extends another 1,000–1,500 miles (1,600 to 2,400 km) to the east where it gives way to Russia's Eastern Highlands, an area of Alpine orogeny. Both the Central Siberian Plateau and the Eastern Highlands are predominantly permafrost regions, where much of the ground is frozen all year round underneath its surface. The permafrost is more extensive and continuous in the tundra areas in the northernmost areas of Siberia. Permafrost in many parts of Siberia (as well as in Canada and Alaska) is starting to melt because of climate change, which makes travel increasingly difficult from spring to fall as the hard permafrost was easier for vehicles and the region's nomadic herders to travel on.

Trans-Siberian Railway

Population centers and human activity in Siberia are concentrated along the Trans-Siberian Railway. This is one of the greatest railway adventures in the world, connecting Moscow to the city of Vladivostok on the Sea of Japan. The railway crosses Siberia's southernmost territories, which are quite mountainous east of the West Siberian Plain along the borders with Mongolia and China. The Kamchatka Peninsula to the far east of Siberia is a spectacular highly volcanic and glaciated landscape. There are 29 active volcanoes on the peninsula, including the Klychevskoy Volcano, which is the highest volcano in Eurasia.

Siberia's Climate

The climate of Siberia is continental in the extreme. Winters are long and cold. Summers are short and cool to hot. In the city of Verkhoyansk in eastern Siberia, the average winter temperature is -58°F (-50°C), while the average summer temperature is 60°F (15.5°C). This difference of almost 120°F (65.5°C) from the winter average to the summer average is the greatest in the world. The coldest city temperature ever recorded in the world was also in Siberia, in the town of Bherkliask, which once reached -84°F (-64°C).

▶ *The Taiga forest in Siberia is the world's largest continuous forest. Meandering stream and remnant water bodies in the Taiga reflect the impact of continental glaciation during the ice ages.*

Nature and Adventure Tourism

Siberia holds considerable opportunity for ecotourism development, although the costs of these experiences tend to be high owing to the region's remoteness and limited infrastructure. Wilderness activities (hiking and wildlife viewing), sporting activities (fishing and hunting) and arctic activities (dog and reindeer sledding and skiing) are the mainstay of the fledgling tourism industry in interior Siberia. Whale and other sea mammals, birds and volcanoes are attractions in coastal and island areas, especially on the Kamchatka Peninsula. Inuit and other nomadic peoples of the north comprise the cultural attraction of the area. Unfortunately, it is estimated that 10 percent of the wildlife in northern Eurasia (including Central Asia) is seriously in danger of extinction. The region is in need of more concerted conservation efforts, which ecotourism has some potential to facilitate.

Lake Baikal

Lake Baikal, Eurasia's largest freshwater lake and the world's deepest, is located in southern Siberia just north of Mongolia. It serves as a focus for the region's ecotourism because of its wildlife and accessible location on the Trans-Siberia railroad. Lake Baikal is 3.15 million hectares in size and is the oldest (25 million years) and deepest (5,580 ft; 1,700 m) freshwater lake in the world, containing 20 percent of the world's unfrozen fresh water. Although there is increasing concern over lake pollution from timber and paper mills, it still has very high scenic value and is home to many endemic species.

Siberia's Economy

Although Siberia is very isolated and sparsely populated, it is currently undergoing substantial economic development based on its major resources of timber, oil, natural gas and minerals. The Sea of Okhotsk is also one of the richest fisheries in the world owing to its cold currents and extremely large tidal ranges. It produces over 10 percent of the world's annual fish catch, although fishing there has actually declined since the collapse of the Soviet Union. The same has happened to much of the manufacturing base that

◄ *The unique cultural landscape of Georgia in Transcaucasia is appealing as a crossover between human landscape features and natural settings. Much of Georgia's tourism appeal lies in its religious built heritage, which is illustrated here against a dramatic mountain backdrop.*

moved to Siberia during the Soviet period to encourage regional development. Nevertheless, the development of the region's oil and gas fields has given parts of it a major economic boost.

Kamchatka Peninsula

The physical landscape is also very important in providing the resources for developing tourism. Nowhere is this clearer than on the Kamchatka Peninsula. Although the Peninsula has some mineral deposits as well as timber resources, the landscape is increasingly being recognized as an important economic resource. The Peninsula is similar to Alaska in terms of its abundance of wildlife and nature; however, only very small numbers of tourists currently visit there.

Much of the Peninsula is either national park or nature reserve, and a number of reserves are part of the Volcanoes of Kamchatka World Heritage Site, regarded as one of the most outstanding volcanic landscapes of the world. Because of this, the Peninsula is attractive for photography as well as for ecotourism. Several of the national parks and reserves are also occupied by some of the local indigenous people who are developing commercial handicrafts as part of cultural tourism. The area also has many geysers and hot springs that feature strongly in the promotion of the area to travelers. As with Alaska, the area is becoming attractive for hunting and fishing tourism, and it is now possible to take direct flights between Alaska and Kamchatka.

Transcaucasia

In the southern part of the European portion of the CIS there are three mountain ranges: the eastern edge of the Carpathian Mountains (a small portion of which is in Ukraine); the Crimean Peninsula (a small mountain range extending out from the north shore of the Black Sea); and the Caucasus Mountains (situated between the Black Sea to the west and the Caspian Sea), which is one of the world's largest inland bodies of water, to the east. The Caucasus Mountains are a high alpine system of recent geological origin that is rich in mineral resources and lush with semitropical vegetation (due to the neighboring seas) on its lower slopes. Over the past 400 yrs, this area has served as a shifting boundary between the Ottoman Empire, the Russian Empire/Soviet Union and the Persian Empire.

Many different peoples have passed through the Transcaucasia, some as conquerors and others as traders. Transcaucasia refers to the mountains and the lowlands to their immediate north and south. It is known as the land of languages because over 50 different languages and dialects are spoken in this small territory. The major languages are Azerbaijani (a Turkic language); Ossetian and Tatic (Persian-Iranian languages); Armenian (an ancient Indo-European language); the Caucasian languages of Georgian, Chechen, Dagestani, Abkhasian; pockets of Greek; and the Slavic languages of Russian and Ukrainian. On the northern side of the Caucasus are various regions belonging to Russia, including Dagestan, which has some 40 different ethnic groups, many with distinctive dialects.

Chechnya, & the Transcaucasian States

The Russian territory of Chechnya is also in this region and has been the site of a bloody independence effort by the Muslim Chechens against the Russians at various times since the break-up of the Soviet Union. On the southern slopes of the Caucasus range are the independent countries of Georgia, Armenia and Azerbaijan. Georgia is the site of an ongoing dispute over the sovereignty of two regions, Abkhazia and South Ossetia, which are claimed by Georgia but instead regard themselves as independent states aligned with Russia.

Azerbaijan has assumed great economic importance in recent years because of its oil reserves, which it has been seeking to expand through an active drilling and exploration program in the Caspian Sea. However, there are some concerns about the potential pollution effects of oil exploitation, as the Caspian Sea is already suffering from increased levels of industrial and agricultural pollutants and declines in fresh water entering the sea due to water diversion for irrigation schemes. Georgia has been developing strong ties with Western countries and is strategically important because it carries a major oil pipeline from the CIS countries.

The Western Caucasus of the Russian Federation is one of the few large mountain areas of Europe that has not been significantly impacted by humans; only wild animals graze its alpine pastures. There are also substantial areas of relatively undisturbed forest, and it is one of the remaining homes of the European bison (buffalo). Like the American buffalo, it once roamed over the plains and woodlands of Eastern Europe, but now it is limited to a small number of protected areas. This area is increasingly attracting heritage and ecotourists. However, an important issue facing tourism development in the region is that many international tourists associate the political instability in the southern Caucasus with the Russian regions that are relatively unstable.

As already noted, Armenia, Georgia and Azerbaijan share the Caucasus region with Russia. During Soviet times, the region was an extremely favored domestic destination for Russians from the north, who could return home with boxes and suitcases stuffed with fresh fruits and vegetables that were difficult to acquire in other parts of the Soviet Union; the Transcaucasia republics were known as the fruit basket of the USSR.

Today, the three countries boast some of Eurasia's most impressive mountain scenery, and all three are heavily dependent upon religious heritage in their tourist offerings. Armenia and Georgia are overwhelmingly Christian nations with most of their heritage focusing on ancient churches and cathedrals. Azerbaijan is a largely Muslim country, with its religious landscape comprised more of picturesque mosques and other Islamic features. Cuisine, wine, mountain scenery and village life are salient parts of the tourist product, although all three capital cities are important gateways into the region and centers for business tourism. Georgia boasts an affiliation with the ancient saga Jason and the Argonauts, and Azerbaijan boasts its position on the ancient Silk Route, which has become an important UNESCO-endorsed, multi-national heritage corridor through Asia and the Middle East.

ISSUES AND INSIGHTS   Siberia and Asia's Monsoon Climate

There are no moderating marine air influences in Siberia, which makes it, along with Central Asia, much drier than areas closer to Europe. Because land masses heat much more quickly than water, Siberia is dominated in winter by a high air pressure system and in summer by a low air pressure system. Rapid heating of the ground and the air rising into the upper atmosphere cause the summer low. The ground's cooling and the subsequent descent of air from the upper atmosphere create the winter high.

High pressure systems are always caused by air that is descending from the upper atmosphere to the ground. Usually this air is very dry, which is why high pressure systems are associated with clear skies. Conversely, low pressure occurs when air near the ground rises into the upper atmosphere. If a large source of water is available, this will bring moisture into the upper atmosphere, causing clouds to form, which can then bring precipitation (rain or snow). Another feature of low air pressure and high air pressure is that their winds move in opposite directions. The low pressure winds circulate around the center of low pressure in a counterclockwise direction. High pressure winds circulate in a clockwise direction.

Because the Siberian High in winter and the Siberian Low in summer are so intense, they affect the direction of wind patterns for almost the entire continent of Asia. In the summer, the Siberian Low circulates in a counter clockwise direction, pushing air off the Asian landmass across the Indian Ocean toward India and Southeast Asia, across the South China Sea, and up toward Japan. In the winter months, the Siberian high circulates in a clockwise direction, pushing extremely cold air from Siberia down across northern China and into Japan, and circulating winds from the South China Sea across Southeast Asia and then across the Indian Ocean toward India and Southwest Asia.

This change in wind direction is known as a monsoon wind pattern—with a summer monsoon coming from one direction, and a winter monsoon coming from an opposite direction. Because the summer monsoon is so pronounced when it arrives in India and Southeast Asia, most people refer to the summer monsoon simply as the *monsoon*.

Other European Former Soviet Republics

Moldova

As already noted, Moldova is a Romanian-speaking nation, located just east of Romania. Following its departure from the USSR, a referendum was held in Romania to determine if it should absorb Moldova into its territory, given their common heritage and linguistic roots. The result was no, owing partly to the financial and social burden on Romania to upgrade Moldova's infrastructure, economy, social system, education system and health care scheme to meet national standards. Similar problems were encountered by West Germany regarding East Germany when the two states were reunited in 1990. Tourism in Moldova is limited, but growing, and focuses mainly on rural landscapes, scenic villages and inexpensive wines.

Moldova is struggling economically but surviving in its new capitalist environment. However, even this relatively small country is experiencing some political conflict and concerns for its future have been exacerbated by events in Ukraine in 2014 and the activities of pro-Russian paramilitaries. The thin wedge of territory between the Dniester River and the border of Ukraine declared independence from Moldova with support from Russia when the USSR dissolved. The territory, known as Transdniestria or Transnistria (although its self-appointed name is Pridnestrovie), has close relations with Russia and contains a largely Russian and Ukrainian population. It has established its own government offices, customs and immigration, money, and trade relations with Ukraine and Russia. It is not recognized by the international community as a sovereign nation, but it has become a tourist curiosity, even though visas are sometimes difficult to acquire.

Belarus

At the time of the USSR's disintegration, *Belarus* (White Russia) achieved independence as one of 15 equal Soviet republics. However, it is unique in the sense that this was not the country's desire. Following economic and political turmoil in the earliest days of independence, Belarus desired to be united with

Russia, which never came to fruition. Today, Belarus is functioning independently and has some interest in tourism, although there are influential factions in the country trying to dissuade tourism development.

Ukraine

Ukraine has sought closer ties with the European Union, which has antagonized Russia to its east. Despite these tensions, it has become a fashionable tourist destination for personal heritage tourists because of the widespread Ukrainian *diaspora* throughout the world (one of the largest of which is in Canada). Ukraine's tourism promotional efforts spotlight this form of roots tourism, wherein people of Ukrainian descent travel to their homeland to do genealogy, meet relatives and see native villages.

Like Russia, Ukraine is a large country with an array of natural and cultural environments that exude tourist appeal. Wildlife viewing, mountain trekking and visiting national parks are important elements of tourism, as are golf courses, health spas, Black Sea resorts, religious sites and historic castles. Its cultural and historical connections with Russia also mean that Ukraine is among the top ten countries in the world with respect to international visitor arrivals, although visitor numbers have been substantially affected by the conflict with Russia and pro-Russian independence groups in 2014. In 2014, the Crimea, which had a long history as a Russian tourism destination, again became part (functionally at least) of Russia through unilateral annexation.

Kiev

Kiev, the capital of Ukraine founded in 862 CE, has a rich architectural and cultural heritage available to the visitor. To visitors, it contains an interesting mix of Soviet and neo-baroque architecture, although a number of buildings need urgent renovation. Ukrainian attempts to get politically and economically closer to the West have made the country attractive to tourists, although the political relationship with Russia continues to divide the country and is a major source of tension. Nevertheless, its relative ease of access from the West as well as its cheapness in comparison with Moscow make it an increasingly attractive tourist destination for the more adventurous traveler who is seeking something different. Odessa, a major seaport on the northwest coast of the Black Sea, is a major tourist destination with a long tradition of spa and health tourism, as are Yalta and Sevastopol in the Crimea.

Arguably the most unique tourist attraction in Ukraine is the possibility of being able to take a day trip from Kiev to Chernobyl, the site of the nuclear power plant meltdown and explosion that occurred in 1986 and sent clouds of radiation across Europe. A number of tour companies run day tours through the 30km exclusion zone around the power station. Tours usually include sightseeing of the Chernobyl reactor 4 (from an observation point 200m from the reactor sarcophagus), which also includes a visitor center, the abandoned city of Pripyat and abandoned villages. Day passes are only available through tour operators. Belarus also offers day trips into the Chernobyl restricted area, although these are not nearly as comprehensive or popular as those from the Ukrainian side.

Name _____

Name Human Mobilities

1. Identify and describe two examples of human mobility that shaped the historical development of one or more of the subregions of Western Europe and Eastern Europe.

_____

_____

_____

_____

2. Identify and describe the role of human mobility in the contemporary societies of one or more of the subregions of Western Europe and Eastern Europe.

_____

_____

_____

_____

Tourism Destinations

3. Describe the contemporary international image of one or more of the subregions of Western Europe and Eastern Europe. How do these images impact the potential of the regions as tourist destinations?

_____

_____

_____

_____

4. Selecting one or more of the subregions of Western Europe and Eastern Europe, how might the countries in the sub-region use their cultural and physical resources to improve their global image and attractiveness?

_____

_____

_____

_____

Sustainable Environments

5. Identify and describe tine major environmental challenges of one or more of the subregions of Western Europe and Eastern Europe. What challenges do the countries in the region face in addressing these issues?

_____

_____

_____

_____

6. Identify and describe the major social and economic development challenges of one or more of the subregions of Eastern Europe and Western Europe. What challenges do the countries in the region face in addressing these issues?

_____

_____

_____

_____

Instructor Questions (answer any additional questions from your course instructor in the space below)

_____

_____

_____

_____

CHAPTER | 3

The Old World: Central Asia Through Africa

3.1 CENTRAL ASIA

3.2 SOUTHWEST ASIA AND NORTH AFRICA (THE MIDDLE EAST)

3.3 SUB-SAHARAN AFRICA

CHAPTER 3 OVERVIEW QUESTIONS

HUMAN MOBILITIES
- In what ways do the regions covered in this chapter reflect the geographic concepts of "crossroad" and "cultural hearth" in their origin and modern geographies?

TOURISM DESTINATIONS
- What is the role of heritage in the modern tourism landscapes in, and tourist motivations to, the countries from Central Asia through Africa?

SUSTAINABLE ENVIRONMENTS
- What challenges do the countries and cultures of these regions face in conserving their cultural and natural resources for future generations?

INTRODUCTION

This chapter covers a vast territory that is called the *Old World*. It extends from Central Asia, through Southwest Asia (also known as the *Middle East*), across North Africa, and across the Sahara to Sub-Saharan Africa. We call this the Old World because it includes the political and *cultural hearth* of some of the greatest empires of Eurasia (the Mongol Empire, the Arab Empire and the Ottoman Empire, among others), as well as being the probable source of modern humans (in Africa). Africa is geologically the oldest continent and over 200 million years ago was at the center of the *Pangaea* supercontinent. The areas north of the Sahara (from North Africa to Central Asia), in particular, share many characteristics, including history, the Islamic religion and largely arid environments. As noted in Chapter 1, the term Old World as used in this book differs from standard usage, which refers to the *Eastern Hemisphere* and includes all of Europe and Asia, in addition to Africa.

The Old World, as defined in this chapter, is rich in physical, cultural and historical landscapes, though much of the region's recent emergence from dominance by European powers has resulted in political turmoil and limited tourism development. Ecotourism and heritage tourism have great potential in almost every corner of the Old World. International tourist arrivals are relatively low throughout the Old World, though year-on-year growth rates have been

▶ *The Old World Region from space, showing its major physiographic landforms and relative location to Europe and Asia. (Note: This image exaggerates elevations to make them more prominent than they would appear on the actual planet.)*

higher than the global average in recent years. This is despite wars and political turmoil that stand in the way of tourism development in many corners of this vast region. There also remains a great need to increase environmental awareness and to address major environmental problems, which is especially challenging in more impoverished countries. As these issues are addressed, the Old World of Central Asia, Southwest Asia, North Africa and Sub-Saharan Africa could blossom as major international tourism destinations.

▶ *Regions and Subregions of the Old World.*

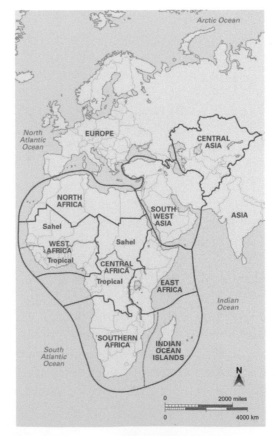

3.1 | CENTRAL ASIA

Current Central Asia Issues and Resources (http://wrgeography.com/centralasia.html)

Central Asia includes the countries of Afghanistan, Kazakhstan, Kyrgyzstan, Tajikistan, Turkmenistan and Uzbekistan. The suffix *-stan* means land, and so these are the land of the Afghans, the land of the Kazaks, the land of the Kyrgyz and so on. Central Asia was formerly known as *Soviet Central Asia*. It is a predominantly *Turkic*-speaking Muslim cultural region, with large minority Russian populations in some locations, usually urban areas. Its population is the fastest growing among the territories that were once part of the Soviet Union. (The former Soviet Union today is sometimes referred to as the Commonwealth of Independent States, or the CIS, although not all former Soviet republics are members of the CIS.) Central Asia has a mostly *arid* climate, though rivers from its high mountain areas allow for irrigated agriculture. Ethnic tensions have also been significant in Central Asia, often in the form of anti-Russian sentiments following many years of Soviet efforts to *Russify* the region (to make it Russian) or as a result of non-participatory and autocratic regimes.

Physical Geography of Central Asia

Central Asia's physical landscape can basically be divided into two subregions: (1) the high mountain areas in its southeast, where Central Asia meets the *Tibetan Plateau*, and (2) the flat plains surrounding the Aral and Caspian Seas in the south and west of the region. The two mountain countries of Central Asia are Tajikistan and Kyrgyzstan.

The Roof of Asia

Tajikistan is the home of the *Pamir Mountains* (bordering India and Pakistan) and has Central Asia's highest mountain peaks, reaching over 23,000 ft (7,000 m) in elevation in the *Pamir Knot* (also known as the *Roof of Asia*), These mountains are a major tourist attraction, though political instability and intense winter blizzards, which can last for days, make travel to them extremely challenging.

© Nikm, 2011. Used under license from Shutterstock, inc.

◀ *In the middle of the Pamir Mountains on the borders of Tajikistan, China, Afghanistan and Pakistan. Most of the major mountain ranges of Central Asia extend out from the Pamir Range, which is also known as the Pamir Knot.*

Kyrgyzstan is north of Tajikistan and includes the slightly lower *Tian Shan* mountain range, which borders China. The Tian Shan (also spelled "Tien Shan") is an extension of the Pamir range and extends eastward through the middle of the Asian land mass. The Tian Shan range forms the northern border of the massive land area that was uplifted by the collision of the Indian and Asian plates. This uplifted area includes the Tibetan Plateau, the Pamir Knot and the Himalaya mountain range to the south. The *Tarim Basin* (containing the Taklimakan Desert), between the Tian Shan range and the Tibetan Plateau, is the remains of an ancient micro-continent that was surrounded and partially covered by the younger uplifted mountains to its north, west and south.

The Tian Shan Range is a major potential tourism area for both Kyrgyzstan and China and is easier for tourists to access owing to the relatively greater degree of political stability in Kyrgyzstan. The *Ala-Archa Canyon* is a national nature park not far from the Kyrgyz capital of Bishkek, where numerous trekking trails lead visitors to glaciers on the country's highest peaks. Kyrgyzstan's *Issyk-Kul Lake* is the second largest mountain lake in the world (after Lake Titicaca in the Andes Mountains of South America). Surrounded by snow-capped peaks, it was a major resort destination during the Soviet era and is a growing tourism destination today. Historically, some researchers believe that the Issyk-Kul Lake area was where the *Black Plague* originated in the early fourteenth century CE, killing 75 million people across China, South Asia, Europe and North Africa.

Both Tajikistan and Kyrgyzstan have rolling hills and some lower-lying plains in their western portions that border the other three countries of Central Asia. Kazakhstan, Uzbekistan and Turkmenistan encompass the lower plains that comprise most of the Central Asia region. *Kazakhstan* is the ninth largest country in the world in land area. The country slopes from its higher mountains in the east, through rolling hills, to vast sand deserts and marshes in its western and southern portions. Uzbekistan and Turkmenistan similarly slope toward lower-lying areas in their western reaches as they approach the Caspian Sea.

Caspian and Aral Seas

The Caspian Sea and the Aral Sea are both internal drainage systems, meaning that while rivers flow into them, they do not flow out to an ocean. They were once much larger than they are today, but have shrunk considerably since the early 20th century because of climate change and the diversion of water for agricultural irrigation. In the case of the Aral Sea, the water has become so salty that saltwater fish have now replaced the former freshwater fish varieties. Much of the Aral Sea is also polluted with pesticides and herbicides from cotton farming, while the dust from the South Aral Sea is a major health problem in the region. However, the northern Aral Sea has been partly restored as a result of the efforts of the Kazakh government to the point that limited commercial fishing has been able to resume.

The northern portions of Central Asia reach toward Siberia and the very center of the Asian continent, and therefore they experience the extremes of summer heat and winter cold, though not as extremely as Siberia to the north. Arid conditions are most prominent in the southern lowland areas, with higher precipitation in the mountains in northern Kazakhstan, which borders Siberia.

Arid Climate

Even with the high eastern mountains, virtually the entire region of Central Asia receives less than 20 inches (51 cm) of rainfall each year, creating grassland, desert and drought conditions. The arid climate creates highly sensitive aquatic ecosystems in all of the seas and lakes of Central Asia, the largest of which are the Black Sea, the Caspian Sea and the Aral Sea. These water bodies concentrate drainage from their surrounding watershed and serve as sensitive measures of regional environmental degradation. Most all of them are severely polluted, though concerted international efforts in the 1990s to address

◀ *Issyk-kul Lake in Kygyzstan is the second largest high mountain lake in the world. The high peaks of the Tian Shan Mountains can be seen in the distance.*

© And Andreev, 2011. Used under license from Shutterstock, Inc.

these problems have helped to some degree. Ecotourism offers an alternative to the large-scale and heavy industries and some forms of intensive agriculture that have caused this widespread damage.

The intersection of desert, grasslands and mountains also creates a rich diversity of ecosystems that attract hunters and fishers, though some argue that they are starting to threaten some of the region's endangered large game. Migratory bird watching on wetland areas between the sand dunes and drier plains is a major niche tourist attraction in the lowland areas of Central Asia. However, the greater difference between summer and winter temperatures in lowland Central Asia results in large numbers of summer insects, which can be a major distraction for visitors. Because of their location on the former edge of a vast Jurassic sea, the southeastern mountains of Turkmenistan and Uzbekistan have some outstanding dinosaur sites, including the Kugitang Reserve, located on the Turkmen side of their shared border.

Human Geography of Central Asia

Central Asia shares many characteristics in common with Southwest Asia (often referred to as the Middle East) and North Africa. All three have low overall population densities, although higher density centers do exist. All three have mostly arid grassland and desert climates. Most of the people of all three regions share Islamic religious beliefs and historical ties that involve the Arab and Ottoman Empires. Central Asia is different from the other two regions in that it has much colder winter temperatures, all six of the countries of region are *landlocked* (having no border access to an ocean), and they all have a history of being part of the USSR.

Central Asian Empires

What we see today in the former Soviet Socialist Republics of Central Asia is the result of the disintegration of one of the world's last great empires. An empire is a group of countries and nations ruled by a single supreme authority. This is very different from the nation-state concept, which was a formative historical feature of Europe (Chapter 2). Some of the world's major empires originated in Central Asia, including that of the Huns from the northern portion of the region, whose empire was concurrent with that of the Romans.

The Huns attacked China in the fourth century BCE, and conquered the Volga region, including Moscow, in the fourth century CE. Under their leader, Attila, the Huns ransacked Roman Italy in the

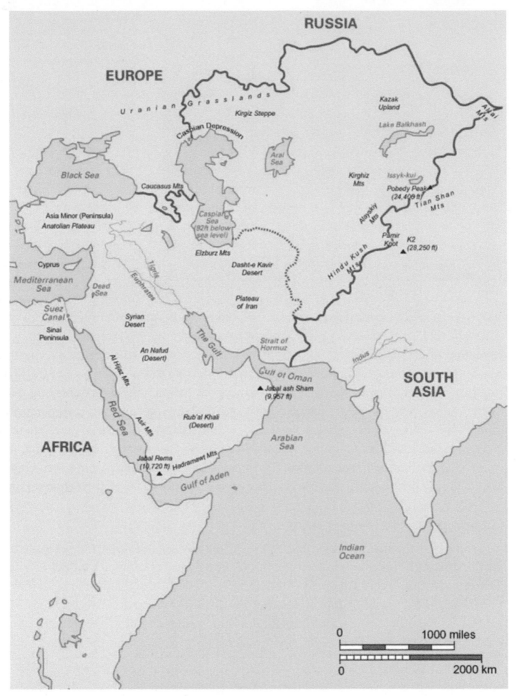

▲ *Central and Southwest Asia Physical Geography Features.*

fifth century CE, marking the final end of the Roman Empire. Genghis Kahn, from Mongolia, unified the Central Asian lands around 1200 CE. The Mongol Empire expanded to control most of Asia and Eastern Europe by 1260 CE. (This was the same time that the European Renaissance was beginning in Southern Europe.) In Europe the Mongols were known as the Tartars; they ruled the European portions of their empire from the Moscow area. In the late fourteenth century, they were removed from power in China; in 1401, they were removed from the Moscow area, and by the mid-fifteenth century they were completely removed from Russia.

The Silk Road

The Silk Road actually is comprised of many different routes that connected the Chinese city of Chang'an (modern day Xi'an) to cultures of Asia Minor and the Mediterranean Sea. Although there is evidence of ancient Egyptian trade with Southwest Asia, and probably Central Asia as early as 4000 BCE, the earliest Mediterranean connections to China are found in jade dating back to about 1500 BCE. Evidence of Chinese silk in Egypt dates back to approximately 1000 BCE. Some silk trade may have existed earlier than that, but its evidence has completely decomposed over time.

The Greek conqueror Alexander the Great is credited with opening the silk route as the major historical trading route that it came to be, starting in the fourth century BCE. China's Han Dynasty greatly expanded trade along the route in the second century BCE, after which the Romans came to develop a near-insatiable desire for Asian goods, especially silk. This resulted in a considerable flow of gold to China and the establishment of a Roman Embassy in Chang'an in 166 CE. Although disrupted by the rise of the Turkish Empire in Southwest Asia, the Silk Road trade routes were re-established under the Mongol Empire (thirteenth to fourteenth century CE) until finally declining at the end of the fourteenth century when sea routes came to dominate trade between the East and West. Today, the Silk Road is promoted as a major tourist attraction in several countries, though its isolation and limited infrastructure make traveling the route a challenge without a well-organized tour group.

Ancient cities of Central Asia still maintain a strong sense of cultural continuity with the past. Herat, Afghanistan, whose citadel was built by Alexander the Great, and the Silk Road city of Merv in Turkmenistan (reputed to have been the largest city in the world in the twelfth century) are two examples. Magnificent buildings are found in the Central Asian countries from the pre-European era, many of which were built by Timur (known as Tamerlane in Europe), who conquered much of Central Asia (including the Mongols) in the fourteenth century, establishing the Timurid Empire (1370–1405). His empire was centered in Uzbekistan and covered all of Central Asia, plus large parts of Southwest Asia (southeast Turkey, Syria, Iraq, Kuwait, and Iran), the Caucuses (Azerbaijan and Georgia) and South Asia (Pakistan and India). His dynasty lasted in some form until 1859. Among the well preserved buildings that date back to Timur's influence are the magnificent Hodja Akmed Yassavi Mausoleum in Zhabagaly, Kazakhstan, and Registan Square, which was the center of old Samarkand in Uzbekistan.

European Imperial Expansion

The mid-fifteenth century was the beginning of the age of *European exploration*, discovery and colonization. This era started with the Portuguese sailing to Africa and Asia, and the Spanish, who focused on the Americas following the explorations of Christopher Columbus in 1492. The Dutch soon followed, settling in South Africa and colonizing Indonesia, along with the British, who moved into Africa, India and North America, and the French, who colonized areas of Africa and Indochina (in Southeast Asia). At the same time, the Russian Empire began expanding eastward and westward. Muscovite armies led the Russian push after Russian independence from the Mongols in 1401. By 1815, the Muscovite Russian armies had controlled all of the territory in Eastern Europe and Central Asia that would later become the republics of the USSR, and other areas that did not become part of the Soviet Union, including Poland, Finland and Alaska.

In the late 1800s, the Russian Empire was a largely rural, agrarian society. The lands east of the Urals, including Central Asia, had very low population densities, and few Europeans considered the Russian Empire a major threat. In 1917, however, a revolution occurred in Russia that overthrew the czar (Russian emperor). In the confusion that followed, several territories of the old Russian Empire broke away, including the Transcaucasian countries of Georgia, Armenia and Azerbaijan; the Baltic states of Estonia, Latvia and Lithuania; and Belarus, Ukraine, Moldova, Poland and Finland. The United States had already purchased Alaska from Russia in 1867.

World War II marked the last great push for imperial expansion by the European powers, which now were joined by Germany in Africa, the United States in Asia and the Caribbean, and Japan, which sought control over most of East Asia. By the end of World War II, all of the old Russian Empire, except for Poland, Finland and Alaska, was retaken by the new Russian/Soviet leadership and integrated into the USSR. In addition, the countries of Eastern Europe, from Poland to Yugoslavia, fell within the sphere of Russian/Soviet political domination.

The End of the Empires

Most of the European empires dissolved in the decades following World War II. India and Pakistan gained independence in 1949. In the 1950s, the Philippines, Indonesia and Indochina (Vietnam, Laos and Cambodia) gained independence, along with some African countries. In the 1960s and 1970s, most of the remaining African colonies gained their independence. However, it was not until the end of the 1980s and the beginning of the 1990s that the Russian Empire/USSR finally dissolved (replaced by the decentralized and somewhat smaller CIS organization).

The Central Asian republics have had significant growing pains in establishing legitimate political rule and economic stability following their independence from the Soviet Union in the early 1990s, though their remote locations and poor transportation infrastructure may be even greater barriers for international visitors. Every country in the region has politically sensitive areas, and these issues need to be kept in mind in any discussion of tourism resources in both Central and Southwest Asia (see map on p. 143).

Countries and Tourism of Central Asia

Kyrgyzstan and Tajikistan

The two highly mountainous countries of Kyrgyzstan and Tajikistan represent two extremes of post-Soviet Central Asia. Despite similar features of physical geography, Tajikistan has suffered through civil war for most of the 1990s, resulting in tens of thousands of deaths and hundreds of thousands of refugees. On the other hand, Kyrgyzstan is considered the most democratic and stable of the Central Asian countries.

The word Kyrgyz comes from the Turkic word for "forty" and refers to the 40 tribes that defeated their Uyghur rulers in 840 CE, and that is memorialized in their flag showing a sun with 40 rays. A treaty between the Qing Dynasty of China and the Russian Tsarist Empire ceded the region of Kirgizia to Russia in 1876, and many Kyrgyz people moved to other parts of the Pamir Mountains to escape Russian rule. Tensions between Kyrgyz nationalists and Moscow continued into the Soviet era, which started in 1919 and resulted in large numbers of Russians, Ukrainians and other Slavic peoples coming to dominate the larger cities of the "Kirghiz Soviet Socialist Republic." The dissolution of the Soviet Union resulted in full independence in 1991 and an official name change to the Kyrgyz Republic in 1993, although Kyrgyzstan is still widely used as well.

Bishkek is the capital of Kyrgyzstan with close to a million people. It is located in the north of the country and has a large number of Russian and other Slavic nationals. Osh is the second largest city in the country and is located in the Fergana Valley in the south, which extends into Uzbekistan. Large numbers of Uzbeks live in this part of Kyrgyzstan, and both they and the Russians in the north have been sources of ethnic and political tensions since the country's independence. Of the country's 5.2 million people, 69 percent are Kyrgyz, 15 percent are Uzbeks, and 9 percent are Russians. The Russian language is an official language of the country, in addition to the Kyrgyz language, which reflects a sense that Kyrgyzstan had benefited culturally and economically when it was part of the Soviet Union. In fact, the breakup of the USSR resulted in a major disruption in its mostly agriculture-based economy.

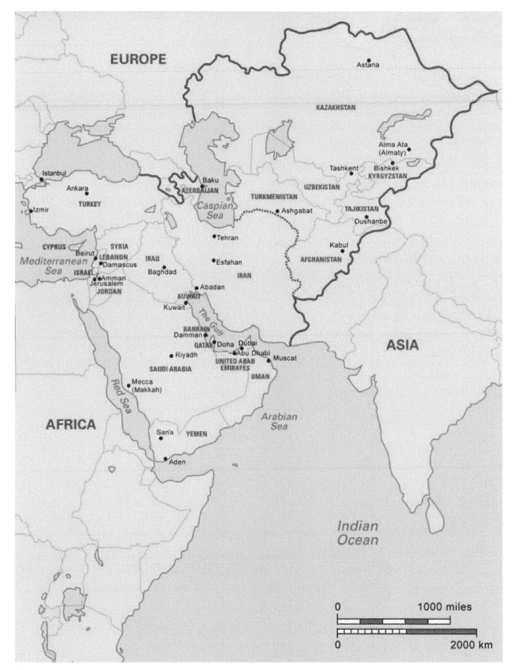

▲ *Countries, Capitals and Major Cities of Central Asia and Southwest Asia.*

While Kyrgyz is a Turkic language, the Tajik language is Persian-based (Iranian or Farsi), and the people of Tajikistan have historically been more closely tied to modern-day Afghanistan and Iran. This reflects the more southern location of the Republic of Tajikistan, which borders northern Afghanistan. The word Tajik was originally used by Turkic peoples to refer to Persian (or Iranian) speakers. Islam was introduced to the Tajik area in the seventh century CE, and into the Kyrgyz area in the eighth century CE.

Unlike Kyrgyzstan, the Tajik Soviet Socialist Republic was the most impoverished and least developed region of the USSR, and following its full independence from the USSR in 1991, the country fell into

ISSUES AND INSIGHTS   The Turkic Peoples

With the exception of Kyrgyzstan, all of the countries of Central Asia have official languages that are of Turkic descent. Turkic languages are also officially recognized in western China (the Xinjiang Uygur Autonomous Region), many of the republics (provinces) in Russia's Siberia, and in the countries of Azerbaijan, Cyprus and Turkey. Turkic language speakers are also found in Iran, large areas of southern Russia, eastern China, northern Iraq, Pakistan and Afghanistan, and throughout the Balkan Peninsula (Romania to the countries of the former Yugoslavia).

The term "Turk" was first used in the West in reference to the Gokturks, who were credited with bringing together the many nomadic tribes from Central Asia to modern day Mongolia into an empire (552–744 CE) that controlled the silk trade route from Europe to China. Chinese historical records talk about earlier Turkic peoples, as early as 200 BCE, and they indicate that the Huns, who invaded Europe in the fifth century CE, may have been a Turkic tribe. Contact with Arabs in the seventh and eighth centuries CE spread Islam to most of the Turkic-speaking peoples of Central Asia. Many different Turkic-speaking groups invaded Southwest Asia, Eastern Europe and South Asia in many waves over the centuries, including the Avars, Karluks, Tartars, Bulgars and Pechangs, who reached Constantinople (and the Eastern Roman Empire) in the year 1091 CE.

The Mongol Empire (1206 to 1324 CE) united the Turkic people and Mongol (East Asian) people in the thirteenth century CE, creating a single empire that extended from the East China Sea into Eastern Europe and South Asia. The dissolution of the Mongol Empire gave rise to the Turkic Ottoman Empire (1299 to 1923 CE), which made Constantinople (Istanbul) its capital.

The Turkic language group is, as one might guess, very diverse and complex due to this long history of intermingling with various other language groups from Asia to Europe. Although a Turkic alphabet (known as runes) once existed, today it has been replaced by the Uyghur alphabet in Central Asia, Arabic script in Southwest Asia and Cyrillic in Eastern Europe. Most of the major Turkish language groups are mutually unintelligible, similar to the difference between the Germanic languages of English, Norwegian and German. The Turkic languages are considered a branch of the Altaic language family, which also includes Mongolian, Tungusic (in Siberia and northeast China), Korean and Japanese (although inclusion of the latter two is open to debate.) (Similarly, the Germanic languages are considered a branch of the Indo-European language family, which also includes the Romance languages of southern Europe, Persian/Iranian, Russian and Hindi.)

a devastating civil war among different clan factions. Most of the non-Muslim Russians fled Tajikistan at that time, as civil strife continued through the 1990s. Over half of the land area in Tajikistan is over 9,800 ft (3,000 m) in elevation. The rugged terrain, combined with the long civil war, has kept poverty levels in the late 2000s below those of the Soviet era. In addition, political freedoms remain very limited under an autocratic, one party rule.

▶ *Yurts are the traditional dwelling for many of the Turkic peoples of Central Asia, especially those on the region's extensive plains. This extended family encampment is on the plains of Kazakhstan and neighboring areas of Central Asia.*

Kazakhstan

The Republic of Kazakhstan covers an area of Central Asia the same size as Western Europe, making it the ninth largest country in the world in land area, and the world's largest landlocked country. Its population of 16.7 million (2014 estimate) is 63 percent Kazakh and 24 percent Russian, with much smaller proportions of Tartars, Ukrainians, Uzbeks and even some Germans who had migrated there from other parts of Russia. In addition to Kyrgyzstan, Kazakhstan is the only other Central Asia country that continues to use Russian as an official language in addition to Kazakh. Russian continues to be the common language of business in the country and reflects the strong Russian presence there since it came under control by Tsarist Russia in the early 1800s. The word Kazakh comes from the Turkic word for "independent" or "free spirit," which refers to the Kazakhs' traditional nomadic horse-based culture. This culture was well suited to the steppe grasslands that cover much of the country, but it was at odds with the settled urban culture of Russian immigrants. Tensions between these two groups over land and water were increased by Soviet efforts to transform pastoral grazing lands into grain production. Despite these issues, Kazakhstan was the last of the Soviet Socialist Republics to declare its independence, which occurred in December 1991.

The name of Kazakhstan's capital, Astana, literally means capital. The largest city in the country is Almaty, which is the commercial center and possibly the most cosmopolitan city in Central Asia. Kazakhstan has an emerging market economy and has benefited from the export of agricultural products (from its steppes), oil

---

ISSUES AND INSIGHTS   Borat's Kazakhstan

Kazakhstan is undergoing substantial economic development based on its enormous oil and gas reserves, and the government is also hoping to promote tourism to the region that focuses on its cultural and natural heritage attractions, including coastal resorts on the Caspian Sea. Unfortunately, the country may have become better known in the West as a result of Sacha Baron Cohen's 2006 film, *Borat* (sub-title: *Cultural Learnings of America for Make Benefit Glorious Nation of Kazakhstan*), a mockumentary about a fictional Kazakh television journalist character who visits the United States of America with a cheap suit and a bad mustache. The film was a satirical perspective of American life and of Kazakhstan, particularly with respect to anti-Semitism. (Cohen is a Jew who is a grandchild of a Holocaust survivor.) Unfortunately, residents of both countries failed to see the comic value of such satire.

The Kazakhstan government was unhappy about the image the film portrayed of their country and culture. It first tried to sue Cohen and then hired a public relations firm to launch a debunking marketing offensive, including a four-page advertisement in *The New York Times*. Both failed and instead gave even more publicity to the film and the character. The satirical Borat character had actually been on British television since the mid-1990s, but it was the success of the film that led to expressions of concern from the Kazakhstan Foreign Ministry and other members of the government.

The film has been banned in Russia and Kazakhstan, and the domain name www.borat.kz was suspended in December 2005. The 2006 United States State Department annual human rights report cited the loss of the .kz website as evidence of the Kazakh government's efforts to curb free speech. Although actually filmed in Romania, the film has nevertheless increased awareness of Kazakhstan in the travel marketplace. For example, since the release of the movie, the respected British newspaper *The Guardian* has provided several stories on the country (which are generally positive), and not on any of the other central Asian republics.

The Kazakhstan government reportedly invested $40 million in the production of the film *Nomad*, a historical epic set in eighteenth century Kazakhstan. A young boy is destined to unite the country's three warring tribes and free them from the Dzungars, a confederation of West Mongolian tribes. The Kazakh language film premiered in 2005, and the English language version was released in the US in 2007. The film, however, was not a box office success.

For serious international tourists, Kazakhstan has become a major destination for adventure and ecotourism. Almaty (population 1.6 million) is the center for trips into the Tian Shan mountain range, bordering Kyrgyzstan and China. The world's largest skating rink (as of 2010) is located near Almaty, as are some of the leading winter ski resorts in Central Asia. Canyons similar to the Canyonlands of the southwestern US are also found in portions of the Tian Shan range. The country's traditional foods are similar to those of the Mediterranean, but with more meat, reflecting their traditional pastoral animal husbandry tradition. The capital of Kazakhstan, Astana, is a very modern city, rebuilt in the 1990s after the country gained independence. The world's tallest tent building, a 490-ft (150 m) entertainment and shopping center, opened there in 2010.

(from the Caspian Sea coast) and minerals (from its mountains). Politics there have been more challenging, as the country's first president, elected in 1991, was made president for life in 2007 by the country's parliament. Periodic elections routinely fall short of international standards according to the Organization for Security and Co-operation in Europe, and intimidation and suppression of political opponents is common. The country also has substantial environmental problems stemming from Soviet nuclear testing, industrial dereliction and the ecological problems of the Aral Sea (see Issues & Insights, p. 146).

Uzbekistan and Turkmenistan

Historically, both Uzbekistan and Turkmenistan, as well as modern-day Tajikistan and Kyrgyzstan, have been ruled by various Asian empires, most notably the Persians/Iranians (different empires from 500 BCE on), the Arabs in the seventh century CE (introducing Islam), the Mongols in the thirteenth century CE, and the Timurid Dynasty (1370–1526 CE). The founder of the Timurid Dynasty, Timur (known to Europeans as Tamerlane), made the Uzbek city of Samarkand his capital and brought to it artists, scholars, scientists and theologians from all the lands he had conquered, including portions of Russia, India and China.

There were also short-lived Uzbek and Turkmen kingdoms, but by the early 1800s both areas came under Russian tsarist control, with Turkmenistan forming an agreed boundary between areas of Russian and British Empire domination in Central Asia. Like the Kazakh to the north, the Uzbek and Turkmen were traditionally horse-dependent nomads comprised of many different clan groups unified mostly by common languages that distinguished them from their neighbors, and unified in their modern day form by Russia and the Soviet Union. While most are Muslims, they freely mix Islam with older animist beliefs.

Uzbekistan is one of only two countries in the world that is *double landlocked*, which means it is a landlocked country surrounded by other landlocked countries. The easternmost portions of the country, where Samarkand is located, borders Tajikistan and has large numbers of Tajik speakers. Although Russian is not an official language in Uzbekistan or Turkmenistan, it is widely used in the urban areas of both countries as a *lingua franca* among different ethnic groups and for commerce. Economically, Uzbekistan is more impoverished, like its mountainous neighbors to the east (Kyrgyzstan and Tajikistan). Turkmenistan, on the other hand, is more like Kazakhstan to the north with which it shares access to rich natural gas and

▶ *The Presidential Palace in Ashgahat (or Ashkhabad), the capital of Turkmenistan. The capital cities of central Asia are a mix of modern and traditional architecture, reflecting their long history as part of the USSR.*

oil reserves on the Caspian Sea. Both countries, however, have been ruled by authoritarian regimes that discriminate against minority groups and severely limit the rights of their populations in general.

Out of the 177 countries ranked on the Transparency International's Corruption Perception Index (see: http://www.transparency.org/cpi2013/), Uzbekistan and Turkmenistan were tied as 168th; Kyrgyzstan was 150th, Tajikistan was 154th and Kazakhstan was perceived as the least corrupt in Central Asia at 140th. By comparison, New Zealand and Denmark were seen to be the least corrupt, Australia was 9th, the U.K. was 14th, and the U.S. was 19th.

Afghanistan

In the past, Afghanistan has been included in the region of South Asia, mostly because it was not part of the former Soviet Union. Today, however, it is usually included in Central Asia because of its common cultural and environmental characteristics. Like its two northern neighbors, Tajikistan and Kyrgyzstan, Afghanistan is a largely mountainous country in its eastern portions, with lower lands in its western and southern portions. The major mountain range in Afghanistan is the Hindu Kush, which is a southern extension of the Pamir Knot, and a western extension of the Himalaya Range.

Afghanistan has always been at the crossroads between Central and South Asia and between Europe and China. Because of its strategic importance, it has long been sought after by imperial powers. The Hindu Kush region is easily accessible from nearby Pakistan, from which adventure travel into the region often originated prior to the Soviet Union's disastrous invasion of Afghanistan in the 1980s. The famed Panjshir Valley was known for its lush vegetation and hospitable population (the Panjshiris), but it also was a major military battlefield in the war against the Soviet armies.

In 1978, the communist party of Afghanistan (the People's Democratic Party of Afghanistan) took control of the government in Kabul, but there were two or three major factions within the party, each with

---

ISSUES AND INSIGHTS   The Aral Sea

The Aral Sea is fed by two rivers that drain from the Pamir Range. The Syr Darya flows through Kazakhstan and enters the lake from the north. The more important river is the Amu Darya, which flows through Uzbekistan and Turkmenistan and enters the lake from the south. In ancient times, the Amu Darya has sometimes changed course and drained into the Caspian Sea instead, leaving the Aral Sea largely dry. In modern times, however, the diversion of river waters for agriculture (especially water-intensive cotton) in Kazakhstan, Uzbekistan and Turkmenistan, along with a severe drought in the mid-2000s, almost drained the entire lake. This is sometimes regarded as the greatest human-induced disaster in the world, as the shrinking lake poisoned the land and people with salt, heavy metals and pesticide. Today the Aral Sea is only about one-tenth the size it was before 1960 when it was the fourth largest lake in the world. Animals and people who were once dependent on the Aral Sea have either migrated to other areas or have died. The North Aral Sea is the only area to have had substantial ecological restoration work conducted on it, which has been reasonably successful. Further efforts are underway.

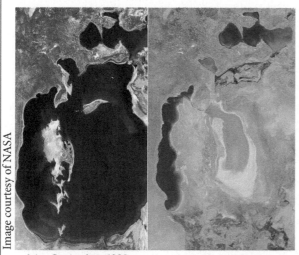

Image courtesy of NASA

July - September, 1989   October 5, 2008

Sources

Image source: NASA Earth Observatory (US Government public domain), see also http://en.wikipedia.Org/wiki/File:Aral_Sea_1989–2008.jpg

Information sources: The Shrinking Aral Sea Recovers http://earthobservatory.nasa.gov/IOTD/view.php?id=46685

its own goals for the future of the country. The Soviet Union staunchly supported the new communist country, but concerns about how matters were being handled by the faction in power led to the invasion of Afghanistan by the Soviet military in December 1979. While the Soviet forces never controlled most of the country, they did control the puppet government in Kabul. For ten years, until its withdrawal in 1989, the Soviet Union fought against various rebel forces that resided in the mountainous north and other isolated parts of the country that did not support the communist government.

Taliban and Al-Qaida

After the USSR withdrew from Afghanistan in 1989, the country fell into civil war, which led to the collapse of the communist government in 1992. The Taliban was one of these factions opposing the government in Kabul. The Taliban was a fundamentalist Muslim group that severely enforced religious piety and the inferior role of women in society. In 1996, they seized power in Kabul and effectively controlled most of Afghanistan until the United States invasion in 2001, when the regime was overthrown for supporting the Al-Qaida terrorist network within its borders.

Given its history of war and otherwise political instability, many observers have questioned whether or not tourism could ever be re-established in Afghanistan. This is particularly the case today. As of 2014, foreign military troops were still stationed there to help control the Taliban and other terrorists who were reportedly living in the rugged mountains along the border with Pakistan. Because of the political instability in the country, most governments continue to issue warnings against their citizens traveling to the country.

Tourism in Afghanistan

During the late 1960s and 1970s, Afghanistan received approximately 90,000 foreign visitors each year, and tourism was considered one of the most important foreign exchange earners. Tourists visited primarily because of the country's wealth of natural and cultural attractions, including rugged mountains, valleys, ancient cities and very old Buddhist cultural sites.

By 2006, Kabul's international airport was receiving international flights from Europe and other parts of Asia, and there was a revival in tourist demand among the more adventurous and intrepid travelers. Several tour companies began to offer adventure tours to Afghanistan around 2007. New hotels are being built and foreign investment, particularly by expatriate Afghanis, is being routed into (re)developing the country's utility and tourism infrastructures. Afghanistan's domestic tourism is growing as well, as middle and upper class Afghanis and limited numbers of foreign aid workers from the cities spend their weekends in places such as the lakes of Band-e Amir and the cultural sites of Bamiyam (a UNESCO World Heritage Site), many of which were destroyed in 2001 by the Taliban with the justification that the area's ancient Buddhist statues represented the prohibited act of idol worship.

In 2005, the government appointed its first tourism minister since the United States invasion to promote Afghanistan as an emerging tourist destination. The ministry emphasizes the country's mountains and opportunities for skiing and trekking, as well as heritage sites. There are even plans to establish a tourist destination at Tora Bora, near Osama Bin-Laden's old headquarters. In recent years, several international aid agencies have helped spearhead a variety of tourism development programs. The Aga Khan Foundation is one such agency. With funding from the Norwegian and German governments, Aga Khan initiated the Wakhan Tourism Development Program, which encourages trekking tours through Afghanistan's Wakhan region (the protruding, elongated territory in the northeastern portion of the country). Multi-day treks to observe wildlife, photograph mountain scenery and meet villagers in the region are encouraged in a part of the world few tourists have ever visited. Despite these new tourism promotional programs, in many parts of the country there are still legitimate concerns about land mines, rival mountain rebels, kidnappings and roadside banditry, but the Ministry of Education, Culture and Tourism has

ISSUES AND INSIGHTS Lonely Planet Afghanistan

Lonely Planet is of the most influential guidebooks in the world. In 2007, the Lonely Planet Library added Afghanistan to its list for the first time. Does this signal a renaissance of tourism in the country and acceptability of travel to and within Afghanistan? Or does it constitute a form of extreme adventure tourism? Arguably tourists are now returning to some parts of Afghanistan anyway, while the country is also host to a number of foreign aid workers, contractors and military personnel already on the ground who would probably appreciate guidance on things to do and see when they are off duty. Such activities may also help bring income to some business people and communities while the information may also help cultural sensitivities.

Nevertheless, the guidebook does urge caution and discretion. There is a whole section on safety and risk assessments on each region at the start of the relevant chapter, and some parts of the country are off-limits to foreigners. For example, the message regarding the south is basically that you should not go there, although the guide is more positive about traveling to Pakistan on the historic route through Jalalabad and the Khyber Pass. A point of debate therefore is will the publication of a guide encourage people to travel there, or will it be used more for armchair travel?

attempted to alleviate people's fears in recent years by providing armed escorts for tours. In 2014, the South Asian Association for Regional Cooperation (SAARC), the main supranational body of South Asia, designated Bamiyam the Cultural Capital of SAARC for 2015. Plans for this title include tourism development efforts through exhibitions, seminars, music and dance performances, and food and cultural shows to be promoted throughout the SAARC member countries.

3.2 | SOUTHWEST ASIA AND NORTH AFRICA (THE MIDDLE EAST)

Current Central Asia Issues and Resources (http://wrgeography.com/middleast.html)

The cultural and geological diversity of Southwest Asia and North Africa (and to a lesser degree *colonialism*) has created many countries in this region, a process that is ongoing to the present day. In recent years, Southwest Asia and North Africa has increasingly been referred to as *MENA* (Middle East and North Africa).

Geographically (based on culture and landforms), the Southwest Asia portion of MENA can be divided into two major subregions. To the south is the Arabian Peninsula, home of the world's largest true desert. It includes the countries of Kuwait, Saudi Arabia, Bahrain, Qatar, the United Arab Emirates, Oman and Yemen. Syria, Lebanon, Israel, the Palestinian Territories, Jordan and Iraq are in a geographical transition zone between the Arabian Peninsula to the south, the Mediterranean Sea to the west, and mountains to the north.

The northern part of Southwest Asia (from Asia Minor to Central Asia) is a highland region that includes the countries of Turkey, Azerbaijan and Iran. Cyprus, Armenia and Georgia are sometimes included in the highland Southwest Asia region, though they are also sometimes included in Europe owing to cultural similarities; this book discusses them in the European context.

Algeria, Egypt, Libya, Morocco and Tunisia constitute the North Africa portion of MENA. Western Sahara is a stateless territory administered by Morocco, while Sudan shares the Nile River with Egypt. Sudan is sometimes included in North Africa, though in this chapter Sudan is placed in the Central Africa region of Sub-Saharan Africa. Egypt is sometimes included in Southwest Asia because of its close physical and historical ties to the Arabian Peninsula, as well as its occupation of the Sinai Peninsula, though we place it in North Africa in this book based on its geographical location.

Defining the Middle East

Southwest Asia and North Africa are together often referred to as the Middle East. The term Middle East was first used to define Southwest Asia by the British India Office in the 1850s, and basically it referred to the region east of the Ottoman Empire (today's country of Turkey was the *Near East*) and centered on the Persian Gulf. This definition, however, was widely criticized for its geographical inaccuracy, and later for its *Eurocentrism*.

Until World War II, confusion continued over the use of Near East and Middle East to refer to Southwest Asia, though it was well established that China and Southeast Asia were the Far East. In the decades after World War II, however, the term Near East has been largely dropped from popular use (although it is still used by historians and archaeologists), and Middle East has come into common acceptance. (The United States government, for example, first used the term in an official document in 1957.)

Despite its wide acceptance by news media organizations, the term Middle East is still not clearly defined. Most of the time it refers to the countries of the Arabian Peninsula, plus Turkey and Iran to the north, and Egypt across the Red Sea. However, it can also include Afghanistan and Pakistan (which are placed in Central Asia and South Asia, respectively, in this book), and the remaining countries of North Africa, west of Egypt.

The Muslim countries in the Horn of Africa (Ethiopia, Eritrea, Djibouti and Somalia) and the country of Sudan are sometimes included as part of the Middle East, but more frequently they are considered Sub-Saharan Africa. In other definitions, the *Maghreb Region*, comprising the northwest part of the African continent (i.e., Morocco, Algeria, Tunisia and sometimes Libya and Mauritania), is not included in the definition of Middle East. Because of this confusion, this book defines Middle East based on the geographic concepts of Southwest Asia and North Africa.

Physical Geography of Southwest Asia and North Africa

Geologically, Southwest Asia is a region of considerable tectonic (earthquake) activity. This is where the African and Eurasian continents collide, creating alpine orogeny mountain systems. These include the Atlas Mountains in Northwest Africa, Asia Minor and the Caucasus Mountains, the Zagros Mountains of Iran and the Hindu Kush Range, which branches out from the Pamir Knot and into Afghanistan. (The Himalayas, the Rocky Mountains, the Andes and the Alps are also part of the Alpine Orogeny, or period of mountain building.) These high mountain ranges provide sources of water in an otherwise very dry region. The active continental movement in this zone results in frequent earthquakes that occasionally cause major damage to human settlements.

Bam

In 2003, a very large earthquake occurred in the ancient city of Bam, Iran, killing 27,000 people and destroying some 75 percent of the city's buildings, including its ancient and imposing mud-citadel (said to be the largest adobe building in the world). The damage caused by the 2003 earthquake in Bam meant that the internationally culturally significant parts of the city were placed onto the *World Heritage in Danger List*. However, there has been an international effort to try and repair the damage that was caused to Bam by the earthquake as much as possible. Bam, the site of an *oasis* in the desert of the Iranian high plateau, dates to the sixth century BCE. However, between the seventh and eleventh centuries, Bam's fortified settlement and citadel (Arg) were developed as the city was at the crossroads of silk and cotton trade routes. The citadel contains 38 watchtowers, government quarters and a historic town with one of the oldest mosques in Iran, dating from the eighth to ninth century CE. The modern city of Bam surrounds the old town and citadel.

From Seas to Oil

South of the mountain regions of Southwest Asia are large areas of land that were formerly under water. The Persian (or Arabian) Gulf is a remnant of one of these large seas. The North Africa plain, east of the Atlas Mountains, was similarly once submerged. As old shallow seas, these areas were rich in sea life and vegetation, which over time have been transformed into oil and natural gas. The sand-covered flat lands of Southwest Asia and North Africa are estimated to contain two-thirds of the world's oil reserves, and they are of strategic global geopolitical importance. The modern economies of some areas, especially on the Arabian Peninsula, have been transformed by the region's oil and gas wealth.

Although some oil fields were initially drilled by British and American oil companies in the first half of the 20th century, it was the development of oil by companies owned by the countries in the region in the 1960s and 1970s that provided both the basis for rapid economic expansion in the region and changes in the world's geopolitical balance. These were both caused by the dependence of western developed countries on the region's oil. While the oil economy has provided the basis for the development of petrochemical industries, the countries of the region have been diversifying their economic base, with tourism being a major target industry. In addition, a number of Gulf countries, such as the United Arab Emirates (UAE) and Qatar, have developed substantial airlines and airline infrastructure in an attempt to position themselves as global transport, business, tourism and leisure hubs.

The Arabian Peninsula

The land area in this southern part of the Middle East and North Africa is comprised of massive platforms (known as *massifs*) of extremely ancient rock. The rock underlying these areas is among the oldest in the world, and it was part of the core of Pangaea, the ancient supercontinent that comprised all the continents we know today. The Arabian Peninsula is one large massif tilted upward on its western side, where

Photo: Dallen J. Timothy

◀ *A low-budget Red Sea resort in Egypt's Sinai Peninsula. The Red Sea divides the continents of Africa and Asia and is situated in one of the driest regions of the world. It is also one of the most important water transportation routes, connecting to the Mediterranean Sea via the Suez Canal, This is an example of Egypt's efforts to develop coastal resorts that cater to different budget type travelers.*

its higher mountains are located, and downward on its eastern side, sloping into the Persian Gulf. (The country of India has a similar geologic structure that is tilted in the same way.) The continent of Africa and the Arabian Peninsula massif are separated by the Red Sea but were once a single geologic structure. However, the movement of molten magma deep inside the earth caused the surface to crack along the Red Sea, the Gulf of Aden and the Great Rift Valley in eastern Africa, all of which today are slowly widening and pushing their adjacent land areas apart from one another.

Sea Sands

These flat land areas are covered by the large sand seas of the Sahara Desert and the Arabian Peninsula. (*Sahara* is often translated as *Great Desert*, so some prefer to just use the word by itself, while others prefer the traditional full name of Sahara Desert.) Together these two seas of sand extend 5,000 miles (8,050 km) from east to west. They were formed by the erosion of the underlying rock by wind, water and time. The area has very little cloud cover, causing extremes of daytime heat and nighttime cold. The Rub' al Khali (Empty Quarter) covers much of the Arabian Peninsula and is the largest true desert in the world. A true desert is a solid expanse of sand seas with no vegetation. The entire region, extending across North Africa and into Central Asia, has a hot and dry climate, although the more mountainous northern portions are colder in the winter months.

Arid Climate Zones

The largest expanse of dry, arid and desert lands on the planet extends from the Atlantic Ocean to the Indian Ocean, and into Central Asia. The reason for this vast arid climate zone is the location of the African continent on the surface of the earth. Climate refers to the average weather conditions in a place over the course of a year. The main variables in climate are seasonal changes in climate and precipitation; arid climates have little or no rain throughout the year. A climate map of the earth shows a belt of tropical, humid climate (high rainfall and high temperature) along the equator.

Drier climates are found just north and south of the tropical equator and are either arid (little rainfall) and generally found on the western sides of continents, or subtropical (summer hot/winter cool) and found on the eastern sides of continents. Further north of these regions is another belt of humid climates, including the Mediterranean and Marine West Coast climates on the west coasts of continents, and increasingly colder humid climates on the east coasts of the continents. Finally, the North and South Polar regions are vast arid regions that receive very little precipitation, though their soils can be quite water logged and frozen with moisture.

Environmental Adaptation

In the hot and dry desert areas, buildings in towns and cities are traditionally built very close together to protect against heat and wind. Although winds from the upper atmosphere are generally not strong, the daytime heat rising from the desert landscape can create very strong convection winds. Making a building very compact is like raising the ground level, with the heat affecting the roofs of the houses while the interior spaces below remain cooler, like a cave in a mountainside. Courtyards provide shade and an opportunity for the circulation of air as winds pass overhead and pull cooler air from below. Round roofs, which are common features of the region's traditional architecture, also contribute to this circulation of air. Some traditional buildings use wind towers, which are about 45 to 140 ft (15 - 45 m) high and collect air at their top on their windward side (facing into the wind), directing it into the house and then exhausting the wind out through the windows on the leeward (back) side of the building. Water channels, wells and water fountains are also used in traditional architecture to keep the indoors cool as the air passes through.

Human Geography of Southwest Asia and North Africa

Bedouins

The traditional (pre-modern) economy of North Africa and Southwest Asia consisted of two non-urban lifestyles. The first is nomadic herding and the second is sedentary agriculture. Nomadic herding in Southwest Asia and North Africa is mostly practiced by the Bedouins (desert dwellers), who are Arab pastoralists and comprise about 10 percent of the population in Southwest Asia and North Africa. They live in a delicate balance with the desert ecosystem because *overgrazing* can permanently destroy a pasture area, turning it into a desert.

Most are not truly nomadic, but practice *transhumance*, which is movement in a prescribed route among pasture areas. They move from one pasture to another to prevent overgrazing, often spending the dry season in higher areas and the wet season in lowland areas. Some only travel short distances within a small area, while others may move up to 600 miles (966 km) a year, which requires almost constant travel. The distance traveled depends on the quality of the pastureland in a region and the seasonal availability of food and water. The area and route traveled is clearly defined and respected by others. Individuals have use and access rights to certain places, even though they may only be there a few days out of the year.

Bedouins travel in family groups, with the sheik, or headman of the tribe or clan, as its leader. They own very little (pots and utensils, jewelry, radios), so they can more easily travel. Their animals are prized possessions and a measure of social and individual value. Sheep are the dominant livestock, producing milk, wool and meat. Goats are often better for drier climates, as they are able to survive on the sparsest vegetation. Camels are kept more for meat and milk, and occasionally for transportation, though that is not their dominant use.

Bedouins have a strong attachment to their traditional lifestyle, and they often feel discriminated against by urban-based governments in the region. Governments throughout Southwest Asia and North Africa have tried to settle the nomads, which has often led to disease and overgrazing, as well as much resistance. Many Bedouin groups in countries such as Egypt, Israel and Jordan have become involved in tourism, offering camel guiding services and overnight experiences (including food and culture) in Bedouin tents.

◀ *A Bedouin extended family encampment in southern Jordan. Despite efforts by some countries to force the Bedouins into permanent settlements, their preferred lifestyle is in a nomadic tent.*

Photo: Alan A. Lew

Ecotourism

Throughout the region, nature-based tourism is poorly developed, with most tours focusing on archaeology, ancient cities, churches and monasteries, traditional markets, and village and urban life. On the Arabian Peninsula, the UAE and Oman have been developing tourism based on their desert resources, including safari tourism and Bedouin ethnic tourism. Educational tourism has also emerged in the more developed areas of the region, such as Israel, based on modern and technological adaptations in these arid lands.

The arid conditions throughout the region from Central Asia to North Africa create highly sensitive and endangered aquatic ecosystems, the largest of which are the Red Sea and the Persian Gulf, and the Black Sea, Caspian Sea and Aral Sea in Central Asia. These water bodies concentrate drainage from their surrounding watersheds and serve as sensitive measures of regional *environmental degradation*. Most of them are severely polluted, though concerted international efforts in the 1990s to address this problem have helped to some degree. In Central Asia, ecotourism offers an alternative to the large-scale and heavy industries that caused some of this widespread damage. However, there remains a great need throughout to increase environmental awareness and to address major environmental problems.

Ecotourism opportunities are actually quite plentiful along the coastal seas and oceans that surround much of the Arabian Peninsula. These include diving in coral areas and sea turtle communities, which are usually away from major cities where pollution has caused considerable coral damage, and coastal wetlands that attract large numbers of birds. The Farasan Islands, for example, support the largest wild gazelle population in Saudi Arabia, in addition to a great variety of birds, dolphins, turtles and whales in marine habitats, which range from coral reefs to mangroves. Israel has many nature reserves that protect a diversity of wetland ecosystems. A geologic depression in western Jordan includes the Jordan Valley, the Wadi (oasis or dry wash) Araba, Lake Tiberias (the Sea of Galilee or Kinneret) and the Dead Sea, the lowest point on any continent. In eastern Jordan, wetland parks are managed by the World Wildlife Fund (WWF) in the oases of the Shaumari and Azraq. The famous archaeological site of the old trade caravan city of Petra, with its many large buildings and tombs carved into rock walls, is located close to the Wadi Araba in Jordan.

▶ *A souk in the Medina of Tunis, Tunisia. A souk is a covered market street, with thick ceilings that keep out the Middle Eastern summer heat. Souks come in many different sizes, the largest being the Grand Bazaar in Istanbul. This is one of several in the Tunis Medina, a UNESCO World Heritage Site, which is the oldest part of the city core. It is characterized by narrow streets and alleys, though it is also the home to important monuments, mosques and a university.*

Photo: Alan A. Lew

The Maghreb

Maghreb means the land where the sun sets in Arabic (*al-maghrib*) and is the name given to the northwest corner of the African continent. By contrast, Southwest Asia and Egypt are known collectively as *Mashriq*, or the place where the sun rises. Traditionally the core of the Maghreb has been Morocco, Algeria and Tunisia, through which the Atlas Mountains run. Today it often includes these three countries, plus Libya and Mauritania, all of which have a large Berber population. In 1989, these five countries formed the l'Union du Maghreb Arabe (UMA), a relatively weak economic trade union.

Between Morocco and Mauritania is the territory of the former Spanish province of Western Sahara, one of the most arid and least populated regions on earth. Today, Moroccan forces control most of Western Sahara, with a small part controlled by the Prolisario Front, which, backed by Mauritania and Algeria, fought for the territory's independence after Spain withdrew in 1975. The Arab League of countries supports Morocco's claims to Western Sahara, while the African Union countries support the Prolisario.

The rugged Atlas Mountains (peaks over 13,000 ft (4000 m)) form a geologic spine through much of the Maghreb region. These mountains are part of the Alpine system that extends through southern Europe and provide great biological diversity to the northwest corner of Africa. Just off the Atlantic coast of the Maghreb region are Spain's Canary Islands and Portugal's Madeira and Azores Islands.

Civilizations and Empires

Southwest Asia has long been a crossroads of travel between Asia, Europe and Africa. It was the center of the earliest known civilizations of humanity, including the Sumerians (3500 to 199 BCE), the ancient Egyptians (3000 to about 30 BCE) and the Babylonians and Assyrians (900 to 500 BCE). As in other parts of the world, sedentary agricultural areas have the largest number of people and are mostly found along the rivers, in more humid highland areas, and in scattered desert oases. The Nile River Valley and the Tigris-Euphrates Valley were home to two of the oldest civilizations on the planet.

Egyptian Civilization

The ancient Egyptian civilization existed along the middle to lower reaches of the Nile River for thousands of years, with the dynastic period lasting from about 3150 BCE to 31 BCE. The Egyptian dynasties reached their largest territorial extent about 2000 BCE, and they ended when they were conquered by the early Roman Empire. The Egyptian civilization was based on a social structure designed to manage the waters of the Nile River to support agriculture. This included the collective development and management of an irrigation system and a semi-divine ruler to influence the unpredictability of nature. From this grew a writing system and literature, trade relations and military domination over neighboring territories, and an elaborate system of religious beliefs. The pyramids and the sphinx of Giza (in Egypt) are some of the great wonders of the ancient world, and they continue to be among the most remarkable architectural accomplishments of human history.

Mesopotamian Civilization

Contemporary with the Egyptian dynasties was the ancient Mesopotamian civilization along the Tigris and Euphrates Rivers (in modern-day Iraq). Ancient Mesopotamia is usually dated from the Bronze Age about 3500 BCE to the arrival of Alexander the Great in the fourth century BCE, at which time Greek culture came to dominate the region. The Tigris and Euphrates Rivers provided similar opportunities to the Nile River for the emergence of highly organized societies. Among the major kingdoms and empires of Mesopotamia were Sumer, Babylon and Assyria.

The Sumerians arose about 3500 BCE, before the Egyptian Dynasties, and lasted to approximately 2800 BCE. The ruins of the ancient city of Ur (birthplace of the Old Testament's Abraham) in southern

ISSUES AND INSIGHTS   Volunteer Tourism

Volunteer tourism combines travel with voluntary work that is often of a social or environmental nature, and usually in a location where government or private sector assistance would not otherwise be forthcoming. Volunteer tourism is therefore seen to provide a more reciprocally beneficial form of travel in which both the volunteer and the host communities are able to gain from the experience. The concept of volunteer tourism has therefore come to be strongly related to concepts of sustainable tourism and sustainable development, especially with respect to *pro-poor tourism* (PPT).

While the concept of international volunteering is by no means a new phenomenon, in recent years there has been a rapid increase in the number of individuals taking part in short term, organized volunteer tourism programs (VTP). The increasing demand for VTPs has been paralleled by an increase in sending organizations that promote, sell, and organize programs for volunteer tourists. Sending organizations now offer a large variety of options depending on volunteer tourists' preferred activity, location and duration.

It has been suggested that the interactions encouraged by this form of tourism between volunteers and their hosts lead to improved cross-cultural understanding, with volunteer tourists gaining a sophisticated understanding of the local culture in which they participate, as well as of the issues facing their host communities. In addition, some group VTPs attract volunteers from all over the world, and the friendships developed between volunteer tourists themselves may also reduce racial, cultural and social boundaries.

However, with the growing number of short-term VTPs, recent studies have begun to question whether volunteer tourism does indeed always result in increased cross-cultural understanding. In particular, research relating to the United Kingdom 'gap year' (a year taken out between high school and university, or post-university, which often includes an element of voluntary work) suggests that international volunteering may in fact reinforce existing stereotypes and deepen dichotomies of *them and us*. It has also been suggested that certain types of VTPs may represent a form of neocolonialism or imperialism, in which volunteer tourists inadvertently reinforce the power inequalities between developed and developing countries.

For example, when volunteer tourists inappropriately take on the roles of *expert* or teacher, regardless of their experience or qualifications, this may be seen to represent the neocolonial construction of the westerner as racially and culturally superior. Therefore, it is essential for programs to be carefully developed and managed to ensure that the development goals are achieved. Where individuals volunteer through sending organizations, these organizations must play an important role in ensuring that stereotypes are broken, rather than reinforced.

Three key recommendations for sending organizations seeking to achieve cross-cultural understanding have been identified. First, they should develop programs that will be of genuine value for the local communities. Second, the importance of approaching VTPs as a learning process rather than simply an *experience* should be recognized through the use of experiential learning techniques. Third, opportunities for interaction with other cultures should be deliberately facilitated.

Sources

Raymond, E. M., and C. M. Hall. 2008. The development of cross-cultural (mis) understanding through volunteer tourism. *Journal of Sustainable Tourism*, 16(5), 530–543.

Iraq date back to Sumerian times. From about 2000 BCE until the dominance of the Greeks, southern Mesopotamia was largely (not always) dominated by Babylonia (capital city of Babylon), while the northern portions of the Tigris and Euphrates Rivers were the home of the Assyrian Empire, which at one time extended their control to the Nile River. The city of Babylonia was famed for the legendary Hanging Gardens of Babylon and the Tower of Babel, though no signs of the city exist today.

Byzantine Empire

Turkey has been the center of two major empires in Southwest Asia. In its early history, it was closely tied to Greece and at times was ruled by its Mediterranean neighbor. It was part of the Roman Empire (31 BCE to 475 CE), and then became the center of the Eastern Roman Empire (also known as the Byzantine Empire, ca. 5th c. CE 1299) after Rome fell to invading groups from other parts of Europe and Central Asia. The Byzantine Empire was based in Constantinople, which was made the Christian capital of the Roman Empire by Constantine the Great in 324. This Greek-speaking empire spread Orthodox Christianity throughout large parts of Eastern Europe, where it is still the dominant religion today. The empire was named after the Greek city of Byzantium, which was named New Rome by Constantine, but soon became Constantinople; it was named Istanbul in 1930, which is a traditional Turkish name for the city.

◀ Physiography of Southwest Asia and Africa.

The Byzantine Empire reached its peak in 550 CE when its empire encompassed much of the Mediterranean Sea. This did not last, however, as within less than a century the Arab Empire began to expand outside the Arabian Peninsula. Little of the Byzantine period remains in the cultural traditions of Southwest Asia, or even of Turkey, with the exception of its architecture. The Hagia Sophia (St. Sophia) in Istanbul was built as a Christian Church in 532 to 537 CE and is considered by many to be one of the most beautiful buildings in the world. It later became a mosque and today is a museum that continues to dominate the skyline of old Istanbul.

The Arab Empire

The most influential empire for most of Southwest Asia and North Africa was the Arab Empire (632–1258 in various forms and sizes). One cultural characteristic that all of Southwest Asia and North Africa (except for Turkey) share is their history of being part of the Arab Empire, resulting in shared cultural traditions that reach back to that time and including religion, language and architecture. Today, Arabic is the dominant language of the region from the Arabian Peninsula to Morocco. It was spoken by the Prophet Mohammed, a trader born in Mecca in the year 570 CE and the founder of Islam. Within some 50 years of his death in 632, Arabs from the Arabian Peninsula had conquered the Persian Empire (Iran) to the east, much of the Byzantine Empire (Turkey) to the north, all of North Africa to the west and most of the Iberian Peninsula (Spain and Portugal). Further to the east, Pakistan in South Asia, and southern portions of Central Asia, also came under Arab Empire control. The Arabs built Baghdad as a new capital city for their empire.

▶ *Ruins of the Roman Baths at Carthage in Tunisia. Carthage was the largest Roman settlement in North Africa, and these baths were the largest outside of Italy. Carthage, which is on the outskirts of the modern city of Tunis, was the main trading city for the rich agricultural hinterlands that surrounded it.*

Photo: Alan A. Lew

The Arab Empire reached its peak in the late eighth century. It was a cosmopolitan empire, blending the wisdom of three continents. Scientific innovation included Arabic numerals (that are still used today), mathematics based on the number zero (which was a relatively new concept), algebra and trigonometry, and medical colleges that produced the first government-certified physicians in the world; free public hospitals were also established in the Arab Empire. Aristotle, Plato and Hippocrates were translated from Greek into Arabic, thereby preserving their writings through Europe's Dark Ages. Baghdad became the richest and one of the largest cities in the world. The dominance of the Arab Empire on the Arabian Peninsula continued until the year 1258 when the Mongols conquered and destroyed Baghdad. The Mongols ruled the city for 200 years, and the Arabs never again were able to establish the unity of former times. The vision of a return to a unified Arab world has, however, continued into recent times; Libya, Syria, Iraq and Egypt have all tried to establish themselves as heirs to the Arab Empire, though none have succeeded.

Sunni and Shiite

The most salient legacy of the Arab Empire is Islam. The countries of Central Asia, Southwest Asia and North Africa share the religion and culture of Islam, and over 90 percent of the people in this region are Muslims. Within Islam, there are two major sects: Shia (or Shiite) Islam and Sunni Islam, as well as some other minor sects not typically associated with the two larger ones. The Sunnis believe that Mohammed was the last prophet of God, while the Shias believe there were other prophets after Mohammed. Sunni Muslims predominate throughout the region, with the exception of Iran, where Shiites are the majority, and Iraq where they comprise the majority population but are not as pervasive.

Iraq was ruled by a minority Sunni government for many decades, which brutally suppressed the majority Shiite population. This resulted in the Shiites tending to become more impoverished and more fundamentalist in their religious beliefs. Differences between Sunnis and Shias are a major source of tension in some parts of Southwest Asia today. (There are other major centers of Islam and Islamic culture that are not part of North Africa, Southwest Asia and Central Asia. These include Pakistan and Bangladesh in South Asia, and Malaysia, Brunei and Indonesia in Southeast Asia. Indonesia has the largest Muslim population of any country in the world.)

Ottoman Empire

After the fall of the Arab Empire, a second Islamic empire arose in Turkey. Asia Minor (the peninsula where Turkey is located) was never part of the Arab Empire, though Islamic religious influences gave rise to the Ottomans, whose empire eventually replaced the Byzantine rulers in Constantinople. The Ottoman Empire (1299–1922) reached its peak in 1526 when it controlled the area from Hungary to the Caucasus, the entire Arabian Peninsula and most of North Africa. The empire lasted until World War I, when it was defeated by Great Britain and Russia. Although they saw themselves as heirs to the Arab Empire, the Ottoman leaders exhibited a great distrust of science, commerce and industry. They isolated non-Islamic peoples, such as Armenians, Jews and Greeks, and political corruption was a major problem throughout the Empire.

These characteristics formed the basis of European and other Western stereotypes of much of Southwest Asia and North Africa today. These perceptions are not helped by the political turmoil and unrest that continue to plague many parts of the region and make this region among the most dangerous in the world to visit. At the same time, strong cultures and traditions, opportunities to see distinctive landscapes, and lifestyles and historic relics give tourism enormous potential.

Non-Islamic Peoples

Most of the region's population today speak Arabic, consider themselves Arabs and practice Islam. The major non-Arab, yet Islamic, areas are located in the high plateau and mountain regions of Turkey, Central Asia (mostly Turkic speaking), Iran, northern Iraq and portions of Afghanistan. Besides Islam, Southwest Asia is also a hub of Judaism and Christianity, two other major world religions. The Jewish state of Israel was created after World War II, in a region that historically had a large Jewish population, but also a large Arabic-speaking Palestinian population. Jerusalem, the capital and largest city of Israel, has long been a *pilgrimage* destination for Jews, Christians and Muslims. It is one of the three most important religious cities in Islam, as Muslims believe that the site atop the Temple Mount is where the Prophet Mohammed ascended to heaven. Unfortunately, competing religious interests in such a small territory have more often served as a source of conflict and division rather than of religious tolerance and unity.

Amazigh/Berber

The southern portions of Algeria, Tunisia and Morocco have large Amazigh (Berber) populations, which have lived in this region for over 2000 years. Similarly, the southern edge of the Sahara Desert is a transition zone where semi-Arab peoples transition to Sub-Saharan African populations. Countries there include Mauritania, Mali, Niger and Chad (which are included in the Sub-Saharan Africa region of this book). Although on the edge of the Middle East region, the highlands of Ethiopia are the center of a civilization that dates back over 2000 years and that has evidence of Greek and Egyptian influences. These include the adoption of the Coptic Christian tradition, which is related to the eastern Orthodox form of Christianity. Smaller early Christian church populations are scattered throughout Southwest Asia and North Africa, including a large Christian population in Lebanon, Syria, Egypt and Israel.

Countries and Tourism of Southwest Asia and North Africa

Israel

Israel's primary tourist appeal is its religious heritage sites. Christians, Jews and Muslims around the world travel to Israel and the adjoining Palestinian territories to view and experience places significant to

► *The Church of the Holy Sepulcher in the Old City of Jerusalem commemorates the site where Roman Catholics and many Orthodox Christians believe Jesus rose from the dead after he was crucified. Other religions believe that this event occurred at other sites. Worshipers here are kissing the Stone of the Anointing, where the body of Jesus is believed to have been placed after he died on the cross.*

Photo: Alan A. Lew

their religious traditions. Christmas and Easter are especially popular seasons for Christians to visit the Holy Land. Bethlehem and Jerusalem, sites of important events in the life of Jesus Christ, are the main destinations in Israel and Palestine for this group. Muslims (some of whom can get permission from their home governments to visit Israel) gravitate to the Haram al-Sharif (Temple Mount to Jews) to see the Dome of the Rock (*Masjid Qubbat As-Sakhrah* in Arabic) and to worship at the al-Aqsa Mosque adjacent to the Dome, located in the Muslim Quarter of Jerusalem. The rock is believed to be the place where the Prophet Mohammed ascended into Heaven to consult with Moses. The existence of the Dome of the Rock renders Jerusalem the third most sacred location for Muslims, following Makkah (Mecca) and Medina in Saudi Arabia. The rock is also venerated by Orthodox Jews as being the location where Abraham was willing to sacrifice his son in the Old Testament.

For Jews, Jerusalem represents the center of their religious and cultural heritage. It was one of the capitals of ancient Israel, and many events took place there, including temple buildings, which have made the city a holy place. Of particular importance is the Western Wall—the last remaining vestige of ancient Solomon's Temple (see photo on page 20). Jews travel from all parts of the world to pray and worship at this site, and many young Israelis and other Jews of the diaspora celebrate their bar mitzvahs (boys), bat mitzvahs (girls) and marriages at this location, adding special significance to these events. Unfortunately, the ongoing conflicts between Israel and its neighbors, and the bitter relations between Palestinians and Jews in Israel, have led to many important sacred sites for Jews, Christians and Muslims being inaccessible and off limits for security reasons.

There are several other types of tourism in Israel beyond those of a religious nature. During winter months, the Red Sea beach resort of Eilat becomes crowded with warmth-seeking tourists from northern Europe, with charter flights arriving directly to Eilat from as far away as Finland and Sweden. The Dead Sea between Israel and Jordan is also an important center for *health and medical tourism*. A unique form of heritage tourism has developed in Israel since 2000—Birthright Israel. This program, which is funded by various Jewish organizations around the world and the Israeli government, provides free trips to Israel for youth of the diaspora to become more familiar with the country's history, culture, language and religion, and to increase support for Israel among the world's Jewish youth.

These trips are not primarily religious in nature but instead focus on building solidarity for the secular state of Israel.

In terms of urban tourism a number of cities such as Jerusalem and Nazareth have a number of non-religious tourist attractions that are connected to the Byzantine period while other more-modern urban heritage attractions, such as museums and art galleries are increasingly significant. Tel Aviv is also notable for its World Heritage Bauhaus streetscapes and its thriving arts and cultural scene.

Modern Turkey

Turkey today probably has the best-developed tourism industry in the entire region from Central Asia to North Africa. Similar to other areas in the region, archaeological and historic sites are a cornerstone of tourism in Turkey, which has historically served as the bridge between Asia and Europe. The higher elevations of the country and other parts of highland Southwest Asia are considerably cooler than the Arabian Peninsula, and many distinct forested regions are protected in Turkey's nature reserves and 21 national parks.

Some of these parks, first established in 1958, were initially designed for archaeological and historical purposes, but they have since become important ecological habitats protecting biological diversity. Wetlands are also more plentiful on Asia Minor and around the Caucasus Mountains. Trekking in the mountains (including ski-mountaineering in Georgia), bicycle touring, and sun and sand resorts along the Mediterranean coast have become popular travel experiences in these areas. Antalya, a province on the Mediterranean coast of southwestern Turkey, has in particular become a center for coastal tourism development that has been boosted by direct charter and budget airline flights from across Europe. The region attracts 30 percent of the foreign tourists visiting Turkey.

Turkey had applied to join the EU with the other ten countries admitted in 2004. However, the country has issues related to human rights and the treatment of its Kurdish minority that must be addressed before membership conditions can be fulfilled. In addition, Turkey is a very large and poor country, with a predominantly Muslim population. This all raises new issues for the EU, and particularly for countries such as Catholic Poland, which has been perceived as seeking to entrench Christianity within the EU framework. The sensitivities of this issue were further highlighted in 2010 when it was revealed that the Holy See, which had unsuccessfully been seeking to secure a reference to Europe's "Christian roots" in the EU constitution, had also been hostile to Turkey joining the EU.

Iran

After a period of turmoil following the overthrow of the Shah of Iran in 1979, tourism returned by the late 1990s, though visas for independent travel were very difficult to obtain. Rampant urban and industrial development, combined with devastation from the Iran-Iraq War in the 1980s, has caused widespread environmental damage, especially along coastal areas of the Persian Gulf and Caspian Sea. The country has established a few national parks, mostly in forested areas of the Alborz mountain range where the rare black bearded and spiral horned Alborz sheep reside. However, these parks are almost entirely unmanaged and unprotected. Iran is a significant Shia Muslim pilgrimage destination, with many sacred mosques and graves of Imams (religious leaders) being significant draws.

Iran has several amazing ancient cities, some of which have been listed by UNESCO as World Heritage Sites. Detentions of Americans and other foreigners in Iran, as well as political unrest related to a desire for popular democracy, had strained efforts to develop international tourism in the Islamic Republic of Iran for many years, although there was significant domestic tourism in coastal areas on the Caspian Sea and the Gulf of Oman. The situation may be slowly changing, however, as there were there signs of significant growth in international arrivals to Iran in 2013 and 2014.

Saudi Arabia and the Hajj Pilgrimage

As already noted, the dominant religion in the region is Islam; the center of Islam is Saudi Arabia. It is mandatory for Muslims to make the *Hajj* pilgrimage to the city of Makkah in Saudi Arabia at least once in their lifetime, if financially and physically possible, making this one of the greatest international tourist movements in the world. (There is some debate as to whether or not religious pilgrimage is a form of tourism. Most pilgrims say it is not. However, from an economic and technical viewpoint, there is little difference between religious travel and leisure travel.)

The Hajj is the fifth of the five pillars (or practices) of Islam. The other four are Shahadah (the declaration of faith), Salat (prayer), Sawm (fasting) and Zakat (charitable giving). The Hajj is the pilgrimage to the Holy City of Makkah (Mecca is its anglicized form) during the Islamic month of Dhu'l-Hijjah. Hajj literally means to set out for a place. The greater Hajj (al-hajj al-akbar) is performed over five or six days, starting on the eighth day of Dhu'l-Hijjah. Together with the lesser Hajj (Umrah), the Hajj comprises a series of religious activities and ritual acts that must be carried out in a particular order at specific religious locations in and around Makkah that are symbolic of the lives of Ibrahim (Abraham) and Hajarah (Hagar), and that are cornerstones of Muslim faith. Prior to the greater Hajj, many pilgrims will visit the city of Medina and the Mosque of the Prophet, where the Prophet Mohammed's tomb is located.

Managing the Hajj

The Hajj attracts over 2 million international and domestic visitors to Makkah and Medina each year (see Table 3.1). It is the second largest religious gathering in the world (after the Kumbh Mela pilgrimage in India, which occurs every twelve years and attracted some 80 million people in 2013). In poorer Islamic nations, governments and international organizations offer grants for the needy to assist in paying for the trip to Mecca, and some TV shows offer Hajj trips as competitive door prizes. Because of the large numbers of people that participate in the Hajj over a short period of time, Saudi authorities have developed an extensive infrastructure to manage the pilgrimage. The Saudi government issues special visas to non-Saudi nationals for the purpose of the pilgrimage, and only a set number of visas are allocated to each country that pilgrims come from.

Guides are often hired to help travelers perform the rites of Hajj, to organize lodging and to provide meals, drinks, transportation and interpreter services. Women are encouraged to undertake the Hajj in

TABLE 3.1 Number of Hajj Pilgrims 2004 to 2012	
Year	**Total Pilgrims (Islamic/International)**
1424/2003–2004	2,012,074
1425/2004–2005	2,164,469
1426/2005–2006	2,130,594
1427/2006–2007	2,378,636
1428/2007–2008	2,454,325
1429/2008	2,408,849
1430/2008–2009	2,313,278
1431/2009–2010	2,789,399
1432/2010–2011	2,927,717
1433/2011–2012	3,161,573

Source: Ministry of the Hajj, Kingdom of Saudi Arabia.

conjunction with a close male relative, although the Saudi government will allow women to travel accompanied by a group of other women if they have written permission from a male relative. Entrance to Makkah is for bidden to non-Muslims as the entire city is regarded as a holy site to Islam.

Saudi Arabia Influence

Makkah and Medina are two of the three most important Islamic religious sites. Their location in Saudi Arabia gives the Saudi government and the Saudi royal family an enormous role and responsibility with respect to Islamic affairs. Together with its significant oil wealth, the Saudi rulers are a major influence in the political affairs in Southwest Asia and North Africa. Nevertheless, the same combination of holding a monopoly of natural resources and offering religious significance have been a source of instability in the region that has often served to harm tourism. Radical Islamists, such as the leaders of al Qaida, are often driven more by their disdain for Saudi government policies than by their hatred of the West.

Saudi Arabia has some of the most restrictive international travel policies in the world. The country receives over 9 million visitors a year, about half of whom are guest workers, followed by religious-related travelers. Until 2004, Saudi Arabia did not issue tourist visas. The only way a person could enter the country was to attend the Hajj (on a Hajj visa), or as an invited guest visiting relatives, a company or a university, or by attending a conference. Tourism is now starting to emerge, though tourist visas are only issued to organized group tours, and visitors are expected to respect the culture and traditions of Saudi Arabia. Homosexuality, drug smuggling, adultery and trying to convert a Muslim from Islam are all punishable by death. Non-Muslims are not allowed to visit the holy cities of Makkah and Medina.

Tourism in the Middle East

As noted in the examples of Israel and Saudi Arabia above, religious pilgrimage is a primary form of tourism in Southwest Asia, followed by heritage tourism based on the archaeology of ancient empires. Compared to other parts of the world, tourism has not been a major economic growth engine in Southwest Asia and North Africa. Part of the reason for this has been the idealistic contentions of religious conservatives who view tourism as promoting hedonism, immoral behavior and cultural degradation.

Religious Influences on Tourism

In the view of many clerics and government officials, the negative religious and social impacts of tourism (e.g., prostitution, nudity, open affection between genders, immodest attire, alcohol use, gambling and irreverence for sacred places and ceremonies) outweigh the economic and social benefits that the industry might offer. Even though countries such as Turkey, Morocco, Tunisia and Egypt have become important leisure travel destinations, especially for Europeans, most of the Muslim-dominated states of the region have been cautious to accept tourism. The possible exception to this has been the United Arab Emirates (UAE), which has been anxious to position itself as an aviation and tourism hub, especially for when revenues from gas and petroleum and related sectors begin to diminish.

Religious beliefs in primarily Muslim nations profoundly affect the growth, operation and management of tourism. Forms of tourism that emphasize culture over recreation tend to be emphasized, and tourist resorts are more likely to be located inland rather than near beaches on the coast. Beach and coast-based tourism, in particular, is generally seen as a major promoter of vices by openly promoting recreation and bodily pleasures. This worldview has led to the encouragement of domestic and intra-regional travel rather than to catering to non-Muslim outsiders, which serves as a protective measure to reduce the influences of Western tourists on destination populations. Of this, Aziz (2001 154) notes," . . . no Islamic country has yet managed to accommodate the needs of Western tourists without compromising the religious and cultural expectations of most of its own people."

ISSUES AND INSIGHTS   Halal Food

Many westerners have heard of *kosher* food and have a rough idea that it refers to Jewish dietary laws. But few know of *halal* food. Given the increasing numbers of non-Islamic visitors to Islamic countries and communities, and the growing number of Muslim tourists to non-Islamic countries, an understanding of halal is therefore extremely important for the tourism and hospitality industry as well as for appreciating Muslim cultures.

Halal is one of the most important aspects of Islamic life. Although it is a standard that Muslims around the world live by, many non-Muslims are still unaware of halal food and its significance for the Muslim community. Halal food and drinks must conform to Islamic dietary laws as specified in the Qur'an, the hadeeth (sayings) and sunnah (tradition) of the Prophet Muhammad, as well as in the fiqh (teachings) of the four Islamic Jurists: Hanafi, Shafi'ie, Maliki and Hambali. Other sources of laws are the Ijima (collective approval) and Qiyas (syllogy) of Islamic scholars.

Halal and haram are the two major terms used in Islamic dietary laws. While halal means permitted and legitimate, haram, on the other hand, means unapproved, forbidden, illegitimate or illicit. Several criteria must be met before food products can be certified as halal. First, the product must be free from any substance or ingredient taken or extracted from a haram animal or ingredient. Halal products are made from naturally halal animals, such as cattle, goats, sheep and chickens that are slaughtered according to the manner prescribed in Islam. Second, halal products should also be processed, produced and/or stored by using utensils and/or equipment that have been cleansed according to Islamic law.

The main idea behind this is that products should be free of contamination and must not come into contact with haram substances during preparation, production and/or storage. In addition, halal ingredients should also not be mixed with objectionable or haram ingredients like enzymes and emulsifiers of porcine origin or other non-halal animals. Any other groups of food, like cheese and meat, may be combined together, as long as no haram or prohibited foods are included in the mixture.

Islam places a very strong emphasis on cleanliness in everything, especially in the context of food and drink. In Islam, eating is regarded as a matter of worship, like prayer and other religious activities. So, just as Muslims perform the ablution as a means of cleansing themselves before their daily prayers, they must also ensure that the food they consume is clean and prepared in the correct manner, starting with the avoidance of items prohibited in Islamic dietary laws. The Qur'an clearly prohibits Muslims from consuming the following categories of food: alcohol, carrion, flowing blood, pork and animals that have been slaughtered with the invocation of a name other than the name of God. (This type of food is associated with the practice of idolatry that Islam strictly opposes, and this prohibition also addresses the issue of halal slaughtering, in which pronouncing the name of God is a required condition while slaughtering an animal.) Islam prescribes that an (halal) animal should first be slaughtered before its consumption, in order to get rid of the blood. The Islamic method of slaughtering an animal is to cut its throat to enable as much blood to run out as possible and not congeal in the veins.

There are also other prohibitions that have arisen, either according to the (fiqh) teachings of Islamic jurists or based on the (hadeeth) sayings and (sunnah) tradition of the Prophet Muhammad. These prohibitions include the consumption of all beasts of prey and birds of prey, amphibians, undesirable insects, and vermin or poisonous animals. In addition, Muslims are strictly prohibited from eating meat that has been cut from a live animal.

According to Islamic law, fish and other sea creatures are exempted from the category of dead animals. All marine animals are therefore halal and can be eaten without the requirement of slaughtering and bleeding, although there is some debate among jurists of different schools of Islamic thought with respect to the status of crustacea and mollusks. Human beings are permitted to catch seafood in any manner they like, using methods that are as humane as possible. It also matters not whether seafood comes out of the water dead or alive, as a whole or in pieces, and if caught by a Muslim or by a non-Muslim.

Sources

Wan Hassan, M., and C. M. Hall. 2003. The Demand for Halal Food among Muslim Travelers in New Zealand, In *Food Tourism Around the World*, eds. C. M. Hall, et al. Oxford: Butterworth-Heinemann.

Syed Marzuki, S.Z., Hall, C.M., & Ballantine. P.W. 2012, Restaurant manager's perspectives on halal certification. *Journal of Islamic Marketing*, 3(1), 47–58

The Gulf States

Although total numbers are relatively low compared to more developed regions of the world, in recent years Southwest Asia and North Africa have been among the fastest growing regions for international tourism in the world. This may seem surprising given the wars and civil unrest that many parts of the region have experienced in recent decades (including Iraq, Iran, Azerbaijan, Egypt, Tunisia, Lebanon, Syria and

Israel/Palestine). Nevertheless, the gulf states of UAE, Qatar, Kuwait and Oman have been reinvesting large amounts of oil money into tourism and related leisure and sports developments as part of a strategy of economic diversification, which may be disproportionately contributing to increases in tourism throughout the entire region. Dubai, one of the seven emirates that comprise the UAE, has become a leader in this area, building a postmodern landscape of human-created islands (in the shape of giant palm trees), indoor ski runs (the world's largest) and the world's tallest building (160 floors, completed in January 2010).

Even Saudi Arabia, among the most religiously conservative countries in the region, is gradually encouraging more international tourists. Investment has been focused in hotel, leisure, sports and shopping developments as well as in airlines. In fact, in the same way that intense competition has occurred in Southeast Asia between Kuala Lumpur and Singapore for hub status, so the Gulf States have also been competing for the growing international tourism and travel market to the gulf region, as well as for the development of stopover tourism between southern Asia and Europe. From the Persian Gulf, where much of this competition is concentrated, London is seven hours away by air, Frankfurt is six hours, Cairo is four hours, and Hong Kong is eight hours by air.

Egypt

Egypt, with over 14.7 million visitors in 2010, is the dominant non-religious international tourist destination for the Old World region (Central Asia, Southwest Asia and all of Africa). Egypt has long been an important group tour destination in the Middle East, particularly in conjunction with visits to Israel. In the late 1970s, a peace treaty was signed between Egypt and Israel, paving the way for the development of tourism between them and opening doors through which Israelis now vacation in Egypt, and Egyptians spend their holidays in Israel. Elsewhere in North Africa, Tunisia has become an important resort destination for Europeans, while Morocco is a well-established adventure and youth travel destination.

Libya

Libya has been largely closed to the world following military clashes with its neighbors and accusations of supporting terrorist attacks against the United States and Europeans in the 1970s and 1980s. That changed in 2006 when the country agreed to settle its international disputes and began to allow its first Western tourists in decades. Political instability by way of popular uprisings against autocratic leaders in Tunisia, Algeria, Bahrain, Yemen, Egypt and Libya in 2011 resulted not only in the overthrow of the Egyptian and Tunisian governments, but it also set back tourism in North Africa considerably. Arrival numbers plummeted; flights to and from Cairo were cancelled for weeks at a time. Tour operators cancelled group packages in Egypt, resort guests in Tunisia can celled their bookings, and the promise shown by Libya in developing an international tourism industry disappeared.

United Arab Emirates (UAE)

Dubai (one of the UAE's emirates) is one of the best examples of a place expanding tourism and seeking to develop a transportation, business and leisure hub. As of July 2010, there are over 6,000 weekly flights operated by 130 airlines to over 220 destinations across every continent except Antarctica operating out of Dubai International Airport (DXB). In 2013, DXB handled 66.4 million passengers, an increase of 15.2 percent from the previous year, making it the seventh busiest airport in the world by passenger traffic. Al Maktoum International Airport opened in 2010. Initially designed for cargo flights as part of new commercial, residential, exhibition and logistics development known as Dubai World Central, it began receiving commercial passenger flights in 2013. If completed as planned, the airport will eventually become the world's largest cargo and passenger hub. A high-speed rail expressway is proposed to link the two airports in Dubai.

▶ *The Burj al Arab Hotel in Dubai, United Arab Emirates. This hotel was the world's first seven-star hotel, a title it gave itself. There are only 202 rooms, each taking up two floors, and rates range from about US$1,000 to $27,000 a night. It was also the world's tallest hotel when it was built, though there are a few others that are taller now, most of which are also in Dubai. The shopping center in the foreground uses the architectural style of traditional wind towers, which were used to capture winds to cool homes.*

Photo: Alan A. Lew

In addition to airport development, Dubai also developed a cruise ship terminal that opened in 2001. The importance attached to tourism development in Dubai is illustrated by the fact that the tourism organization, the Dubai Department of Tourism and Commerce Marketing (DTCM), is chaired by the UAE Vice President and Prime Minister and Ruler of Dubai, His Highness Sheikh Mohammed bin Rashid Al Maktoum. In addition to its head office in Dubai, the DTCM has eighteen overseas offices. They are located in New York (U.S.), London (the U.K. and Ireland), Paris (France), Frankfurt (Germany), Stockholm (Scandinavia), Milan (Italy), Moscow (the Russian Federation, CIS and Baltic States), Sydney (Australia), Johannesburg (South Africa), Mumbai (India), China (Beijing, Guangzhou and Shanghai), Hong Kong (Far East), Tokyo (Japan), Saudi Arabia (Jeddah and Riyadh) and Zurich (Switzerland and Austria).

The growth of tourism in Dubai has gone hand in hand with an aggressive aviation development policy which combines an open skies policy with the development of the national airline, Emirates, the international airline of the UAE, which is wholly owned by the government. Growth of Emirates Airlines has been substantial, doubling in size every 5.5 years in its first eleven years since its establishment in 1985, and every four years between 1996 and 2008. As of late 2014, Emirates serviced some 140 destinations in 78 countries with approximately 3,400 flights per week. Dubai has built this economy on loans from its more oil-rich Emirate neighbors, through creative business acumen and because of an open and global economy. To attract international tourists, Dubai has expanded beyond the traditional tourist attractions of Southwest Asia and North Africa. Those tend to include cultural attractions (Arab culture, historical cities and archaeological sites) and natural attractions (desert caravans, marine tourism and bird-watching). Instead, Dubai has focused on developing sporting facilities, such as golf courses, and a meetings and conventions industry. It is also relatively liberal with respect to the enforcement of regulations on alcohol, dress and the mixing of sexes, which allows Dubai to promote itself as an international pleasure tourist resort that can cater to a wide range of tastes and markets. Likewise, Dubai is one of the most desired shopping destinations in the world.

Other emirates (e.g., Sharjah, Abu Dhabi and Ras al Khaimah) in the UAE have followed Dubai's lead and have started testing the tourism waters. Aside from Dubai's shopping, amusement parks and other commercial attractions, UAE tourism is based largely on business travel (oil) and desert safaris.

Qatar, Bahrain and Kuwait

The three Gulf countries of Qatar, Bahrain and Kuwait are interested in tourism as an economic growth mechanism to supplement their large oil supplies. Like other petroleum-producing countries of the region, business travel comprises a large part of their international market. Qatar and Bahrain are especially interested in developing tourism based on beaches, historic sites and cities. Kuwait became known to the outside world as the center of the Gulf War in the early 1990s and has some tourism based on that legacy. All three countries are well connected throughout the region, and Qatar Airways, like the UAE's Emirates Airline, is working to become a major air transport hub between Europe, Asia and Africa. Bahrain is physically connected to Saudi Arabia by a large bridge, or causeway, which provides another vital form of transportation linking large Saudi Arabia with tiny Bahrain, and there has been talk in recent years about connecting Bahrain and Qatar with an even longer bridge, although soured relations between the two countries and a lack of financial backing has put the project on hold.

Extreme Tourism

Extreme tourism is defined as travel to dangerous places or participating in dangerous activities. Sometimes called shock tourism, it is related to extreme sports and dark tourism, but it may place the tourists in even more life threatening, but adrenaline pumping, danger than either of those. One form of extreme tourism is travel to the most dangerous places of the world—places that most leisure travelers would avoid. War-torn areas of the Old World (from Central Asia through Africa), where civil disorder is the norm, are the types of destinations that draw the small, niche extreme tourism market.

Central Asia, Southwest Asia and North Africa have suffered more than their share of global conflict in recent decades. Some of these have included the Iran-Iraq War (1980–1988); the hostilities between Israel and the Palestinians (since World War II) and between Morocco and the Prolisario Front in Western Sahara (1975–1991); the two United States-led invasions of Iraq (1991 and 2003); Soviet (1979–1989) and American (2001) invasions of Afghanistan; Turkey's invasion and splitting of the island of Cyprus (1974); civil war in Lebanon (1975–1990), Syria (2011–2014), Tajikistan (1992–1997), Sudan (since 1955),); recent genocides in Sudan (since 2003); and political uprisings in Algeria (1992–2002), Tunisia (2011), Egypt (2011–2012), Yemen (2011–2014), and Libya (2011–2012). In addition, the entire region has a long history of powerful rulers for life and insurgent terrorism related to ethnic divisions and the rise of extreme versions of Islamic fundamentalism. Many of these are examples of both the internal struggle for scarce resources in this hostile yet fragile environment, as well as the result of external influence and interference by European, Russian and American interests after centuries of colonialism. It is amazing that despite these challenges, some form of tourism still persists in every country in this region.

Iraq

Iraq is an example of a nation attempting to develop its tourism economy while experiencing internal war. In March 2003, based on suspicions that Iraq's leader, Saddam Hussein, was manufacturing weapons of mass destruction (WMDs) (e.g., nuclear and biological weapons), which he might use against the United States and Israel, the United States invaded Iraq and overthrew its autocratic leader. WMDs were not found, and there was little evidence of any possible links between Iraq and al-Qaida terrorists based in Afghanistan. However, the United States maintained its presence in Iraq with the goals of bringing democracy and human rights to a nation that had been ruled for decades by a despotic dictator. By the end of 2011, U.S. troops had withdrawn from Iraq, but media reports continued to show roadside explosions, murders, kidnappings and general lawlessness on a near daily basis.

Despite the problems in Iraq, there have been nascent efforts to begin promoting and developing tourism in the midst of the 2003–2011 conflict. In the face of social chaos and obvious dangers, the first organized tour of the country since the United States-led invasion took place in September 2003, arranged and operated by Hinterland Travel in the United Kingdom. Since then, there have been many media reports of tourists visiting Iraq on small, organized tours or individually.

In addition, Iraqi tour operators have organized trips for Iraqis to neighboring countries—something they were not permitted to do under Saddam Hussein's regime. As well, not all of Iraq was subject to the frequent violence of Baghdad, and domestic tourism was beginning to thrive in some areas. At least this was the case until 2014, when a new terrorist threat, the Islamic State (ISIL/ISIS), formed from Syria's civil war and spread into Iraq, creating dangerous conditions for Iraqis of all religions and ethnicities. ISIL's despotic control of some of Iraqi and Syrian territory, which has included high-profile kidnappings and murders of western journalists and aid workers, has put an end to any nascent efforts in Iraq to develop tourism.

Lebanon

Lebanon endured a bloody civil war between 1975 and 1990, spurred by religious conflict between the country's Muslims and Christians. Lebanon's internal strife was exacerbated by the country's small size and proximity to Syria and Israel. Prior to the civil war, tourism was a major sector of Lebanon's economy. Beirut, the capital, was often referred to as the Paris of the Middle East due to its extensive French colonial influence and the capital's wide avenues, boutique shops and meticulous landscaping. However, much of Beirut was destroyed during the civil war, and visitation ceased almost entirely. The governments of the United States, Canada and various European countries issued warnings against travel to Lebanon. Between 1990 and 2013, however, Lebanon had a more stable political, social and economic environment with only a few skirmishes, including those associated with the September 2005 withdrawal of Syrian troops from the country.

At the close of Lebanon's civil war, adventurous Europeans—particularly Belgians, Germans, Dutch, Italians, French and expatriate Lebanese—quickly returned to visit the country. The tourism industry once again began to flourish, albeit at a smaller scale than before the war. However, in July 2006, Israel invaded Lebanon, at first to find two Israeli soldiers who had been captured by Islamic Hezbollah fighters. However, as the invasion and attacks on Lebanon's infrastructure continued, the Israeli government's justification was to prevent Hezbollah from attacking Israel with rockets launched from Lebanese territory. Unfortunately, the destruction wrought by the Israeli invasion meant that much of the social, economic and tourism infrastructure that had been rebuilt following the civil war was again destroyed, but perhaps more seriously for the longer-term development of tourism in the region, the destruction served to reinforce the negative image of war and political instability that existed among many potential Western tourists.

Syria

Syria has many desirable attributes: Roman heritage, religious sites (Christian and Muslim), interesting cultures and a well-known cuisine. Suspected support of Hezbollah and other militant factions bent on destroying Israel and the United States has led to souring relations between Syria and the United States, as well as other western nations. The United States government claims Syria is a hotbed of terrorist support and strongly warns Americans against visiting Syria. This has many spillover effects on other United States allies. In an attempt to realize the downfall of the Syrian dictator, Bashar al-Assad, the way Egypt, Tunisia and Libya did, various rebel factions began efforts to oust him in 2011. Their efforts were not as successful as others in the *'Arab Spring'* in deposing their president, and at the time of writing (late 2014),

Photo: Alan A. Lew

◀ *The ancient city of Petra was the capital of the Nabataeans in the sixth century BCE when most the rock cut buildings were created, although it may have been an established trading community as early as 1200 BCE. Today it is a UNESCO World Heritage Site and the most visited attraction in Jordan. Most of the cave buildings in this view were created as tombs. Camel rides and art and souvenir shops can also be seen at the base of this hill.*

their efforts had become an outright civil war, which has killed thousands of Syrians and destroyed countless homes and public buildings, including some extremely important heritage structures. The Syrian civil war has destabilized the entire region, including neighboring Lebanon and Iraq and gave rise to a new face of terror, ISIL/ISIS.

Jordan

Like Egypt, Jordan signed a peace treaty with Israel in 1994, which opened the border to cross-border travel to and from its Jewish state neighbor. Israelis regularly visit Jordan today, and many more progressive-minded Jordanians visit Israel on day trips or longer as business tourists and pilgrims or on pleasure holidays. Jordan is a well-known heritage destination, particularly because of its association with Petra on the old spice trading routes, as well as the Roman ruins scattered throughout the country. As already noted, Bedouin culture is also becoming a new attraction for people willing to spend time in the harsh desert environments (badia) of Jordan. In common with neighboring Israel, the Dead Sea has become an important resort, spa and health destination for domestic and international tourists on the Jordanian side of the border.

3.3 | SUB-SAHARAN AFRICA

Current Sub-Saharan Africa Issues and Resources (http://wrgeography.coni/subsaharan.html)

This section covers the region of Africa located south of the Sahara Desert. Sub-Saharan Africa has historically been among the most impoverished region of the world, and poverty and short life expectancies are continuing challenges for much of the region. It was almost completely controlled by different European powers during the colonial era, and today it remains among the more remote and exotic destinations for tourists from the developed countries of North America, Europe and Japan. Sub-Saharan Africa is believed to be the place of origin for humankind, and it is the most culturally and ethnically diverse region in the world today. At the same time, it has a significantly lower population density than many

▶ *A sign marking the Tropic of Capricorn in the Namib Desert of Namibia. Africa is the second largest landmass on the planet and is the only continent that straddles the Equator, as well as both the Tropic of Cancer in the north and the Tropic of Capricorn, here, in the south. This means that the sun is always directly overhead at some point in Africa at mid-day.*

Photo: Jarkko Saarinen

other areas of the world. The physical geography, however, is not all that complicated for an area that is three times the size of the United States.

Physical Geography of Sub-Saharan Africa

Unlike Europe, and indeed much of the Eurasian landmass, the entire continent of Africa is very compact. It has no large, penetrating bays or seas. It is a relatively flat continent, with a generally uniform elevation and only a few high mountain regions. Most of Africa is a plateau surrounded by a narrow coastal plain. Over 90 percent of the continent is more than 500 ft (150 m) in elevation. The exceptions to this plateau character of most of Africa are the Atlas Mountains in the northwest extreme of the continent, which are an extension of the Alps in Europe; Mount Kilimanjaro, which is an extinct volcano and Africa's highest peak at 19,340 ft (5895 m); the Cape Range in South Africa, which is a major source of gold and diamonds; and the Great Rift Valley, which is the largest *rift valley* region in the world, averaging 30 miles (48 km) wide and up to 3000 ft (915 m) deep. (A rift valley is formed by a continent splitting apart, with each side of the valley moving in an opposite direction.)

Africa's larger landforms are relatively easy to comprehend. Africa has several large river basins, in addition to the major mountainous regions that rise above the continent's plateau-like topography. Starting in Southern Africa, the Kalahari Basin is an arid internal drainage basin. The rivers in this basin flow from higher elevations that receive more precipitation down toward the center of the basin (home to the Kalahari Desert) where they dissipate into the soil. (They do not flow out to the ocean.) North of the Kalahari is the Congo Basin, which is the drainage for the Congo River. The Congo Basin is in the most humid portion of Africa and has dense tropical jungles. The Congo River flows over a large number of waterfalls as it makes its way down from the higher fringes of the basin toward its center and out to the Atlantic Ocean. These waterfalls prevent navigation into the interior of the Congo, and, along with similar ones in other basins, they kept much of interior Africa a mystery until only about 200 years ago.

Northeast of the Congo Basin, the Nile River flows through the Sudan Basin and out to the Mediterranean Sea. To the west of the Sudan Basin is the Chad Basin, another internal drainage system in a very arid landscape (the Sahara Desert). West Africa is the home of the Niger River, which flows through the arid Djouf Basin before it enters the Atlantic Ocean (see map below).

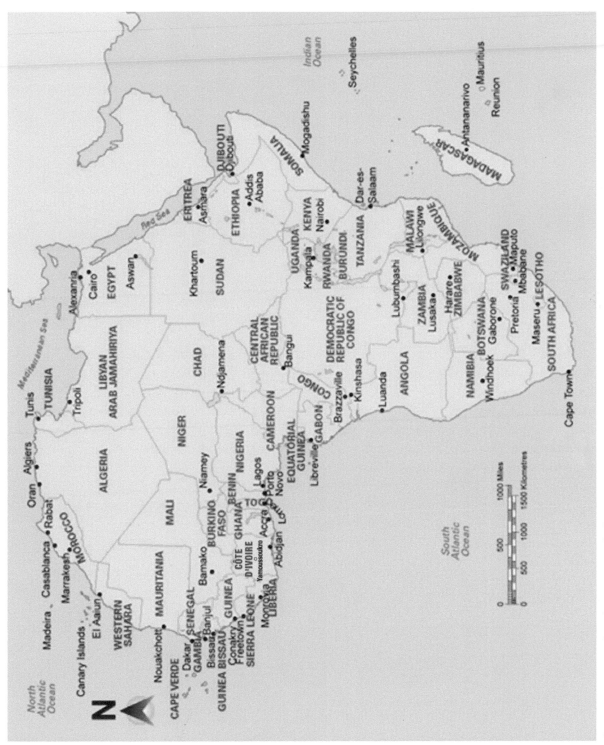

▲ *Countries, Capitals and Major Cities of Africa.*

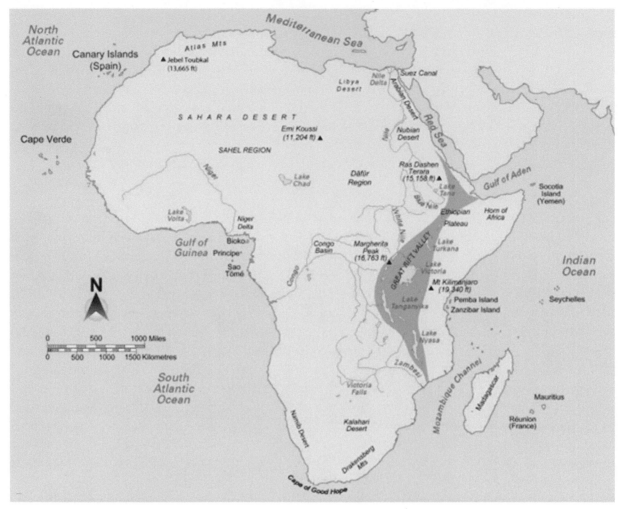

▲ *Physical Geography of Africa.*

Climate

Most people in the more developed countries of the world imagine that the landscapes of Southwest Asia and North Africa are little more than deserts. For Sub-Saharan Africa, the image is mostly that of a *jungle* (tropical rainforest). While these images are accurate for large portions of these regions, each also has a much greater diversity of climatic conditions. Africa is the only continent that sits astride the equator. Africa extends almost the same distance north and south of the equator. The northwest African coast and the eastern portions of the Mediterranean Sea have a Mediterranean climate with rainfall in the winter and drought in the summer. These areas have among the highest population densities in Africa, because their rainfall helps make their soils more productive.

Moving south, the climate of Africa quickly changes to steppe grasslands and then desert. Some desert areas here only receive recorded rainfall once in ten or more years, with recorded temperatures reaching 150°F (65°C). This area has the lowest population density in Africa, mostly because of the lack of water. Where water is available, settlements can be very dense, such as along coastlines and along the Nile River. Elsewhere, nomadic lifestyles predominate.

The Sahel

South of the Sahara Desert is the Sahel region, which is a transition zone to more humid grasslands further to the south. Here Arab and Berber cultures from the north and black African cultures from the south meet and interact with one another, sometimes in violent ways. This region also extends through Ethiopia. Rainfall in the Sahel region is very unpredictable, with some years receiving adequate precipitation for agriculture and grazing, while other years experience severe droughts. *Desertification* (the southward expansion of the Sahara) has become a serious problem due to overgrazing and not allowing vegetation to regenerate. These problems have grown due to increasing population pressures in the region, and all of the countries along the Sahel have experienced famines over the last several decades.

The ITCZ

The unpredictability of rainfall in the Sahel is due to the pattern of wet and dry climatic zones that surround the globe. Along the equator around the entire globe is a zone of tropical climate with year-round high temperatures and high rainfall. This area is known as the *intertropical convergence zone*, or ITCZ. Air from north and south of the equator meets along this zone and rises up (low pressure), creating clouds and rain throughout the year. The fact that most of the equator is ocean contributes to the high rainfall in the ITCZ.

The air that moves into the upper atmosphere along the equator eventually must descend back to the earth. This happens to the north and south of the equator in regions known as the horse latitudes (also known as the Calms of Cancer north of the equator and the Calms of Capricorn to the south, and more formally as the Subtropical High). In these areas, the air that descends from the upper atmosphere (high pressure) is generally dry with calm winds. The Sahara Desert in the northern part of Africa and the Kalahari Desert in the southern part of Africa are both situated in the horse latitudes, as are the Australian Outback and North America's Sonora Desert. The reason these areas are called the horse latitudes is because early sailing ships would sometimes be stuck in these regions of the oceans for many days at a time and sailors are said to have thrown their horses overboard in hopes that a lighter boat would more easily move in the calm winds.

The ITCZ and the Subtropical High follow the sun, moving northward in the northern hemisphere summer, and southward in the northern hemisphere winter. Thus, in the summer months of the northern hemisphere, the ITCZ brings precipitation to the Sahel region and the edges of the Sahara. The problem is just how far north the ITCZ will travel in any year. In good years the ITCZ travels well north and stays there for a long time, bringing plenty of moisture to the Sahel. If this extends over several years, people will migrate to the region. However, it is just as possible for the ITCZ to stay south of the Sahel, causing extended years of drought, famine and out migration.

Humidity increases further southward. In the southern part of West Africa (on the Atlantic Coast) and in the Congo Basin the ITCZ is present most of the year, creating tropical rainforests in the Congo Basin. In between the tropical climate and the transition region of the Sahel are grassland and savanna (sparsely forested grasslands) climates and vegetation. These subtropical grasslands exist both north and south of the Congo Basin and are the homes of Africa's large game animals, including giraffes, elephants, lions and zebras. The very southern part of Africa has a Mediterranean climate, which is the same climate as that found in the very northern tips of the continent.

Human Geography of Sub-Saharan Africa

Poverty

Sub-Saharan Africa has a population of about 800 million (2007), which was approximately 12 percent of the world's population. This is a relatively low population given the large land area that Sub-Saharan

Photo: Jarkko Saarinen

◄   *The Okavango Delta (or Okavango Swamp) in Botswana is the world's largest inland delta. The Okavango River drains into this interior basin, which was a lake 10,000 years ago, instead of the sea. The Okavango is renowned for its diversity of big game wildlife.*

Africa encompasses. However, the current growth rate of 2.3 percent and other demographic change leads the UN to predict that the region will have a population of nearly 1.5 billion in 2050. It is also one of the poorest regions in the world, accounting for little more than 1 percent of the world's gross domestic product (GDP). Over 60 percent of the people live in slums and squatter settlements in urban areas. High infant mortality rates exist in most countries of this region, and Sub-Saharan Africa is the only major region in the world where poverty has increased since 1980. Although the region's economy has been experiencing significant growth since 2000, and it was one of the few areas relatively unaffected by the global financial crisis of 2008 to 2010, the region still contains 70 percent of the world's 50 poorest nations.

AIDS

The biggest challenge today for Sub-Saharan Africa is disease. Perhaps foremost of these is AIDS. In 2010, approximately 68 percent of the world's population suffering from AIDS lived in Sub-Saharan Africa, and some 70 percent of new infections were in the region. The same year, Sub-Saharan African accounted for an estimated 67 percent all AIDS-related deaths. Of the estimated 34.3 million people in the world who have had AIDS, 23.8 million of them live in Sub-Saharan Africa, and 19 million of them have already died from the disease. Approximately 91 percent of all HIV-infected children live in Sub-Saharan Africa. AIDS has reduced the life expectancy in Sub-Saharan Africa from 59 years to 49 years, or less (the global average of 62 years).

In Zimbabwe alone, it has been reduced from 61 to 33 years, and in Chad it is only 32 years. Researchers project that the overall impact of AIDS by 2010 reduced the population of Sub-Saharan Africa by 71 million people. This includes people who would have been born if it were not for the deaths of women of childbearing age. This number is more than twice the 30 million Europeans (one-third of the population at the time) who died of the bubonic plague (1347–1352 CE), which was previously considered the worst outbreak of disease in world history.

Malaria and Ebola

Another disease that severely affects the region is malaria. According to World Bank research, the disease slows growth by about 1.3 percent per year through lost time caused by illness and the cost of treatment and prevention measures. The region's GDP would have been 32 percent higher in 2003 had the disease been eradicated in 1960. Under the support of organizations such as the Gates Foundation, there are substantial international initiatives to control, if not eradicate, malaria.

In 2014, a severe outbreak of Ebola occurred in West Africa. It began in Guinea in 2013, and in less than a year had spread across borders to Liberia and Sierra Leone, with a few isolated cases in Nigeria and Senegal. The disease is spread through contact with bodily fluids, so healthcare workers are especially prone to exposure, and many have become sick or died as a result. The 2014 Ebola outbreak has seen widespread fears among the traveling public, as many people cancelled trips not only to the most affected countries, but also to many other non-infected countries in Sub-Saharan Africa. Several countries instituted travel bans (e.g., Colombia and St Lucia) on arrivals from certain West African countries, and some governments have expressed concern over a possible Ebola pandemic. As of mid-October 2014, there had been more than 9,200 reported cases and 4,500 deaths in a handful of countries.

Economic Development

Despite these difficulties, several countries in Sub-Saharan Africa have thriving economies. Good governance and sound fiscal policies have created bustling and robust economies in Botswana, Ghana, Uganda, Senegal, Gabon, Mauritius and the Seychelles. Ghana, for example, has had one of the strongest stock markets in the world in recent years although a recent decline in exchange rates has created some difficulties for local businesses.

Traditional Economy

Much of Sub-Saharan Africa has been relatively isolated from the rest of the world throughout history, and even in modern times. North Africa and East Africa have been important trading routes and have been closely connected to world trade for a long time. However, West Africa and areas further to the south were cut off from the major European, Asian and North American trade routes due to deserts, seas and jungles.

Vasco de Gama, from Portugal, was the first European explorer to sail around the Cape of Good Hope (the southern tip of Africa), which he accomplished in 1497. This route became the major transportation corridor between Europe and Asia for the following 300 years. During that time, Europeans became very familiar with coastal settlements, but they knew nothing of the interior regions of Africa, which they perceived consisted only of deserts and jungles. The plateau character and river navigation challenges made explorations of the interior difficult.

The traditional economy of Sub-Saharan Africa (also sometimes referred to as Black Africa, although this term is decreasing in use because of its colonial overtones) was primarily subsistence agriculture. Diets consisted mostly of starches, with small amounts of protein. This was true on the savanna and in grassland areas, where maize (corn), sorghum and millet were the staples, as well as in the forests and jungles, where sweet potatoes, cassava and fruit were major foods. These traditional diets were supplemented by the hunting of wild game, which led to the extinction of many large animals.

Shifting Cultivation

Shifting cultivation was a common practice, especially in the savannas and jungles of central and eastern Africa. On the savanna, entire families move as they shift their farms from one field to another. In the jungles, villages usually remained in one place while farmers shifted from one field to another. In both cases, the fields that were left behind were allowed to lie fallow for several years to regain their nutrients. Also, in both areas, land holdings were communal, based on the tribal society, and each family received a sufficient amount of land to provide for a subsistence livelihood. In the colonial period, land came under private ownership. This resulted in the migration of landless farmers to the cities, along with increased incentives to produce surplus cash crops for sale and export. This directly led to overuse of the soil, shorter fallow periods and less productive farmlands.

Pastoral herding lifestyles were found on the savanna and on the steppe or grassland areas. Sheep and cattle were the dominant grazing animals and outnumbered people in most of these herding areas. Nomadic and transhumance herding practices were common in the past, but they are very rare today. Most of the sedentary herding populations today supplement their cattle and sheep-raising with crop production. Cattle, especially, are highly prized symbols of social status in these areas, and they are important in marriage ceremonies and religious sacrifices. African cattle breeds are very hardy and resistant to diseases, but they tend to be poor milk producers.

Dual Economy

Africa today has what is known as a dual economy. Traditional agricultural economies (lower circuit) exist outside of the cities, while urban centers have more modern market economies (upper circuit),

▶ *"Jesus Magnet Carpentry Shop" and "God's Abundant Gift Fashion & Business Centre" are typical of the business names in the mostly Christian town of Koforidua in the highlands of Ghana. This scene also shows the high degree of entrepreneurial spirit that is found in many parts of Sub-Saharan Africa.*

Photo: Alan A. Lew

and there is very little trade between these two groups. In the dual economy model, the upper circuit economy is a money-based system that is tied to international trade. This exists in the coastal urban areas of Sub-Saharan Africa, which have been more influenced by European trade. The lower circuit is a non-monetary barter system that is mostly local. This characterizes many of the non-urban areas of Sub-Saharan Africa.

Ethiopian Empire

Several empires have arisen in Sub-Saharan Africa over the centuries. The Ethiopian empire, which we discussed briefly in relation to the Middle East, is the earliest major civilization of Sub-Saharan Africa, dating back to pre-Christian times. Ethiopia is the dominant country of East Africa. Legends indicate that the Ethiopian Empire was founded by Menelik I, the son of King Solomon of Israel (960–922 BCE), and the Queen of Sheba. The royal line of Menelik I ended in 1974 with the death of King Haile Selassie. In 324 CE, the King of Ethiopia converted to Christianity, and ancient Coptic Christian churches are still found in the highlands of Ethiopia. Beta Jews from Ethiopia, who practiced a very old form of Judaism, have immigrated to Israel, based on an Israeli law that permits Jews from around the world to live in Israel.

ISSUES AND INSIGHTS Pro-Poor Tourism

Poverty reduction has become an important item on the tourism agenda. In 2007, the World Tourism Organization (UNWTO) identified poverty reduction, along with climate change, as a global challenge to the tourism industry. One particularly important aspect of the interest in equity and human wellbeing as a tourism development goal has been the growth of academic and development agency interest in the relationship between tourism development and poverty-reduction strategies, which is often referred to as pro-poor tourism (PPT).

As with the relationship between tourism and sustainable development, the PPT field has developed in great part as a result of interchange between researchers and public and non-government organizations with interests in the less-developed countries. However, as with sustainable development, the PPT field has become increasingly open to contestation and critical debate.

The focus on providing tourism employment to the unemployed or under-employed is not far removed from the goals of many regional economic development programs. PPT advocates though tend to suggest that there are qualitative differences with respect to its approach to the poor. However, critics suggest that it fails to address the structural reasons for the north-south divide, as well as the internal divides within the developing countries, particularly with respect to international trade. For example, in the area of agriculture the European Union (EU) and the United States of America have been strongly against the free trade that developing countries have been seeking. Nevertheless, despite such concerns, the UNWTO has resolved that the liberalization of the conditions governing trade in services is compatible with sustainable tourism development and the protection of social and cultural values and identities, and it advocates the concept of *tourism liberalization with a human face*.

The idea that openness is good for growth and human development has become deeply ingrained in many development institutions, such as the World Bank. However, other agencies, such as the United Nations Development Program (UNDP), have commented that, in practice, the relationship between trade and growth is determined by a complex array of domestic and external factors, and cross-country evidence provides little foundation for the use of loan conditions or world trade rules to promote rapid liberalization.

The UNDP as is the case with many other commentators, does argue that liberalization of trade in services, such as tourism, offers potential benefits to developing countries. Yet they believe that industrial countries have focused on areas that threaten to undermine human development prospects, while failing to liberalize areas that could generate gains for poor countries. However, at a local scale there are a number of PPT initiatives being carried out in Africa, with South Africa and Namibia in particular being a laboratory for the testing and evolution of new approaches towards the planning of local economic development.

Sources

Hall, C. M. (ed.) 2007. *Pro-poor tourism: Who benefits? Perspectives on tourism and poverty reduction*. Clevedon: Channel View Publications.

Rogerson, C. M. (2014). Strengthening Tourism–Poverty Linkages. In A. Lew, C.M. Hall and A. Williams (Eds.), *The Wiley Blackwell Companion to Tourism*, 600–610. Oxford: Wiley-Blackwell.

The Ethiopian empire prospered through its trade with Arabs, supplying gold, ivory and spices from inner Sub-Saharan Africa. During the nineteenth century, invading Italian forces defeated the Ethiopians and colonized the country in 1936. It was freed by a British invasion in 1941 in the early years of World War II. Today, the highlands of Ethiopia are considered the Switzerland of Africa and a Christian enclave surrounded by Islamic peoples; the lowland areas have been troubled by famine, war and civil unrest.

West African Empires

West Africa is a major cultural hearth for much of black Africa. Early civilizations arose along the Niger River. The Ghana Empire existed from the eighth to the twelfth century, at the same time that the Arab Empire and Chinese Empires were at their peaks. They controlled all trans-Saharan African trade during that time. The Mali Empire replaced the Ghana Empire from the thirteenth to the fifteenth century. In 1324, its emperor, Mansa Musa, made a pilgrimage to Mecca and visited Arab leaders. Upon his return, he made Mali one of the major Muslim countries in the region.

Zimbabwe/Karanga and the Congo Kingdom

In southern Africa, Zimbabwe was the center of the Kingdom of Zimbabwe/Karanga Empire in the eleventh century. Through trade with Arabs to the north, they became very advanced in mining industries, agriculture and trade with South Asia. The word Zimbabwe was the Shona name for the venerated houses of chiefs or of gravesites. The ruined city of Great Zimbabwe that was built in the twelfth century by the Karanga people, which demonstrated their high level of architectural skill, is the largest of these zimbabwes. Great Zimbabwe is also a World Heritage Site. When the white-dominated government of Rhodesia was voted out of office in 1979, the new leaders of Rhodesia renamed their country Zimbabwe after the former capital of the Kingdom of Zimbabwe.

In the fifteenth century, Bantu-speaking peoples from the tropical savannas and rainforests of West Africa spread into the Congo Basin, pushing the indigenous pygmy tribes into the Kalahari Desert to the south. The Bantus established the Congo Kingdom, which covered a wide area and prospered through trade with its neighbors. In the sixteenth century, its king adopted Christianity and began a process to modernize the Congo Kingdom, using Portugal as a model. However, in the seventeenth century, the kingdom declined, following wars with its neighbors and the demoralizing impacts of the slave trade. In 1665, it was invaded and colonized by Portugal.

Slavery

All of Sub-Saharan Africa's empires and kingdoms were in decline when the Europeans arrived in the sixteenth century. The major product that the Europeans wanted from Africa was slaves. *Slavery* has existed in Sub-Saharan Africa since Egyptian times. Arab traders sold black slaves to the Greeks and the Romans. (Early Greece had more slaves than free men.) Slavery was also common in Sub-Saharan Africa itself, and it existed there long after it was abolished elsewhere in the world. Some say that slavery still exists in pockets of Sub-Saharan Africa to this day.

In the late sixteenth century, the Portuguese brought African slaves to work on the first plantations in the New World, which they had established in Brazil. Brazil was an area with a very low native population density, so the Portuguese relied on imported labor to work their agricultural lands. In 1602, the Dutch brought the first African slaves to North America, putting them to work in their Caribbean colonies. The first black Africans came to the British colonies as indentured servants in 1619, a year before the Mayflower brought the Pilgrims to New England. By the late 1600s, most black Africans were brought to the colonies as slaves. Large numbers of West African blacks came as slaves to the Americas in the early 1700s to work on the plantations of the British colonial South, the Caribbean and Brazil.

◀ *Ghana's Cape Coast Castle and Dungeon was a major center for the trans-Atlantic slave trade in the late 1600s and early 1700s. It later became the administrative center for the British Gold Coast colonies, and today it is a museum. The ancestors of most of the Africans in the Americas came from West Africa, including Ghana, though they are less able to identify specific countries and regions.*

Photo: Lauren A. Hall-Lew

The British were the main source of African slaves to their colonies. They brought fabrics, costume jewelry and other manufactured goods to West Africa, which they traded for slaves. They also brought corn, peanuts, sweet potatoes, coconuts, bananas and citrus fruit (from other tropical lands) to Africa, all of which greatly improved the traditional diet of Sub-Saharan Africa. From West Africa, they brought slaves to the West Indies (the Caribbean), and to the colonial southern USA. Only about 85 percent survived the trans-Atlantic journey. (This number was much higher than in East Africa though, where only about 10 percent survived the Arab slave trade.) The peak of the African slave trade to North and South America was in 1791 when the British, Portuguese, Dutch, French and Danish brought about 75,000 slaves. In total, an estimated 11 million slaves were brought to the Americas.

In 1807, Great Britain banned slavery within the British Isles. In 1834, the British banned slavery throughout the British Empire. The United States banned the slave trade in 1808, though it was not enforced on a regular basis. There were several reasons why the trans-Atlantic slave trade lasted more than 200 years. First, there was the need for inexpensive labor in the plantations of the New World, especially in areas that had low native populations or where natives were uninterested in working for Europeans.

Second, there was the competitive nature of African tribes. Tribal groups that relied on hunting required vast areas of land to meet their resource needs, and they often fought violently with intruders. Herding tribes needed access to grasslands and savannas, and they resented being enclosed by other groups that prevented them from practicing transhumance. Sedentary tribes needed access to the best agricultural lands, and, along with shifting agriculturalists, they struggled to keep cattle and game animals out of their crops. Competition for scarce resources among these lifestyles has long been an issue in Sub-Saharan Africa. Rather than killing one's enemies, the tradition arose to enslave them, and sometimes to sell them as slaves to others.

Tribalism

The word tribe is often used to describe the social organization in Sub-Saharan Africa. The definition of *tribe* is any group of people who share the same customs and languages, and who believe they descended from a common ancestor. There are an estimated 600 to 1000 tribes in Sub-Saharan Africa, many of which speak their own separate language. (There are an estimated 900 to 1500 different languages spoken in Sub-Saharan Africa, and most Africans speak more than one language.) Tribes range in size from a few thousand to several million members. Tribes form out of communal association. They support individuals in times of need, and they expect conformity to customs, traditions and other basic laws governing behavior. Most tribes also include some form of shared work among their members. In Sub-Saharan Africa, there are often tribal marks that people carry, such as a cut on the face of some kind, which were used in the past to identify one's tribal affiliation.

In Anglo North America, the rights of the individual are often considered more important than the groups or social organizations that they belong to. Examples include the right to free speech and a general belief that an individual has the right to do as one wishes. For much of the rest of the world, however, the group is considered more important than the individual. To many Africans, the family, the clan and the tribe are more important than either the individual or the country in which they live. In part, this is because Europeans created the African countries that exist today—they did not exist prior to colonization, and they have little relation to traditional tribal boundaries. Most are not nation-states, and their citizens have not had a long history of identification with today's African countries.

Political Boundaries

The political boundaries of most of the countries of Africa today were created at the 1884 Berlin Conference where the European colonial powers divided up the African continent among themselves. This was before most of the interior of Africa was colonized or even explored. Because the older empires and kingdoms of Africa were almost non-existent by this time, the European powers had little difficulty in dividing them up indiscriminately. (They had done the same in the Middle East, but they were unable to do this in the rest of Asia.) The result was that the former empires of Africa, along with all of the major tribes, were split or joined in a manner that made little sense with respect to ethnic territories. This was the complete opposite of the nation-state form of countries that had evolved in Europe. Traditional political systems that were based on tribes were shattered, and to this day separatist movements and intertribal warfare exist in many Sub-Saharan African countries as a result of European control.

Tribal Conflict

One of the worst outbreaks of tribal tensions occurred in Burundi and Rwanda, two former Belgian colonies. In 1994, militias from the majority population Hutu tribe in Rwanda slaughtered 937,000 members of the ruling Tutsi tribe and moderate Hutus. This event is known as the *Rwanda Genocide*. The genocide was sparked by events in neighboring Burundi, where Hutus and Tutsis had been in a state of civil war since 1993. These tensions have a long history. In 1972, for example, the Rwanda Tutsis killed some 200,000 Hutus, including all educated Hutus, in an effort to suppress any attempt to overthrow the Tutsi rulers.

Other countries that have experienced large-scale tribal conflicts in recent decades include Nigeria (between Muslims in the north and Christians in the south), Somalia, and more recently Sudan (both in its southern region and western region), Zimbabwe, Eritrea (which fought a successful civil war against Ethiopia and gained its independence), Liberia (which is a country created by the United States as a place for former black slaves to return to Africa) and elsewhere.

While a logical approach to this problem may be to redraw the map of Africa, this is no longer a viable option. Although tribal allegiances remain strong across political boundaries, new political and economic interest groups have been formed based on the boundaries of the countries that exist today. Contemporary politics in Africa require the appeasement of various ethnic groups within a country's boundaries by giving them some form of limited autonomy and providing funding for economic development.

Most of the major conflicts in Sub-Saharan Africa are largely ignored in the developed Western countries. This is so for several reasons: (1) There is often limited television coverage of events there by the news media in the West, (2) there are seldom any political or economic consequences of these conflicts for Western countries, (3) the cultural and social ties of Sub-Saharan Africa to the West are limited, and (4) the West expects things like this to happen in Sub-Saharan Africa. This last explanation is probably tinged with a degree of racism.

Coups d'état and Political Discord

Largely owing to the role of colonialism, autocratic power brokers and poverty, Sub-Saharan Africa has experienced considerable political unrest since the mid-1900s. Since independence, nearly all countries of Africa have been directly affected by warfare or other forms of political instability. A common phenomenon that has plagued several African countries is *coups d'état*, or military overthrows of the government. In many cases, coups occur repeatedly in various countries, such as Ghana throughout the 1960s to the 1990s. These events have major implications for tourism. The immediate effects include closing of international borders and airports, which impedes the flow of people into and out of the country. Coups also result in questionable safety and security for tourists, a lack of investment in tourism infrastructure and frequently strained relations with other countries. Africa has also experienced wars and acts of genocide, which clearly create negative images and thwart tourism development plans.

The ethnic tensions in Rwanda and Burundi during the 1990s, for example, essentially ceased all tourism operations in those countries for international media reports scared off would-be visitors by stories of mass murders and widespread kidnappings and other crimes. Likewise, the present political chaos and lack of functional government in Somalia, and the current conflicts in South Sudan, Mauritania, Mali and Nigeria present an image of Africa that is not conducive to the development and growth of tourism.

Countries and Tourism of Sub-Saharan Africa

Regions of Sub-Saharan Africa

There are about 45 different countries and territories in Sub-Saharan Africa, which is about 20 percent of all the countries in the world. The major regions of Sub-Saharan Africa include West Africa, East Africa, Central Africa and Southern Africa. The regional affiliation of the various countries can, in great part, be identified from their memberships in various African economic unions and regional development organizations, although there is some degree of overlap.

West Africa

West Africa is the region lying south of the Sahara Desert and bordering the Atlantic Ocean. It is geographically divided into two distinct subregions, each of which has distinct environmental and cultural characteristics. The first region consists of countries in the Sahel region that border the semi-arid environment along the southern Sahara desert. These countries include:

Burkina Faso	Cape Verde	The Gambia
Mali	Mauritania	Niger
Senegal		

The second region consists of countries that line the Atlantic Ocean on the humid and tropical Guinean Coast. These countries include:

Benin	Cameroon	Cote d'Ivoire (Ivory Coast)
Equatorial Guinea	Ghana	Guinea
Guinea-Bissau	Liberia	Nigeria
Sierra Leone	Togo	

The Niger River flows through much of the region, although numerous smaller rivers flow from the interior highlands to the ocean. The Guinean Coast, in the southern portions of this region, is a humid tropical rainforest that receives year-round precipitation. Mangrove swamps are common along the coast near the mouths of rivers. Rainfall decreases considerably as one moves northward toward the Sahara desert. Some countries span both the humid and arid regions of West Africa. Nigeria, for example, is both the wealthiest country in West Africa (due to its large coastal oil reserves) and the most populated country on the African continent. Political tensions between Nigeria's more Muslim north and more Christian south, however, have been a contributing factor in the country's political instability and continuing individual poverty. Economically, there is a different split of countries that has its origins primarily in previous colonial relationships, with the Economic Community of West African States (ECOWAS) having links primarily to the former British Empire and the West African Economic and Monetary Union (the UEMOA has links to France). This division is also partly a linguistic divide as well. West Africa is

TABLE 3.2   Country Membership of Regional Economic Unions

Union—Member Countries
ECOWAS (Economic Community of West African States)—Benin, Burkina Faso, Cape Verde, The Gambia, Ghana, Guinea, Guinea-Bissau, Liberia, Mali, Niger, Nigeria, Senegal, Sierra Leone, Togo (Cote d'Ivoire was suspended after the 2010 elections but was later reinstated)
UEMOA (West African Economic and Monetary Union)—Benin, Burkina Faso, Cote d'Ivoire, Guinea-Bissau, Mali, Niger, Senegal, Togo
ECCAS (Economic Community of Central African States)—Angola, Burundi, Cameroon, Central African Republic, Chad, Democratic Republic of the Congo, Equatorial Guinea, Gabon, Republic of the Congo, Rwanda, Sao Tome and Principe
CEMAC (Economic and Monetary Community of Central Africa)—Cameroon, Central African Republic, Chad, Republic of the Congo, Equatorial Guinea, Gabon
IGAD (International Authority on Development)—Djibouti, Ethiopia, Kenya, Somalia, Sudan, Uganda (Eritrea suspended its membership in 2007)
EAC (East African Community)—Burundi, Kenya, Rwanda, Tanzania, Uganda
COMESA (Common Market for Eastern and Southern Africa)—Burundi, Comoros, Democratic Republic of the Congo, Djibouti, Egypt, Eritrea, Ethiopia, Kenya, Libya, Madagascar, Malawi, Mauritius, Rwanda, Seychelles, Sudan, Swaziland, Uganda, Zambia, Zimbabwe (Former members: Angola, Lesotho, Mozambique, Tanzania, Namibia)
SACU (Southern African Customs Union)—South Africa, Botswana, Lesotho, Namibia, Swaziland
SADC (Southern African Development Community)—Angola, Botswana, Democratic Republic of the Congo, Lesotho, Madagascar, Malawi, Mauritius, Mozambique, Namibia, South Africa, Swaziland, Tanzania, Zambia, Zimbabwe

sometimes referred to as Western Africa to avoid confusion with ECOWAS, which is sometimes short-ened to West Africa. We use the term West Africa in here because its more general definition is widely used outside of the region.

West Africa is among the less visited international destination regions of the world. Nigeria has long received the most international visitors, due to its large population and economic importance, although the ongoing religious conflict there has reduced the number of arrivals. Senegal is the second most visited country in West Africa. This designation, however, may reflect the fact that The Gambia is surrounded on three sides by Senegal, and cross-border travel between the two countries constitutes international travel. The Gambia has also seen rapid tourism growth in recent years.

Ghana

Ghana is the third most visited country in the region, especially for English-speaking tourists, followed by Cote d'Ivoire (Ivory Coast). Nigeria and Ghana are English-speaking countries, while French is the official language of Senegal and Cote d'Ivoire. Despite the region's tropical climate and diversity of cultural groups (from interior nomads to coastal agriculturalists and urban settlers), it is the colonial history that forms the basis of Western Africa's international tourism today.

West African Diaspora

Western Africa was the principal source of slaves brought to the Americas between the sixteenth and early nineteenth centuries. The slave trade resulted in the widespread (forced) migration of people from their African homelands to many other parts of the world. Today this African diaspora has significant implications for tourism, primarily because of a large and growing worldwide desire among people of African descent to visit the lands of their ancestors. This form of tourism is sometimes referred to as existential tourism, roots tourism or ancestral tourism, defined as travel to a place that holds special meaning for the tourist's personal identity. Existential tourism by diaspora populations is widespread today as international transportation has become easier and both voluntary and forced migrations have become commonplace.

Because of its historical role, as dark as that may be, international tourists of African descent have become an extremely important tourist market for Western Africa. This is especially so for tourists who come from regions with higher standards of living, better levels of education and better paying jobs, such as in North America, Europe and the Caribbean. Many travel agencies and tour operators in the United States, for example, have started to specialize in ancestral travel to Africa. This is happening particularly in cities with large African-American populations, such as New York, Chicago, Los Angeles, Philadelphia, Washington DC, San Francisco, Atlanta and Miami. These agencies arrange international trips for African-Americans, typically selling them as "going home" or mother-land tours.

Ghana has a high concentration of slave-related visitor attractions. Because of its British colonial history, it is the primary destination of choice in Western Africa for African-American and African-British tourists. This is because it has a higher likelihood of being the origin of many slaves from Western Africa who were brought to the English-speaking countries of the Americas. In addition, the widespread use of English as the country's lingua franca makes Ghana an easy place to visit for American and British tourists. With this realization, Ghana has focused considerable attention on roots tourism, and its most recent fifteen-year National Tourism Development Plan places slave heritage at the center of the country's tourism development efforts.

▶ *Most of the people in Accra, the capital of Ghana, cannot afford a car, so they carry goods on their heads. This is common throughout Sub-Saharan Africa, but especially in West Africa. The individuals on the roadway median are carrying goods to sell to automobile drivers and passengers they approach when cars are stopped.*

Photo: Alan A. Lew

Also related to the historical slave trade of Western Africa, UNESCO has put considerable effort into developing and designating a Slave Route through the region. The purpose of the Slave Route is to conserve, commemorate and educate the world about the atrocities associated with the slave trade. However, the UNESCO designation, which is similar to the designation of a World Heritage Site, also adds to the tourist appeal of the region and is a good example of multi-national cooperation in developing a large-scale tourist route.

East Africa

East Africa includes the area known as the Horn of Africa, which contains the countries of Djibouti, Ethiopia, Eritrea and Somalia. East Africa also includes the coastal countries of Kenya and Tanzania, and the East African Rift countries of Uganda, Rwanda and Burundi. The Horn of Africa, along with Rwanda and Burundi, has been the site of some of the most troubling political instability on the African continent in recent years. The lack of a central governing authority in Somalia occasionally spills over into the tourist regions of neighboring countries, while Uganda has the longest running civil war on the African continent. Lawlessness, war, terrorism, kidnappings and piracy during the past two decades have prevented nearly all forms of tourism from developing in countries such as Eritrea and Somalia. Increased Somali piracy since about 2005 against cargo ships, as well as against private vessels and cruise ships, has received considerable media attention. The piracy, which extends as far south as Kenya and Tanzania and eastward deep into the Arabian Sea and Indian Ocean, has ended the once-lucrative cruise sector along the east coast of Africa and dampened much international trade in oil and fishing. Despite these challenges, East Africa is the second most visited region in Sub-Saharan Africa, after Southern Africa.

East African Rift System

The East African Rift System winds its way from north to south through East Africa, forming its western boundary in the south. It is here that the famous lakes of Victoria, Albert, Tanganyika and Nyasa (Malawi) are found, along with some of the most fertile lands on the African continent. The area has a

ISSUES AND INSIGHTS | Managing the Savanna Landscape

The special appeal of the African savanna landscape for western tourists is well recognized. The African savanna is widely viewed as the archetypal wild environment, a place of unspoiled nature. Countless wildlife documentaries, tourist brochures, films and web advertising reinforce the appeal of this *safari* landscape-scattered thorn bushes, wide vistas, yellow grass and herds of animals. Yet such images tend to obscure the fact that these are environments over which bitter struggles have been fought. Whether or not visitors and tour operators recognize their presence, people that inhabit these environments are often clinging to a precarious livelihood in marginal economic and environmental conditions.

In east and southern Africa, for example, the colonial and post-colonial state has had a major role in reshaping the landscape by creating huge national parks in conflict with the interests of both settled local communities and nomadic pastoralists who find themselves in competition with wildlife for the use of savanna plains. As a result, local people in extreme environments often have contested relationships with agencies of the national government. They have often been moved around at its behest and are unlikely to trust power emanating from a distant capital city. Commercial tourism agencies, also often closely linked to the very states that created this crisis of marginalization, may fail to grasp the highly political nature of questions of environmental sustainability in dry environments.

The devastating social effects of forced removals and the reduction of grazing lands through the creation of national parks and other protected areas is now widely recognized. It is increasingly acknowledged that in the process of expanding conservation areas and tourist operations, local property rights need to be protected as far as possible, particularly given that the appropriation of all resource rights by governments and the central management of them has not worked to the benefit of either conservation or local economic development. Since the 1990s, various experiments in community-based tourism and attempts to give communities greater local rights over wildlife and land have been pioneered in dry environments. In fact, such an approach is now widely viewed as the only hope for sustainable wildlife tourism.

Source

Preston-Whyte, R., S. Brooks, and W. Ellery. 2006. Deserts and savanna regions. In *Tourism and Global Environmental Change*, eds. S. Gössling and C. M. Hall. London: Routledge.

very complex ethnic makeup, with many different languages and language groups. There are a large number of Europeans who were attracted to the agricultural potential during the colonial period, and there are also remnants of the Arab empire that once ruled a large part of East Africa.

Kenya

Kenya has traditionally been the most visited country in East Africa, and one of the three most visited countries in all of Sub-Saharan Africa. However, recent political issues and security problems have plagued Kenya, landing it on several countries' *travel warning lists*. Tanzania and Uganda each receive between 700,000 and 800,000 international visitors a year. Safari tourism, to see big game animals in the wild, forms the basis of tourism in East Africa. In Kenya, there are 60 designated wildlife protection areas, which are divided into parks administered and financed by the country's central government, and reserves that are set aside by local authorities, though they are partially financed and administered by the central government. In total, between 6 percent and 12 percent of Kenya's land area is protected in these ways. (The land area varies as the boundaries of the protected areas fluctuate from year to year according to management needs.)

The Big Five

A goal of many safari tourists is to see the Big Five animals. Big game hunters originally coined the term to designate the five most difficult animals to hunt, but now it is used by big game tourists with their cameras. The Big Five include the lion, the African elephant, the African buffalo (or Cape buffalo in South Africa), the leopard and the black rhinoceros. While these animals are also among the most dangerous in Africa, the list does not include the hippopotamus, which is possibly the most dangerous animal when it

▶ *Mount Kilimanjaro in Tanzania is seen here from the savanna grasslands of Amboseli National Park in Kenya. This extinct volcanic mountain is the highest peak in Africa at 19,340ft (5,895 m), and is considered the tallest freestanding mountain in the world, rising 15,100ft (4,600 m) from its base. It is also a popular trekking destination. Unfortunately, Mt. Kilimanjaro's famous snow-capped peak will probably lose its snow due to climate warming by the 2020s.*

© Graeme Shannon/Shutterstock.com

is in water. Wildebeests, gazelles and zebras are among the other big game animals that are distinctive to the safari lands of East Africa and Southern Africa.

Tanzania

Tanzania is home to Mount Kilimanjaro (although part of it lies in Kenya), the Ngorongoro and the Serengeti. It had, prior to the 1970s, been in direct competition with Kenya for leadership in safari tourism in East Africa. However, the country fell behind during two decades of ineffective government policies. Mount Kilimanjaro is the highest peak in Africa (19,340 ft, 5,895 m). It is an inactive *stratovolcano* that is geologically related to the Great Rift Valley, and although it is famous for its snow-capped peak, global warming is threatening to make that a scene from the past. The Ngorongoro Conservation Area is in the Great Rift Valley, and the Serengeti National Park is located on the plains adjacent to this in the northern part of Tanzania. The Serengeti is also home to the Maasai people, who are heavily involved in providing tourism services and demonstrating their living culture to visitors. It is also famous for being home to an estimated 1.5 million large game herbivores and predators.

Lake Victoria

The Great Rift Valley in East Africa has created a series of large lakes known collectively as the Great Lakes of Africa. The most famous of these are Lake Victoria and Lake Tanganyika. Lake Victoria (or Victoria Nyanza) is the world's second largest freshwater lake and the largest lake located in a tropical region of the world. Its oval shape lies prominently within a large plateau on the east side of the Great Rift Valley.

It is bordered by the three major tourism countries of East Africa: Tanzania, Uganda and Kenya. Uganda's Ssese Islands, a large collection of islands in the northwest of the lake, have become a popular tourist destination in recent years. Lake Tanganyika is the world's second deepest lake and the world's second largest in total water volume. It fills the Great Rift Valley towards its southern end, forming a long (418 miles, 673 km), relatively narrow (about 60 km, 40 miles) and very deep (4820 ft, 1,470 m) body of water. While Lake Victoria is the source of the White Nile River, which flows north to the Mediterranean Sea, Lake Tanganyika's waters flow into the Congo Basin and out to the Atlantic Ocean.

Central Africa

Central Africa is a large region that straddles the equator and is largely drained by the Congo River system flowing out of the Congo Basin, in the north. Central Africa includes the southern Saharan countries of Chad and Sudan, as well as the recently independent South Sudan. Instances of *mass genocide*, forced migration and civil war between the northern and southern Sudanese, who are racially, culturally and religiously different, have caused considerable human suffering, death and huge population shifts from Sudan to neighboring countries during the past three decades.

Based on these tensions, and realizing their socio-cultural differences, a referendum was held in southern Sudan in January 2011. In the referendum, 99 percent of the population voted in favor of independence from Sudan, and a new independent nation was established on July 9, 2011, with the name the Republic of South Sudan. The independent South Sudan aims to create jobs and improve the national economy through various economic efforts, including tourism. Unfortunately, shortly after independence, various political factions in South Sudan began fighting one another, resulting in an ongoing civil war.

The humid, tropical climate zone begins in Cameroon and the Central African Republic, which form the northern part of the *Congo Basin*. Further to the south are the countries of Gabon, the Republic of the Congo (Brazzaville) and the Democratic Republic of the Congo (Kinshasa, formerly known as Zaire). The island country of Sao Tome and Principe is also included in this region. Most of these countries were once part of French Equatorial Africa and have had long administrative ties to the former Belgian Congo.

Central Africa is characterized by a hot and wet climate. The ITCZ hovers over this region twelve months out of the year, and there is no significant dry season. Although rainforests predominate in the lowlands of the Congo Basin, flat savannas and open forests on the higher plateau lands (up to 3,000 ft, 900 m) surround these to the north and south and around the higher elevations that encircle the Congo Basin.

Unlike the Great Rift Valley region to the east, the population density is very low throughout central Africa. Large areas of tropical rainforests are completely devoid of people, and the first cities in this region were not established until the Europeans arrived in the fifteenth century. Most of the people today still live in rural communities that are very traditional. The countries of Central Africa are among the least-developed economies in Africa, and the region overall is the least visited in Sub-Saharan Africa. Gabon is

---

ISSUES AND INSIGHTS   Holidaymakers and the Souvenir Wildlife Trade

In August 2007, the WWF-UK (World Wide Fund – UK) released its annual league table of the plants and animals most commonly brought illegally into the United Kingdom. Many of the seized items derived from species protected under the Convention on International Trade in Endangered Species of Wild Fauna and Flora (CITES), the worldwide agreement that bans trade in 827 species and strictly controls the movement of more than 32,000 others. The items confiscated included snakeskin goods, elephant ivory carvings and, at the top of the list with the highest number of seizures, 605 kg (more than half a ton) of traditional Chinese medicines that contained endangered species such as tiger, rhinoceros, seahorse and deer musk.

There were also 221 seizures of elephant ivory and skin products, 44 seizures of snake and lizard products such as handbags and shoes, and 39 seizures of similar crocodile and alligator products. The United Kingdom's authorities also seized almost 1,000 live reptiles, such as snakes, chameleons, tortoises and terrapins. In total, the United Kingdom Customs confiscated more than 163,000 illegal wildlife trade items, many made from highly endangered species, and more than 158,000 plants, including orchids and cycads. More than 1,270kg of coral was also stopped in a total of 23 seizures. According to Heather Sohl, wildlife trade officer at WWF many tourists could be unwittingly helping to push some of the world's most endangered species to the brink of extinction for the sake of exotic souvenirs. The majority of seizures are not brought in by criminals but by holidaymakers as souvenirs. The message of the WWF is therefore *if in doubt don't buy it.*

Source

WWF, Endangered Species (also see Wildlife trade link on same page): http://www.wwf.org.uk/researcher/issues/rarespecies/

the main exception to this. Due to its coastal oil reserves (similar to Nigeria), Gabon's capital city, Libreville, has very modern buildings, roads and casinos. It is also one of the most expensive cities in the world, and one of the safer capital cities of Sub-Saharan Africa.

The Democratic Republic of the Congo

By contrast, Kinshasa, the capital of the Democratic Republic of the Congo, is a sprawling and dusty city of colorful, yet impoverished, people. The Congo (Kinshasa), as the country is also known, has long been acknowledged to have some of the richest natural resources on the African continent, not the least of which are diamonds and coltan (used in all mobile phones). Corrupt and ruthless governments and civil wars have, however, devastated the country and caused widespread malnutrition. They have also threatened the great biodiversity of the country's tropical rainforests, which include many rare animal species such as common and pygmy chimpanzees, mountain gorillas, okapis and white rhinos. The country has five national parks that are listed as World Heritage Sites, all of which have been listed as In Danger in recent years because the poor economic conditions of the over 63 million people in Congo (Kinshasa) have pushed many of them to turn to bush meat (wild animal meat) as a food source. Because of these problems, the Congo Basin has come to symbolize the dark side of the African continent.

Southern Africa

Southern Africa includes the countries of Angola, Botswana, Lesotho, Malawi, Mozambique, Namibia, South Africa, Swaziland, Zambia and Zimbabwe. Southern Africa transitions from the more humid Congo Basin in the north to the dry Kalahari Desert in its central region. Milder Mediterranean type climates are found further toward the Cape of Good Hope (at the southern tip of the African continent). Southern Africa attracted many European colonialists, and as a result has experienced a long history of political and economic struggles between the white elite and emerging black nationalism. South Africa is the dominant industrial power of Southern Africa and the wealthiest country on the African continent. Mozambique, on the other hand, is by some measures considered the poorest country in the world.

International Tourists

Southern Africa as a whole receives the most international tourists of any of the regions in Sub-Saharan Africa. With 9.5 million tourists in 2014, the country of South Africa received more than Egypt (9.2 million). However, Egypt's arrivals have been affected by substantial political instability and were 11.2 million in 2012) in visitor arrivals for the entire African continent. Botswana (2.2 million in 2010), Mozambique (2.1 million arrivals in 2012) and Zimbabwe (1.8 million in 2013) are the next most important destinations in Sub-Saharan Africa. Namibia (a former German colony) and Zambia each receive close to a million visitors a year. Zambia and Zimbabwe share one of the most famous attractions in Southern Africa: Victoria Falls. At Victoria Falls, the expansive Zambezi River (Africa's fourth longest river) flows into a series of deep chasms. Zambia and Zimbabwe also both share large areas of pristine *nature preserves* that have considerable tourism potential. Zambia, however, suffers from poor transportation and accommodation infrastructure to support international tourism.

In contrast to Zambia, Zimbabwe has a well-developed transportation network and tourist resorts, much of which was built by the white minority that ruled the country under the former name of Rhodesia. In addition to Victoria Falls, Zimbabwe has four other World Heritage Sites, and all of them are threatened by the social unrest and instability brought on by the country's dictatorial leadership. Rhodesia (the former name of Zimbabwe) was once the breadbasket of Africa, but Zimbabwe today is a country of

◀ *View of Cape Town and Table Mountain, South Africa, from Robben Island. Nelson Mandela was imprisoned on Robben Island for 27 years during the country's apartheid era. He was South Africa's first president (1993–1999) to be elected in a fully democratic election that included all of the citizens of South Africa. Robben Island has held political and other prisoners since the late 1600s. The island today is a museum and popular tourist attraction.*

Photo: Lauren A. Hall-Lew

malnutrition, low education rates, high crime rates and one of the lowest average life expectancies in the world (an average life expectancy of only 36 years if born in 2006).

South Africa

South Africa is probably the most diverse country in Sub-Saharan Africa, both culturally and environmentally. It has snow-capped winter peaks, the Atlantic, Antarctic and Indian Oceans, subtropical forests and grasslands, and high plateaus. Its *wildlife reserves* are the most diverse in Africa and include the Greater St. Lucia Wetland Park, which is the only estuary in the world where both ocean sharks and the freshwater hippopotamus share the same habitat. Just as diverse is the human population, which consists of several African tribes and Dutch Boer, British, and South Asian immigrants.

Apartheid

Between 1948 and 1994, the country of South Africa practiced a form of political, racial and social segregation under a policy known as *apartheid*. Apartheid affected all aspects of life in South Africa. It dictated what types of transportation people could use, what educational institutions they could attend, where they could live and work and where they could eat and sleep. Black South Africans were essentially relegated to the marginal areas of the country and had few legal rights. As a result of its racial policies, the global community constantly criticized South Africa, and sanctions limited international trade and tourism.

Apartheid policies resulted in the formation of four black *independent homelands*, which were set aside as places for African natives who were not a part of South Africa. Like American Indian reservations, Ciskei, Transkei, Venda and Bophuthatswana were established in poor resource areas for the sole abode of black Africans. The South African government treated these lands as independent countries, although no other countries in the world recognized them as such. Owing to their unique sovereign status and limited natural resources, all of these homelands pursued tourism as a form of economic activity. They especially developed forms of tourism that were not permitted in the conservative South Africa. This included large-scale resort development with an emphasis on gambling, prostitution and other illicit activities that appealed to the white and colored (mixed race) peoples of South Africa. As occurs in the state of Nevada in the United States, South African tourists frequently traveled to the homelands to indulge in these activities, which were strictly forbidden in the white controlled areas. With the dissolution of apartheid policies and the election of a primarily black government in 1994, the homelands were dissolved and reintegrated back into the Republic of South Africa, though their resorts continued to thrive.

The almost half a century of apartheid policies in South Africa had a significant impact on the country's heritage tourism. The black African past had essentially been written out of the country's official history, and tourism development (museums and monuments) focused primarily on the history of the white minority. In 1994, however, this also changed, and the country is beginning to experience a rebirth of its indigenous heritage, focusing on their struggle for political and social recognition. Black African museums and heritage sites have opened up and now surpass the number of white establishments. Tour itineraries are now more balanced and objective in their treatment of both black and white heritages in the country. The hosting of the 2010 FIFA World Cup was also seen as an opportunity to promote a more modern perspective of South Africa on the world stage.

Madagascar and the Indian Ocean Islands

The islands off the southern Indian Ocean coast of Africa are considered part of the African continent, though culturally they share much in common with the larger Indian Ocean region that extends to South Asia and Southeast Asia. Madagascar is the fourth largest island in the world, and the largest, by far, of the Indian Ocean island countries. It has a very distinct language (Malagasy) and culture that are more related to Indonesia and the South Pacific than to Africa. Geologically, it was part of India until 70 million years ago when it wandered off on its own. Because it has been isolated for so long, Madagascar has a unique biology, including being the home of lemurs, an early cat-like monkey, and many other plants and animals found only on the island. This unique biology has significant potential as an ecotourism resource for the country, if it can survive threats of extinction caused by the destruction of natural habitats.

The eastern side of the island of Madagascar is the windward side, and the natural vegetation consists of dense tropical rainforests. However, this is also where most of the population lives, and much of the natural vegetation has been cut for fuel wood and agricultural purposes. Traditionally, shifting cultivation was a sustainable form of agriculture in which farmers would work a plot of land until the nutrients were depleted, then leave it to regenerate while working on a different plot of land. The growing populations of eastern Madagascar do not give the land sufficient time to regenerate, however, resulting in a massive loss of topsoil and downstream flooding.

Mauritius, Reunion & the Comoros

International tourist arrivals to Madagascar are only a fraction of those to the small island of Mauritius, which received 990,000 international tourists in 2013 and aims to have well over a million arrivals by mid-decade. Significantly in terms of strategy Mauritius is seeking to develop mid- to higher end tourism products, often using upscale international resorts aimed at the South African, German and northern European, Middle East, Indian, Australian and Singapore markets. The tourism product has a strong heritage component, including two World Heritage Sites, as well as being positioned for family holidays and couples.

The Republic of Mauritius consists of several small, formerly volcanic islands, where South Asians comprise just over 50 percent of the population. It has over a million people and, in addition to traditional agriculture, its economy is based on beach tourism, offshore banking, a growing technology sector,

ISSUES AND INSIGHTS   **The Environmental Impact of Tourism: The Case of the Seychelles**

The Seychelles is a republic of just over 100 islands in the southwest Indian Ocean. Tourism is the second most important source of foreign exchange for the islands. Because the environment is an extremely significant component of its attractiveness, especially for water-based tourism such as scuba diving and snorkeling, the country has made extensive conservation efforts. However, one of the biggest issues in the study of the environmental impact of tourism is the area of study. Do we just focus on what happens at the destination or do we try and incorporate the environmental effects of traveling to and from the destination? In a study of the 118,000 international leisure tourists who visited the country in 2000, it was found that 97 percent of the environmental impact (in terms of energy consumption and the production of greenhouse gases) was a result of tourists traveling to and from the Seychelles from their homes in Europe, South Africa and elsewhere.

The environmental impacts of a typical holiday journey to the Seychelles (10.4 days) were the equivalent of the environmental impacts of the average human being in a year. Such a finding therefore presents a significant challenge to our understanding of how to evaluate the environmental impacts of tourism, as well as how to manage such impacts. For long-distance destinations such as the Seychelles, it also presents multiple problems. Are the conservation efforts at the local level that are often funded through tourism being overwhelmed in the longer term by the environmental impacts of *getting there*? In addition, a second issue is that, what other economic alternative is there for island microstates that are often distant from their major markets?

Sources

Gössling, S., C. B. Hansson, O. Horstmeier, and S. Saggel. 2002. Ecological footprint analysis as a tool to assess tourism sustainability. *Ecological Economics*, 43: 199–211.

Gössling, S., and C. M. Hall. 2006. *Tourism and global environmental change.* London: Routledge.

higher education and small-scale manufacturing. Increasingly, Mauritius wants to position itself as the 'Singapore' of East Africa. As a result, conference and business travel is becoming increasingly important. Reunion Island is similar in these characteristics, except that it is still governed directly by France. The Comoros islands are an Islamic republic and one of the poorest countries in the world. The nearby Seychelles are a popular beach tourism destination that is both the smallest independent country in Africa, and the continent's wealthiest on a per capita basis.

Name _____

Human Mobilities

1. Identify and describe thee examples of human mobility that shaped the historical development of one or more of the subregions of Central Asia, the Middle East and Sub-Saharan Africa.

   _____

   _____

   _____

   _____

2. Identify and describe the role of human mobility in the contemporary societies of one or more of the subregions of Central Asia, the Middle East and Sub-Saharan Africa.

   _____

   _____

   _____

   _____

Tourism Destinations

3. Describe the contemporary international image of one or more of the subregions of Central Asia, the Middle East and Sub-Saharan Africa. How do these images impact the potential of the regions as tourist destinations?

   _____

   _____

   _____

   _____

4. Selecting one or more of the subregions of Central Asia, the Middle East and Sub-Saharan Africa, describe how might the countries in those subregion(s) use their cultural and physical resources to improve their global image and attractiveness?

_____

_____

_____

_____

Sustainable Environments

5. Identify and describe tine major environmental challenges of one or more of the subregions of Central Asia, the Middle East and Sub-Saharan Africa. What challenges do the countries in the region face in addressing these issues?

_____

_____

_____

_____

6. Identify and describe the major social and economic development challenges of one or more of the subregions of Central Asia, the Middle East and Sub-Saharan Africa. What challenges do the countries in the region face in addressing these issues?

_____

_____

_____

_____

Instructor Questions (answer any additional questions from your course instructor in the space below)

_____

_____

_____

_____

CHAPTER 4 OVERVIEW QUESTIONS

HUMAN MOBILITIES
- How has human mobility contributed the contemporary globalization seen throughout Asia and Oceania?

TOURISM DESTINATIONS
- How does the concept of geographic accessibility impact the tourism development options in different parts of Asia and Oceania?

SUSTAINABLE ENVIRONMENTS
- What are the challenges in balancing the need for economic development with the desire for sustainable environments in the emerging countries and regions of Asia and Oceania?

INTRODUCTION

This chapter covers about half of the world, both in total area and in total population, and more than half the world when the planet's ocean issues are included. The chapter includes the countries of East Asia, Southeast Asia and South Asia (from Japan to Pakistan), the continents of Australia and Antarctica, most of the island nations and territories of the Indian and Pacific Oceans, and the world's other oceans from a global perspective. Asia, Antarctica and Australia comprise about 35 percent of the earth's land area, and over half of the planet's total surface (including both land and water) when the Pacific and Indian Oceans are included.

The Asia and Oceania realms have about 60 percent of the world's population, with India and China having populations of over one billion, and Indonesia, Pakistan, Japan and Bangladesh all having over 100 million. Given the vast territory covered, it is not surprising that each of the major regions of the Asia and Oceania realm has distinct cultural and environmental characteristics. East Asia has relatively homogeneous cultures that share sinitic (Chinese) influences. South Asia and Southeast Asia are culturally more diverse, though the giants of India

▶ *Asia and Oceania from space, showing their major physiographic landforms and waters. (Note: This image exaggerates elevations to make them more prominent than they would appear on the actual planet.)*

© Graeme Shannon

and Indonesia still manage to keep their many ethnic groups united under one government. Southeast Asia sits as a culture and economic crossroad between East Asia, South Asia and Oceania.

Oceania is divided between Australia and New Zealand, and the widely dispersed islands of the Indian Ocean and the South Pacific. Although a separate continent and possibly the most isolated land of all, Antarctica has been included in this chapter. Because of the importance of the Indian Ocean, Pacific Ocean and Southern Ocean throughout this chapter, issues related to the world's oceans overall are also covered in the last part of this chapter. With the equator running through its middle, most of the Asia and Oceania realm is tropical and subtropical, though cold extremes are found in its far north (China and Mongolia) and in Antarctica.

4.1 | SOUTH ASIA

Current South Asia Issues and Resources (http://wrgeography.com/southasia.html)

Physical Geography of South Asia

South Asia includes the countries of India, Pakistan and Bangladesh, the island nations of Sri Lanka and the Maldives, and the Himalayan countries of Nepal and Bhutan. South Asia is sometimes referred to as the Indian Subcontinent because it is dominated by the country and peninsula of India. It is a very diverse region, but, like Africa, its physical geography is fairly straightforward and easy to comprehend. The Indian peninsula is a large tilted block of land that is geologically connected to the Australian plate,

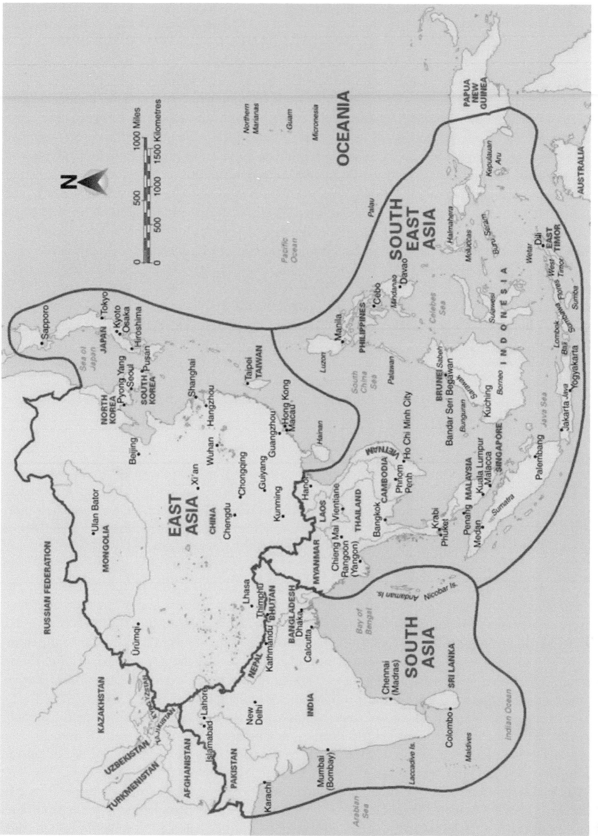

▲ *Subregions, Countries, Capitals and Major Cities of Asia.*

and which has moved northward and collided into the larger Asian continent. On the western side of this peninsula are the Western Ghats Mountains, and on the eastern side are the lower Eastern Ghats Mountains. In between these two mountain systems is the Deccan Plateau.

The Himalaya

High mountains surround the north end of the South Asian subcontinent. In the west, the dominant range is the Sulaiman Mountains, found in Pakistan and along its border with Afghanistan. This comparatively low mountain range extends northward to the Pamir Knot, which is where the Himalayan Mountain Range starts. The Sulaimans were the gateway to South Asia for invading groups from Central Asia and the Middle East. The Himalaya extend across most of the north of South Asia from west to east, and they include most of the highest mountain peaks in the world, such as Mount Everest (29,028 ft, 8,864 m) and K2 (28,250 ft, 8,611 m). North of the Himalayas is the Tibetan Plateau, with a base elevation of 15,000 ft (4,570 m) above sea level. At the eastern end of the Himalayan Range the Indian Subcontinent is bounded by the Chin Hills and the Arakan Range, which form India's border with Myanmar in Southeast Asia.

Major Rivers

South Asia has three major rivers: the Indus, the Ganges and the Brahmaputra. All three flow out of the Himalayan Mountains and the Tibetan Plateau. The Indus River flows through Pakistan to the Arabian Sea. The Ganges River begins near the origin of the Indus but flows eastward across northern India and out to the Bay of Bengal. The vast flat area these two rivers traverse is known as the Indo-Gangetic Plain. The Brahmaputra River also begins near the origin of the Indus and Ganges but flows eastward across the Tibetan Plateau side of the Himalayas and down through Bangladesh, where it joins the Ganges River flowing into the Bay of Bengal.

Monsoon

The term *monsoon* refers to a climate where winds shift direction between the summer and winter seasons. Properly used, there is a winter monsoon wind and a summer monsoon wind, although most people around the world use the term monsoon simply to refer to summer rainstorms, which are often more intense than winter rains. South Asia has the most distinct monsoonal climate in the world caused by the large mass of the Asian continent. In the summer season, the Asian continent heats up, causing air to rise and forming a low pressure air system over Siberia.

The opposite occurs in the winter when a very cold Siberia forms a high-pressure air system. The summer Siberian low-pressure system circulates in a counterclockwise direction, while the winter high pressure circulates in a clockwise direction. The change in direction of these winds in Siberia changes the monsoonal wind patterns in South Asia. The counterclockwise summer wind directions over the continent of Asia pull air across the Arabian Sea and the Indian Ocean, where the air picks up moisture. The air then moves to the Indian Peninsula, bringing the heaviest rainfall to the Western Ghats and Bangladesh, and the mountains that form India's border with Myanmar (Burma).

When the air system is working well, these winds bring rain throughout much of the Indian Subcontinent. When the summer monsoon fails to arrive, as it occasionally does, the result is drought, crop failures and possibly famine. When it is too strong or too concentrated, the result can be widespread flooding and crop and property damage.

The winter monsoon moves in the opposite direction, bringing moist air from the Bay of Bengal to the Eastern Ghats of the Indian Peninsula. However, it tends to be drier than the summer monsoon, and, while it brings rain, it has less of an impact on the entire South Asian region. (The winter monsoon tends to have a greater impact on East and Southeast Asia.) Because of the summer and winter monsoons, the natural vegetation of large parts of South Asia is either *tropical rainforest* or semitropical rainforest. This vegetation is particularly pronounced on the island of Sri Lanka, on the western coast of India, in the Western Ghats, in Bangladesh and on the southern slopes of the Himalayas.

The year-round rainfall in most of these areas, along with high temperatures, also makes them major rice-growing regions. Deciduous trees and shrub lands are the most common vegetation in the rest of India. These are found on the central Deccan Plateau, which lies in the *rain shadow* of the Western Ghats, and on the Indo-Gangetic Plain. Drier grasslands and deserts are also found in South Asia, especially in the lower reaches of the Indus River Valley, where the Thar Desert is located, and in the Sulaiman Range of Pakistan. The Tibetan Plateau is also mostly grassland.

Human Geography of South Asia
Languages
Boundaries within India are mostly based on ethno-linguistic lines. South Asia contains one of the greatest diversities of ethnicity in the world. The region is dominated by India, which has a population of just over a billion people (about one-sixth of the world's population) and is incredibly diverse. The national language of India is Hindi, which is only spoken by about 35 percent of the population. In addition, English is widely used as a lingua franca, which allows all educated people in the country to communicate in a common language. In addition to Hindi, there are 14 other official languages in India, each of which is associated with a different ethnic group and several of which have their own writing systems. Altogether there are at least 1,650 languages and dialects spoken in India. Political borders in India are often based on ethno-linguistic lines.

Dravidians
The major racial groups include the *Indo-Aryan* peoples who live predominantly in the north, and the Dravidian peoples, who are mostly in the east and south. *Australoid-Negrito* people reside in smaller numbers in the central hills of the Deccan Plateau, and Mongoloid peoples (related to Tibetans and Chinese) predominate in the northern mountains and in Assam in the northeastern portion of the country. The other countries of South Asia are less diverse than India, but they still face the challenge of ethnic diversity within their borders. Pakistan, for example, has four major ethnic groups, each of which is associated with a different province. Its capital, Islamabad, was built as a new city in a central and neutral location.

The island country of Sri Lanka is 72 percent Sinhalese (which is related to the Indo-Aryan race), 20 percent Tamil (a Dravidian racial group mostly found in southern India) and 6 percent Moors (who came to the island as traders from North Africa). Between 1983 and 2009, a civil war was fought by the Tamils of Sri Lanka against the dominant Sinhalese rulers of the island. Nepal is mostly Hindu in its southern portions bordering India, and mostly Buddhist in its mountains bordering Tibet. The smallest countries of South Asia are the Himalayan Kingdom of Bhutan and the Islamic country of the Maldives in the Indian Ocean. Afghanistan was considered part of South Asia during colonial times, but it is often grouped with the Middle East or Central Asia today.

Indo-Aryan Invasions
One of the distinct characteristics of South Asia is the influence of the Hindu caste system. The caste system is believed to have originated with the Indo-Aryan invasions of the ancient Indus River Valley civilization—one of the oldest on Earth. It dated to about 3000 BCE, which made it concurrent with the

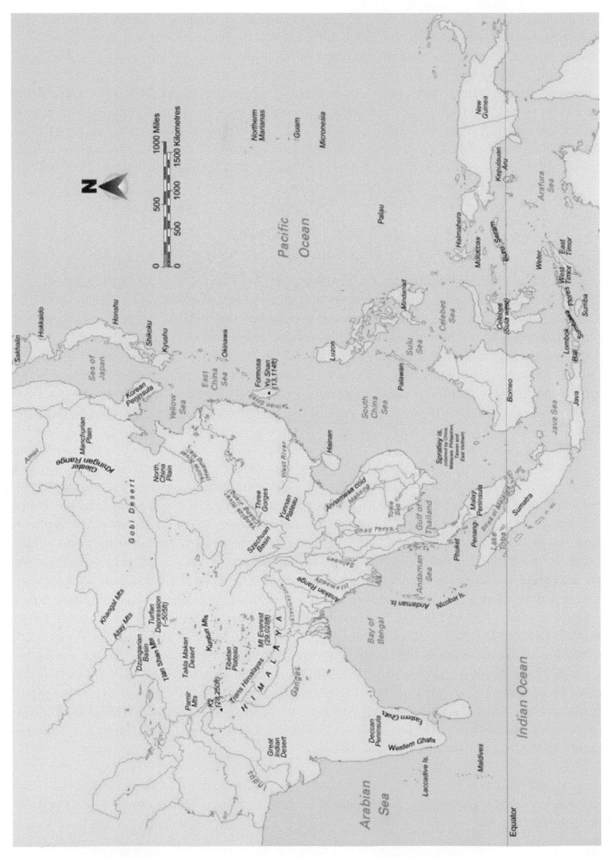

▲ *Physical Geography of Asia*

Photo: Alan A. Lew

◀ *Buddhist stupa at Mong La Pass (13,117ft; 3978 m) in the Khumbu region of Nepal, on the trail to Mt. Everest, which is the snow-capped peak in far distance in the left of this scene. The Khumbu is home to the Sherpa people, whose language and Buddhist religion are similar to that of Tibet. International trekking, which has been popular here since the 1960s, has made this region one of the wealthiest in rural Nepal.*

early Egyptians and Mesopotamians. Much of the Indus River civilization of that time has been buried underground by the waters of the Indus and Ganges Rivers. Archaeologists believe that the Indus River civilization was a matriarchal society, based on evidence that they worshiped a female fertility goddess. They were dark skinned and short in stature, and were probably vegetarians.

The Indo-Aryans probably originated somewhere in Central Asia. They invaded Europe, Southwest Asia and South Asia between 2000 and 1500 BCE (although some argue that this was more of a migration than a military invasion). Common linguistic features can be found in the languages of all three regions dating back to this time period. The Indo-Aryans were cattle herders with highly developed military capabilities. They were probably light skinned and taller in stature than the people of the Indus River, and they were meat eaters. They never united the entire Indian subcontinent, or even the entire Indo-Gangetic Plain, mostly due to constant warfare among themselves. These battles formed the basis of the Mahabharata Epic, which is a classical and semi-religious set of stories of early India. The Indo-Aryan peoples arrived by sea in Sri Lanka in the sixth and fifth centuries BCE, skipping over much of central India.

Hindu Caste System

The Indo-Aryan social structure consisted of three major groups: priests and teachers, warriors and rulers, and herders and manual laborers. In Europe, these became the three social groups of the Middle Ages: priests, nobles and peasants. (Some claim that in the United States this later became the basis for common references to upper, middle and lower socioeconomic classes.) In India, the three groups became the Brahman *caste*, the Kshatriya caste and the Vaisya caste, respectively. In addition to these three principal castes, two other caste groups were developed: the Shudras, who were Dravidians living in the Indus Valley, and the Dalits, who included everyone else. Thus, the five basic caste groups of Hinduism were formed in India. Some scholars argue that these castes were not a clearly fixed hierarchy in India until the late 1800s when the British shaped them to more resemble the class system they were accustomed to:

- *Brahmans* are the priests, scholars and educated caste; they eat no meat, a trait that they picked up from the Dravidians. They live in temples and in the town centers where the temples are located.

They can be very poor, having given up attachments to the material world. Sometimes they work as cooks in restaurants because everyone else can eat the food they cook, including other Brahmans.

- *Kshatriyas* are warriors, landowners and rulers. They are also the merchants in northern India, and they live in the center of the commercial districts. They eat chicken, but no pork or beef. In southern India they often form the highest caste group, because very few Indo-Aryans migrated south, with the exception of Sri Lanka.
- *Vaisyas* are herders, peasants and farmers. In southern India, they are often the merchant caste. They eat chicken, and they can eat pork on special occasions, but no beef.
- *Shudras* are the laborers, crafters and peasants. They eat chicken and pork, but no beef. The original Shudras were Dravidians, who were vegetarians, whereas the original Indo-Aryans were meat eaters. Over time these roles have completely reversed themselves. The Shudras traditionally lived outside of the city and were responsible for slaughtering animals. Sometimes they carry out animal sacrifices in the temples of the Goddess Durga (the demon killer).
- *Dalits* are also known as the Harijans and Untouchables. They are not considered part of Hindu religious society and social structure, and therefore they were not considered human. For many years they were not counted in the local population census in some parts of India. All non-Hindus are automatically Dalit, which is technically not a caste. Because they are outside of the caste system, they can eat any kind of meat, including beef. Their traditional roles are butchers, tanners and fishers, all of whom must kill animals as part of their livelihood. They also slaughter cows for certain sacred festivals, but they cannot cook for anyone in a higher caste group. In Japanese society, the Eta social group serves a similar role, and it also is considered at the bottom of the social classes.

The caste system was officially abolished in the 1950 Indian constitution, and many urban Indians today pay little heed to it. They eat what they want and socialize with anyone, without regard to traditional caste rules. The caste system, however, continues to be a very important and influential force within the more traditional rural areas of India. Although the caste system is frequently criticized, it has had the influence of uniting the great ethnic diversity of India. Caste designation transcends ethnic and language differences. For example, in traditional India a Kshatriya in Tamil Nadu Province in southernmost India could more easily marry a Kshatriya from northern India than someone of another caste in her or his hometown.

Buddhism

Buddhism is another major world religion that originated in South Asia. It is sometimes called Hinduism without the caste system. However, there are some other significant differences between Hinduism and Buddhism. Unlike Hinduism, Buddhism teaches that people can change their individual situation if they take the initiative. Buddhism also considers itself superior to other religions, and it seeks converts, which Hinduism generally does not. This makes Buddhism a proselytizing religion, like Christianity and Islam. Hinduism and Judaism are not proselytizing religions. Buddhism was founded by Siddhartha Gautama (563-483 BCE). He was a prince in the foothills of the Himalayas and gave up his royal life to seek the reasons for so much suffering in the world. His answer was that attachment to worldly goods was the cause of suffering, and renunciation and detachment from the world would lead to freedom from suffering and ultimately to Buddhahood.

Buddhism became the dominant religion in the Mauryan Empire, which was founded in 325 BCE. Asoka, its second Emperor, united all of the Indo-Gangetic Plain for the first time, including today's Afghanistan and a good part of the Deccan Plateau. In 275 BCE, Asoka adopted Buddhism and sent Buddhist missionaries as far as Greece and Egypt. This was the height of Buddhism in South Asia, and

◄ *Varanasi, also known as Benares and Kashi, is situated on the Ganges River in the northern Indian state of Uttar Pradesh. It is one of the seven holiest sites in Hinduism, and there is a general belief that death in Varanasi will significantly advance the devotee's reincarnation cycle. Varanasi is also where Siddhartha Gautama, the Buddha, gave his first sermon and is therefore considered the birthplace of Buddhism. In addition, it is a major historical center for Jainism and Sikhism.*

Photo: Dallen J. Timothy

the arts and sciences flourished under Asoka's rule. Hindus saw Buddha as the tenth incarnation of the Hindu god Vishnu, one part of the Hindu trilogy of supreme gods; Brahma and Shiva are the other two. Buddhism, therefore, was absorbed back into Hinduism, and it almost completely disappeared in South Asia after Asoka's rule. It continued to be followed in Tibet and Nepal in the Himalayas, and on the island of Sri Lanka, where it is the dominant religion today.

Islam in South Asia

Islam is the third major religion of South Asia. The Prophet Mohammed founded it in the sixth century CE in what is now Saudi Arabia. He viewed it as an extension of Jewish and Christian traditions. It rapidly spread across North Africa and into Central Asia after Mohammed's death. Islam did not enter South Asia until the tenth century CE, when Arab traders brought it with them. Dalits were more open to this new religion because of their exclusion from Hinduism and because Islam had no caste system. Islam was especially popular in the Bengal region of India and today's country of Bangladesh. Some of the remaining Buddhists in India were also attracted to Islam.

The Mughal Empire and Taj Mahal

The Mughal Empire was the peak of Islamic rule in the region, and the empire covered much of the same area as the Buddhist Mauryan Empire. The Mughals ruled from 1526 to 1857, when the British took control of a highly disorganized India. The Mughal Empire's leaders were Muslims, although most of the land they ruled remained Hindu. The Taj Mahal was built during this time (1632–1648) and is considered the masterpiece of Mughal art and architecture. The Taj Mahal is actually a mausoleum located in Agra, south of India's capital city, New Delhi. The Mughal emperor, Shah Jahan, built it after the death of his favorite wife, Mumtaz Mahal. The mausoleum building combines elements of Persian, Turkish, Indian and Islamic architecture, and it includes a large garden and a variety of out buildings.

Hinduism and Islam

Hinduism was able to absorb most other religions that entered South Asia, but was unable to absorb Islam because of some major differences between the Hindu-Buddhist tradition and the Judeo-Christian-Islamic tradition. The following comparisons apply to these major religious traditions in their broadest sense. The actual religious teachings and practices of any individual church, temple, teacher or minister may vary considerably.

- The Jewish-Christian-Islamic tradition believes in human equality before God, whereas the Hindu tradition (but not Buddhism) is based on the caste hierarchy, in which a higher caste person is closer to God than a lower caste person.
- The Jewish-Christian-Islamic tradition believes that humans have a free will and can change their lives; Hinduism (but not Buddhism) believes that there is no free will and that all is based on fate (karma). In fact, trying to change one's position in society is wrong because it increases one's karma (resulting in more reincarnations) and is against the will of the gods.
- Hinduism believes that one can only improve one's next life by fulfilling the duties of this life. Buddhism, on the other hand, teaches that the only way out is a total rejection of all material things, which is the Brahman caste role in Hinduism. Buddhists basically believe that everyone can become a Brahman.
- The Jewish-Christian-Islamic tradition believes that the world is real and can be changed; the Hinduism-Buddhism tradition believes that the world is an illusion and that the spiritual world is the only reality.
- The Jewish-Christian-Islamic tradition believes that the world and life can be enjoyed; the Hindu-Buddhist tradition believes that the world and life only bring pain and suffering—enjoyment is only possible in the spiritual world.
- The Christian and Islamic traditions (but not Judaism) believe that their religion (and their God) is the only true religion. Hinduism (but not Buddhism) believes in millions of gods that are constantly changing, and that all religions are equally valid. Traditional Buddhism believes that there are no gods at all; enlightenment is entirely up to the individual. Later forms of Buddhism introduced the concept of Bodhisattvas, which are somewhat similar to saints in the Catholic tradition.
- Islam (but not Judaism and Christianity) teaches that the use of pictures and statues is idolatry and therefore forbidden; Hinduism (but not traditional Buddhism) uses pictures and statues profusely.
- Judaism-Christianity-Islam teaches that there is only one life per person in which one can prove oneself; Hinduism and Buddhism believe in reincarnations that can last forever.
- Judaism-Christianity-Islam teaches that the world will end, and that this is very important; Hinduism-Buddhism believes that the world and spiritual reality are both infinite and eternal, and therefore time has no real meaning. Hinduism measures time in cycles that last thousands and millions of years. The current cycle, the Kali Yuga, is considered an evil time in which to be reincarnated.

These differences between Islam and Hinduism were simply too great to be reconciled and became the basis of the partition of British India into Hindu-dominated India and Muslim-dominated Pakistan, which included East and West Pakistan, at India's 1947 independence.

Sikhs

The Sikhs are a major religious and ethnic group in the Punjab Province of India. The Punjab regions of both India and Pakistan are at the top of the Indo-Gangetic Plain, between the drainage of the Indus and Ganges Rivers. Sikhism was founded by Guru Nanak (1469–1539 CE), who sought to combine the

◀ *The Taj Mahal in Agra, India. This mausoleum was completed in 1648 by the Muslim Mughal Emperor Shah Jahan for his favorite wife. It is considered one of the most beautiful buildings in the world and the best example of Mughal Empire architecture, which combined styles of Islamic, Persian and Indian design. Today, the Taj Mahal has become the most iconic and best-known symbol of India.*

Photo: Dallen J. Timothy

best of Islam and Hinduism. Sikhism is a monotheistic religion that opposes idolatry (like Islam), yet it teaches that all religions are equally valid. Like Hinduism, it also incorporates meditation. The Sikhs were militarized under Govind Singh (1666–1708), who was the tenth and last guru (master teacher) of the Sikh religion. He created a warrior fraternity within Sikhism, introducing the wearing of turbans, the non-cutting of hair and the tradition of always carrying a knife.

The Sikhs were famous for their military skills as mercenary soldiers for the British Empire. In 1947, 2.5 million Sikhs left Pakistan's Punjab region and migrated to Indian Punjab as part of the partition of South Asia, because they were promised complete autonomy under the Indian government. Sikh separatists have been fighting a war of independence to create a separate country (which they call Kalistan), although they only comprise about half of the population of India's Punjab Province.

British Legacy and Colonialism

The British have long been blamed for many of the problems of India in the 20th century, especially the conflicts between Hindus and Muslims. The major impact of British rule, however, has been the *modernization* of the Indian subcontinent. The British built the most expansive and heavily used railway network in the world in India, which extends to many of the smaller and more isolated districts of the country. The British also built a comprehensive road network. The British built industries, introduced a European-style education system, a legal system based on English Common Law (similar to the United States, Canada and Australia) and a parliamentary political system.

In addition, the British united and integrated all of South Asia for the first and only time in its history. India's transportation system has helped enhance the movement of people from one region to another for employment and tourism purposes. Modernization has also had negative impacts. It has increased ethnic tensions and forces that push to divide the Indian subcontinent. Greater education has resulted in greater demand for the use of local languages instead of the national language, which has reinforced local identity over the national identity. Knowledge of other places within South Asia made people realize the great differences in levels of economic development and standards of living, resulting in accusations of favoritism rather than a shared sense of similarity and goals. Muslims in India increasingly identified with Muslims in the Middle East where they see more similarities to their values and traditions, rather than with the Hindus who are their neighbors.

Partition

The one thing that brought the Hindus and Muslims of South Asia together was their desire to rid themselves of British colonial rule. This shared goal, however, did not bring about a lasting unity between the two groups, as trust was never fully achieved. In 1947, India was granted independence, following almost 50 years of anti-British protests and the weakening of the United Kingdom in World War II. The British sought to keep India as a single country, which was also the goal of the Hindu leaders (including Mahatma Gandhi). Muslims, however, wanted an independent Pakistan. The partition of South Asia resulted in about 5 million deaths from ethnic violence, and the migration of 13 million people between India and Pakistan. Even after partition, there were as many Muslims in India as there were in the new country of Pakistan—about 100 million.

West and East Pakistan

Partition created the new country of Pakistan, which originally comprised West Pakistan and East Pakistan (Bangladesh). The only similarity between these two territories was that they were both predominantly Muslim. Almost everything else about them was different. West Pakistan had a dry climate where drought was a constant threat, whereas East Pakistan had a humid tropical climate, with an overabundance of water. Agriculture in West Pakistan was primarily grain production (mainly wheat), while agriculture was primarily wet rice production in East Pakistan. At partition, West Pakistan had an average annual per capita income of US$350 (£175); in East Pakistan it was less than half that at US$150 (£75).

The West Pakistanis were light skinned and more of the Indo-Aryan racial stock; the East Pakistanis were darker skinned and of the Dravidian racial stock. Each of these territories had more than 100 million people, but East Pakistan was only one fourth the size of West Pakistan, resulting in a much higher population density. West Pakistan was much closer to the Middle East, and much more closely integrated with Middle Eastern economies; East Pakistan was economically and culturally closer to India. The Indian state of Bengal was adjacent to East Pakistan, and its major city of Calcutta was the central industrial center for the region. The conflicts that these differences engendered resulted in East Pakistan declaring its independence from West Pakistan in March 1971. A civil war was fought in which one million East Pakistan Bangladeshis died. India intervened with its army on the side of the Bangladeshis and independence was finally achieved in December 1971, followed by the adoption of the new country's name of Bangladesh.

Irredentism

It is known as *irredentism* when people in one country desire to separate and join with people in an adjacent country to form a new country. Irredentism is a major problem for Pakistan with its neighbors to the west, and it is also a major issue for India in the Jammu and Kashmir region in the far north, where 75 percent of the population is Muslim but the rulers have traditionally been Hindu. At partition, the Hindu rulers of Jammu and Kashmir sided with India. However, Pakistan has long supported the desires of the Muslim population in Jammu and Kashmir to separate from India and join Pakistan. Because of this, India and Pakistan have been in an almost continuous state of war since 1947. This situation is especially tense because both countries have nuclear weapons (India since 1974 and Pakistan since 1998) and because of the rise of militant Islam in the border areas of Pakistan and Afghanistan.

Supranationalism

Supranational organization exists in South Asia in the form of the South Asian Association for Regional Co-operation (SAARC). Member countries include Bangladesh, Bhutan, India, the Maldives, Nepal, Pakistan and Sri Lanka. While the goals of SAARC are to promote the socio-economic interests of the region and the cultural development of the people, in part through tourism, several obstacles prevent its successful implementation. The ethnic conflict/civil war in Sri Lanka (1983–2009) and the dispute over Kashmir between India and Pakistan are two of the most challenging problems facing the region in terms of tourism and economic cooperation. These and other conflicts have tainted the region's image and impacted the growth of intra-regional tourism.

Poverty

SAARC has also had a limited effect in improving the impoverished conditions that many of the people of South Asia live in. After Sub-Saharan Africa, South Asia has the second lowest average incomes of any major region of the world. Poor populations are widely seen in the densely populated larger cities of South Asia, and Western visitors often respond to this life on the street with culture shock. It is common for Western-oriented hotels to emphasize how modern they are, providing a sheltered enclave in which to recover from the culture shock of some parts of South Asia. Rural areas are also very impoverished, with many remote areas still lacking electricity and running water.

By Western standards, however, the cost of living in South Asia is extremely inexpensive, and there is a vast network of alternative accommodations that cater to backpacker tourists from throughout the world. Many younger backpack travelers from Europe, North America and Israel (on what is colloquially referred to as the 'hummus trail') spend months and years traveling on very little money through South Asia.

Climbing the Himalayas

The Himalayas are one of the major tourist attractions of South Asia. In Nepal, Mount Everest (also known as *Sagarmatha* in Nepali and *Chomolungma* in Tibetan) and the Khumbu Valley that leads to it are one of the most trekked regions in South Asia. This is the home of the Sherpas, who originally migrated to Nepal from Tibet and continue to practice Tibetan Buddhism. Edmund Hillary and Tenzing Norgay first climbed Everest in 1953. Today some 500 people a year typically attempt the peak, though harsh climate prevents many from achieving their goal, and several die each year in the attempt. Annapurna, which has five peaks over 24,000 ft (7,400 m), is another major *trekking* region in Nepal.

Although it only reaches some 3,510 ft in elevation, a trip over the Khyber Pass in northern Pakistan is one of the top adventures of South Asia. Historically, the pass has been one of the most important trade and transportation routes between Central Asia and South Asia. The pass cuts through an extension of the Hindu Kush range that was used by the Indo-Aryans who invaded South Asia some 3,000 years ago— Alexander the Great from Greece, who conquered the Indus River region in 326 BCE, and the Muslim Mughal Empire, which controlled northern South Asia from 1526. The British also used the Khyber Pass to invade Afghanistan in the 1800s and 1900s. Today the Khyber Pass has significant tourism potential, but it is known as a haven for bandits and as an epicenter for conflict between Afghanistan and Pakistan, and most tourists are encouraged to avoid it.

Kashmir

Between Nepal and Pakistan is the Kashmir region of India. Kashmir has been a popular tourist destination for Europeans since the late 1800s, and a pilgrimage site for Hindus and Buddhists long before that. The main attractions of the Kashmir region are its mountainous landscape, with its many trekking and winter skiing opportunities, and its mild summer climate, distinctive gardens and houseboat rentals on Dal Lake, which is located next to the city of Srinagar. Unregulated residential and tourism development on the banks of Dal Lake have caused it to shrink to a fraction of its original size. This, along with tensions between Pakistan and India over the Kashmir region that cause periodic terrorist and military actions, have affected visitation in recent years.

Pilgrimage and Religious Tourism

Religious *pilgrimage* is a unique form of tourism and shares many characteristics in common with recreational and leisure travel. Infrastructure needs, including transportation, accommodations and meals, are almost identical. The sites visited are often the same for both religious pilgrims and secular tourists, and pilgrims typically include visits to non-religious sites, as well. Managers of both religious and secular destinations must develop policies that protect the site while ensuring consistent experiences for large numbers of visitors over time. Some argue that religious and leisure travelers have similar motivations and experiences, as well. These include motivations to experience social shared values and places of authenticity, and a search for meaning beyond that found in the home place.

There are major religious sites for Buddhists, Hindus, Sikhs and Muslims throughout South Asia. Many of these are in the form of temples that commemorate sacred sites and attract large numbers of devotees from within the region and from abroad. The Kumbh Mela, the largest pilgrimage/religious tourism gathering in the world, is held four times every twelve years rotating among four cities in India. In 2013, the Great Kumbh Mela event (once every 12 yrs) attracted up to 80 million pilgrims who bathed in the Ganges River at its confluence with the Yamuna River (precise numbers of Kumbh Mela are impossible to gather). Similarly, hundreds of thousands of pilgrims travel each year to Lumbini, Nepal, the birthplace of Buddha. Many of these pilgrims come from Sri Lanka, Southeast Asia and East Asia where Buddhism is more widely practiced than in India. Japanese and Southeast Asian Buddhist organizations have specially built accommodations facilities that cater to their different cultural needs.

In India, the Golden Temple of the Sikhs and the Hindu holy city of Varanasi are among the most important pilgrimage destinations that are also major tourist sites. The Golden Temple is located in the Indian Punjab city of Amritsar and is the most sacred site for Sikhs, who come from throughout the world to pray there. The gold-colored temple, officially named Harmandir Sahib, sits in the middle of a small lake and symbolizes the spiritual and political independence of Sikhism.

Countries and Tourism of South Asia

India

As noted above, India is a significant religious tourism destination, both for domestic and international pilgrims of Hindu, Sikh and Buddhist faiths. What the Golden Temple is to Sikhs, the entire city of Varanasi is to the Hindus. Varanasi (also known as Benaras and Kashi) sits on the sacred Ganges (Ganga) River in the Indian state of Uttar Pradesh, one of India's poorest regions. It is considered one of the oldest places of continuous habitation on the planet, dating back to before 3000 BCE, and it is the single most important pilgrimage destination for all sects of Hinduism. More than one million Hindu pilgrims come to the city each year to bathe in the Ganges, to cleanse their sins and to burn their dead on funeral pyres. It is believed that dying in Varanasi will end the continual suffering of reincarnation. Varanasi is also where Gautama Buddha is believed to have given his first sermon on the principles of Buddhism, making it one of the four pilgrimage sites that all Buddhists should visit. The others are Lumbini, Nepal (where Gautama Buddha was born), Bodh Gaya, India (where he gained enlightenment) and Kushinagar, India (where he died).

Indian Diaspora

Another significant form of tourism in India is *visiting friends and relatives (VFR) tourism*. This may be the most significant form of tourism in India in actual numbers, although there are no data to substantiate such a claim. The Indian diaspora (spread of an ethnic group abroad from the homeland) is scattered throughout the entire world, with places such as Fiji, South Africa, Guyana, the United States, Canada and Australia having substantial populations of Indian origin. These people are known for traveling back to their homeland to visit relatives and friends, to experience the land of their ancestors and to participate in religious and other festivals.

India's 28 states and seven territories have unique tourism products, and most tourism centers on large cities and other urban areas. Jungle tours involving tigers and elephants are popular, while the Ministry of Tourism is actively promoting adventure travel and ecotourism in the country's many national parks and protected areas. Desert safaris and the cooler hill stations of the northern mountains are popular destinations and activities. The hill stations were popular holiday spots during colonial times for the British, who needed a respite from the hot interior lowlands; now they are again popular among wealthy Indians and foreign visitors alike.

Former Colonial Territories

In addition to the British, the French, the Portuguese, the Dutch and the Danish also occupied India's soil at different points in history. However, only the British, French and Portuguese had colonies into the 20th century. The French transferred sovereignty of its four small territories in eastern India (Puducherry, formerly Pondicherry) in 1954, in response to India's 1947 independence from the United Kingdom. Portugal was the last to leave India. It gave up its territories of Goa, Daman, Diu, Dadra and Nagar Haveli in 1961 under military pressure from India. Puducherry and the former Portuguese colonies have become important tourist destinations and centers of Christianity in South Asia. Goa and Diu are important beach resorts, and the European architecture of all the territories appeals to people interested in cultural heritage. Many of the older people in these areas still speak French or Portuguese.

Bangladesh

Since the partition of the subcontinent, more than 200 million Hindus have migrated out of Bangladesh, while a similar number of Muslims have migrated into the country. Bangladesh is 95 percent rural. Its major industry is the growing of jute, which is used to make burlap bags and rope. It is grown in Bangladesh and sent to factories in Calcutta for processing. The warm climate and ready availability of water allow the growing of three rice crops a year. Almost half of the country is flooded every year by the Ganges and Brahmaputra rivers, as well as by tidal floods caused by typhoons from the Bay of Bengal.

Because much of the country is low lying, it is regarded as being extremely susceptible to climate change and sea-level rise. Transportation is a problem because water is everywhere, and it can take up to 20 hours to get from Dhaka, the capital of Bangladesh, to the mouth of the Ganges River. Although Bangladesh has some tourism based on beaches, jungles and architectural heritage, severe climatological disturbances have prevented the development of large-scale tourism in the country. In addition, because the country's population is over 80 percent Muslim, it has suffered as much as other largely Islamic populated nations from perceived security concerns.

Pakistan

Pakistan was the first Islamic Republic in the world. When it was founded, the Muslims there shared a high level of religious intensity because of finally gaining independence after many decades of living under Hindu rule. Islam was a unifying force for the people of Pakistan (which means land of the pure), who named their new capital city Islamabad (City of Islam).

Pakistan is made up of four major ethnic groups. Balochistan is where the Baloch reside. There are also large numbers of Baloch in southern Afghanistan and eastern Iran who have long desired their own country. The Sindhis dominate the Sindh Province; they mostly migrated from northern India and the Ganges Plain. This is where Karachi, the largest city in Pakistan and the country's main seaport, is located. The Sindhis speak Urdu—the national language of Pakistan, which is closely related to Hindi, the national language of India. Punjabis live in the Punjab Province of Pakistan, which adjoins the Punjab Province of India, homeland of the Sikhs.

Pakistan's Punjab

The Punjab province is the wealthiest and most populated in Pakistan. It shares many similar climatic and economic conditions to India's Punjab, which is also a relatively wealthy region. The fourth major ethnic group in Pakistan is that of the Pashtuns, who live in the Northwest Frontier Province, although most ethnic Pashtuns are located in neighboring Afghanistan. Like the Baloch, they were divided into two different countries by the European colonial powers so they could be more easily controlled. The Pashtuns also have long desired their own country. The former Taliban rulers of Afghanistan were Pashtuns who went into hiding in the Pashtun area of Pakistan after the United States government overthrew their leadership in the Afghanistan War of 2001.

As with India, VFR tourism is a very important part of the tourism economy of Pakistan. This is critical in a country that has suffered considerably from the conflict in neighboring Afghanistan and from its own internal *ethnic and religious divisions*. Little other tourism presently exists, not for a lack of natural and cultural resources, but because of the broader conflict in the region and the resultant fact that most developed countries have placed Pakistan on their travel warning lists.

Border Tourism

There is promise of growing cross-border travel between India and Pakistan, but presently heavy restrictions have been placed on Indians crossing into Pakistan and vice versa. This is due to the conflict between the two nations, which many world observers fear has escalated to nuclear capabilities in recent years. Indian visas for Pakistanis and Pakistani visas for Indians are difficult to acquire, but citizens of other countries can pass over the border relatively easily. One interesting tourism phenomenon popular among both nationalities is to visit the border itself. Every evening at the Wagah crossing just east of Lahore, the changing of the border guards, a fancy ceremonial affair, takes place. Bleachers and souvenir shops have been set up on both sides, and it is one of the most popular attractions in the region.

The Himalayan Kingdoms

The two other major regions of South Asia are the Himalayan kingdoms of Nepal and Bhutan and the Indian Ocean island countries of Sri Lanka and the Maldives. Nepal and Bhutan are buffer countries between the larger countries of China (and Tibet) to the north and India to the south. Like India, they formerly were under British colonial influence. Both were ruled by hereditary monarchs and have very traditional agricultural societies. Nepal is a mostly Hindu country (although Buddhism dominates in its

◀ *A dzong building in Bhutan. This traditional fortress architectural style is found throughout Bhutan and Tibet. The large outside walls protect a courtyard and may contain religious, military, social and government administrative activities. Their imposing style and social function make dzongs an important part of the historical and contemporary heritage of Bhutan.*

Photo: Dallen J. Timothy

northern mountain regions), while Bhutan is a Buddhist country. Buddhist practices in both countries are closely associated with Tibetan cultural influences. Bhutan and Nepal are among the poorest countries in the world, with tourism as their main export industry. Both are also dependent on India for political and economic support, as well as for their major source of tourists.

Nepal

Nepal has existed as a kingdom centered in the Kathmandu Valley for more than 1500 years. It has nine of the 14 highest peaks in the world, including Mount Everest and Annapurna I. In recent decades, Nepal has suffered from a Maoist-based (a form of Chinese communism) insurgency that had come to control most of the rural countryside, with the central government controlling the cities. Many thousands of Nepalese died in the struggle between the two groups, though most tourists have not been directly affected by this conflict. Peace agreements in 2006 brought Maoists into the government and significantly reduced the influence of the monarchy.

Trekking Tourism

Nepal's world-famous mountain trekking on and near Mount Everest has received a great deal of tourism industry and academic attention. The region is viewed as an ultimate goal for mountain climbing enthusiasts, who typically employ members of an ethnic group known as Sherpas to guide them. Sherpas, who are culturally related to Tibetans, also operate accommodations and restaurants. The success of trekking tourism in the Khumbu Valley has drawn migrants from more impoverished parts of Nepal, many of whom work as poorly paid porters, carry camping and hiking equipment for international trekkers.

Trekking tourism has come under scrutiny by researchers for its culturally and environmentally questionable practices. The utilization of impoverished Nepalis as "pack animals" has been criticized as a neo-colonialist and Eurocentric practice. Likewise, while it is common practice for supplies to be carried into the mountains, not all of it is carried out. The resultant litter that accumulates along trails, streams and scenic valleys, creating an unappealing site for tourists and ecologically unsustainable conditions for the environment, and is periodically cleaned up by volunteer hikers and local residents.

Kathmandu

The Kathmandu Valley is on UNESCO's World Heritage List due to seven major sites, monuments and buildings that demonstrate the historical continuity of the kingdom of Nepal. These sites include Durbar Squares of Hanuman Dhoka (Kathmandu), Patan and Bhaktapur from the days when each was a separate city-state trying to outdo the other in architectural splendor. In addition, the Buddhist stupas of Swayambhu and Bauddhanath, and the Hindu temples of Pashupati and Changu Narayan all continue to be important South Asian pilgrim destinations. As noted above, Lumbini, the birth place of Buddha, is a very important tourist site and a vital pilgrimage goal for Buddhists from around the world.

Bhutan

Bhutan is about one-third the size of Nepal, and it has a population of approximately 2 million. The precursor of Bhutan was the country of Lhomon or Monyul, which existed from 500 BCE to 600 CE, when Buddhism was introduced to the region. Various Buddhist sects fought for political control for many centuries thereafter (with additional influences from Tibet) until the 17th century when a theocratic government united the country. From that time until 1907, the Kingdom of Bhutan (or Druk Yul—Land of the Thunder Dragon) had a system of shared civil and spiritual (Buddhist) rule under an absolute monarchy (the Dragon King). The monarchy has traditionally been absolute, although democratic elections occurred for the first time in history in December 2007 and March 2008. The king is admired and respected, unlike the King of Nepal, who was largely despised by the people.

Bhutan has four major ethnic groups: Ngalop (Tibetan origin); Sharchop (Indo-Mongoloid origin); aboriginal or indigenous tribal peoples; and Nepalese. Fear of the increasing number of Nepalese in the population has prompted the government to adopt policies in the 1980s to support and enhance a Bhutanese cultural identity that is different from that of Nepal. This has caused an increase in ethnic tensions and large numbers of Nepalese left Bhutan for refugee camps in Nepal in the 1990s. Bhutan has more Buddhist monks than soldiers. It has no traffic lights, and the King only allowed television into the country in 1999. In December 2004, Bhutan became the first country in the world to ban the smoking of cigarettes in public and the sale of cigarettes within its borders. Individuals can import them if they pay a 100 percent tax, however.

Bhutan Tourism

Bhutan's primary tourism resources are its well-preserved culture (lifestyles, dress, village life, festivals and celebrations, unique architecture and religion) and its natural bounties in its Himalayan setting. Bhutan has long been closed to mass tourism; even today only small numbers of high-paying tourists are allowed into the country, and only some of the country's regions are open to foreign visitors. While there is no visa quota, there are other mechanisms in place that limit the number and types of tourists that enter the country each year. For example, tourists must pay a minimum of $200 per day. People can, however, pay higher daily tariffs for higher-quality accommodations.

The required daily cost, which must be pre-paid before arrival, includes all transportation to and from Bhutan and within the country, as well as food and accommodations. Each tour is required to have an

ISSUES AND INSIGHTS   The Dangers of Climbing Everest

Since the growth of commercial guided expeditions in the 1990s, Mount Everest has become a playground for those with time and money seeking high-altitude adventure experiences. In the northern hemisphere spring of 2007, more than six hundred people reached the summit, and there were six deaths. Growth in visitor numbers has continued although there are increased concerns over the carrying capacity of the mountain for safe climbing and increasing questions are being asked about the amount of expertise climbers should have before being allowed to participate in climbs.

Most climbers made the ascent using commercial operations that charge between $10,000 and $65,000 (£5,000 to £32,500) for a guide, who is usually a Sherpa, equipment and backup support. There has been a significant change in the profile of climbers on the mountain since the first successful climb. In the early decades, less than 20 percent were 40 years of age or over, but between 2000 and 2005 almost half the climbers were in this bracket. In the same period, those over 60 years old have increased from 0.3 percent to 3.6 percent. Since 2000, an average of 13 sexagenarians have attempted the climb each year. Edmund Hillary and Tenzing Norgay were 33 and 39, respectively, when they became the first men to climb Everest in 1953. Around a third of those under 40 make it to the summit, but by age 60 only around 13 percent succeed. This lowered success rate does not seem to be due to older climbers using their experience to back off from difficult situations. One in 20 is killed on the mountain, which is around three times the overall average. If they do make it to the summit, they have an even higher chance of death-a quarter do not make it back down, more than ten times the average. Gender, by the way, makes no difference to success or death rates.

Sources

Huey, R. B., and R. Salisbury. 2003. Success and death on Everest: How the main routes and seasons compare. *American Alpine Journal*, 2-9.

Huey, R. B., R. Salisbury, J.-L. Wang, and M. Mao. 2007. Effects of age and gender on success and death of mountaineers on Mount Everest. *Biology Letters*, Vol. 3 (No. 5, October 22) 498-500.

attendant guide/interpreter and driver; *independent travel* is not permitted, as even trips for individuals must be organized and pre-planned by the Bhutan Tourism Corporation. Each site visit or activity is planned in advance, and the guide is required to check in with the police or security at each attraction. Visas are not easy to get. These must be acquired in the traveler's home country through a Bhutanese government-approved travel agent, who then works directly with agents in Bhutan to secure the visa once the mandatory fees have been paid. Visitors must fly at least one way into or out of the country. If one enters Bhutan by land border, he or she is required to fly out on the only airline servicing the country, the government-owned Druk Air.

Despite these strict regulations, the Tourism Council of Bhutan is working to increase the number of yearly arrivals. From Bhutan's perspective, even if these policies keep fewer people from visiting than if its borders were completely open, smaller numbers of tourists paying high daily tariffs eliminate the need for mass tourism and helps to avoid the negative socio-cultural and environmental changes that typically accompany tourism development.

The Indian Ocean Insular Countries

Both the Maldives and Sri Lanka have long been major sun, sand and sea tourist destinations for Europeans. They are also island countries. Beyond that, however, they are very different from one another in size, ethnicity and contemporary issues.

Sri Lanka

Sri Lanka lies off the southern tip of India, almost in the center of the Indian Ocean. Its summer monsoon rains bring lush vegetation on the southern half of the island's mountain slopes, the highest peak of which is Pidurutalagala at 8,281 ft (2,524 m). However, the northern half lies in the summer monsoon's rain shadow and is a relatively dry zone. The island is dominated by the Buddhist Sinhalese, who originally arrived from north India in about 500 BCE. They created a high civilization from the third century BCE to the twelfth century CE that was known for its engineering prowess. Sri Lanka came under Portuguese and Dutch control until finally becoming a crown colony of Britain in 1802, known as Ceylon. Sri Lanka has had a democratic tradition dating back to the 1930s, and in 1960 the Sri Lankans elected the world's first female prime minister.

Ethnic Tensions

Today the Sinhalese comprise about 74 percent of the country's population, while 13 percent are Hindu Tamils, who are related to the Hindus in India's nearby state of Tamil Nadu. Ethnic tensions between these two groups date back to the 1950s, and the conflict was only somewhat calmed in the aftermath of the December 26, 2004, Indian Ocean *tsunami* disaster that killed more than 30,000 Sri Lankans. Fortunately for Sri Lankans and tourists, the civil war, which began in 1983, officially ended in 2009, with the government forces subduing the Tamil rebels (Tamil Tigers). The longstanding ethnic conflict between the nation's two main population groups has curtailed tourism development and concentrated it in the southern end of the island. The end of the civil conflict has brought about considerable speculation regarding the potential for tourism development to the northern and northeastern portions of the island, which were effectively outside government control for 26 years.

Sri Lanka receives many accolades as an enjoyable tourist destination, largely because of its remarkable beaches and cultural heritage. Its forests are home to leopards and exotic birds, and its human past affords many opportunities for interactions with ancient temples, villages and colonial buildings. Sri Lanka was one of the hardest hit countries by the Asian tsunami of December 2004, which virtually devastated the beach resorts on the island. Recovery has been quick, and many of them have since been rebuilt, although there are many areas still struggling to recuperate from the disaster.

The Maldives

The Maldives consists of some 1200 coral islands, grouped in a chain of 27 atolls. These coral atolls comprise live coral and sand bars that sit atop an undersea mountain range, which runs north to south for 600 miles (960 km) southeast of the Indian Peninsula. The rising ocean level is threatening the future of the Maldives, as it threatens other low-lying islands throughout the world, which is particularly poignant in this small country because the highest elevation is only 2.4 m above sea level. About 200 of the islands are inhabited, though many have very small numbers of people on them. The Maldives has a population of about 300,000 and is a primarily Muslim country and another former British colony.

Island Resorts

Like Bhutan, the Maldives has a unique tourism system designed to minimize the negative social impacts of tourism. Maldivian tourism primarily caters to affluent Europeans and is the most important sector of the economy, so the government and population want tourism, but they do not want the behavior of the tourists to clash with their own conservative Islamic values. Therefore, many of the islands are used exclusively as self-contained resorts, with Maldivians working, but not living permanently on them. When tourists travel to the Maldives, they essentially book into a single-island resort, where they have only limited interaction with locals—resort employees.

ISSUES AND INSIGHTS The Maldives: A Disappearing Country?

In November 2008, the Maldives made international headlines because the country had, for the first time in its history, democratically elected a president. The new president announced that the country would likely disappear because of the rise of the sea level associated with climate change. As a survival strategy, the Maldives announced that it would use money derived from its tourism industry to buy land in Sri Lanka, India or Australia.

As a tropical island state, the Maldives can be seen as representative of many Small Island Developing States (SIDS). According to the CIA World Factbook, the country has just 300 sq km of land with a maximum elevation of 8 ft (2.4 m) above sea level (asl), and with 80 percent of the area lying below 3 ft (1 m) asl. Only 200 of the 1,190 islands and coral atolls are inhabited, and another 80 are tourist resorts. The population of 393,500 (2013 estimate) had an estimated annual growth rate of 5.6 percent, which would double the country's population in less than 15 years.

Tourism is the most important industry, accounting for 28 percent of GDP and more than 60 percent of the Maldives foreign exchange earnings. More than 90 percent of the government tax revenue is derived from tourism- related taxes, and from related import duties. Fishing, agriculture and manufacturing are of importance for the economy apart from tourism. Most staple foods nevertheless have to be imported. Real GDP growth averaged 7.5 percent per year since the late 1990s, leading to a GDP of almost $1.6 billion in 2007, but high oil prices have led to a growing trade deficit.

With regard to tourism, statistics show a high volatility in arrivals. Tourist numbers went up from about 461,000 in 2001 to almost 617,000 in 2004, but they fell to 395,000 in 2005 (UNWTO 2007), possibly a result of the December 2004 tsunami. Tourists are mostly coming from Europe by air. In 2005, arrivals from Europe accounted for almost 78 percent of all arrivals, entailing high-energy use for flights and transport in the islands (helicopter, speedboats), as well as for accommodation (generator-driven electricity production). Clearly, most of the oil consumption of 210,000 gallons (795,000 liters) per day (2005 estimate) is likely to be used for tourism. Calculated over the year, energy consumption in the Maldives would thus amount to about 229,000 t of fuel and emissions of about 840,000 t CO. This corresponds to about 2 tons of CO_2 per person per year. Even though no official information is available, energy use can be assumed to have been substantially higher in 2004, when tourist arrivals were 56 percent higher than in 2005.

A number of key points can be learned from the Maldives. As in other SIDS, dependence on a single economic sector (tourism) is high, particularly with regard to foreign exchange earnings. Income derived from tourism can boost average income—even though not to levels comparable with developed countries—but the distribution of money may be highly skewed. Furthermore, development is financed at high energy and emission costs, which should be seen in the light of recent demands by SIDS for industrialized countries to cut emissions of greenhouse gases dramatically. Yet, implementing such cuts would clearly also affect long-haul travel by air, and thus the economies of SIDS.

The possibility of achieving the rapid and deep emission reductions needed to limit climate change and sea level rise to levels that would allow at least some of the Maldives islands to remain habitable, has perhaps already been lost. So, while some may question whether it is prudent or even ethical to support further development of energy-intense tourism in the Maldives, there is no other development alternative and few international resources are forthcoming to support the type of adaptation the nation will require. Tourism development is the only way the country can secure the resources needed to support the eventual relocation of its citizens. The nation's historic and current emissions are not the source of its vulnerability, and the reality is that energy-intense tourism development is its best option to build adaptive capacity for its population. This is the major policy dilemma for the coming decades for the Maldives, as well as many other of the world's smaller countries.

Sources

CIA World Factbook. 2008. Maldives, https://www.cia.gov/library/publications/the-world-factbook/geos/mv.html

Gössling, S., C. M. Hall, and D. Scott. 2009. "The challenges of tourism as a development strategy in an era of global climate change." In *Rethinking development in a carbon-constrained world. Development cooperation and climate change*, ed. E. Palosuo, 100-119. Helsinki: Ministry of Foreign Affairs.

Tourists are not encouraged to visit Male, the capital and most populated island, although they may if they dress and behave appropriately. The international airport is on a small island, and inter-island transportation is done via boat. The Maldives is often perceived as a paradise with white sandy beaches and clear blue water, but the country has significant problems with rubbish disposal and is extremely concerned about the impacts of climate change. In addition to beaches, the islands have become popular for scuba diving in recent years.

4.2 | SOUTHEAST ASIA

Current Southeast Asia Issues and Resources (http://wrgeography.com/southeastasia.html)

Southeast Asia is the large area of land and water situated between India to the west and China to the north. Older terms for this region include "Indochina" and "Further India", both of which referred mostly to the peninsular part of Southeast Asia (from Myanmar to Vietnam, and excluding Indonesia and the Philippines). The island realm of Southeast Asia was historically known as the East Indies (the West Indies being in the Caribbean) and the Spice Islands, though the latter term also referred to the Moluccas, a group of islands that are part of Indonesia.

Compared to South Asia and East Asia, Southeast Asia has a relatively low population density. Its estimated population in 2010 was about 593 million, which is half the total populations of India and of China. About 20 percent of the population of Southeast Asia lives on the Indonesian island of Java (136 million), which is among the most densely populated places in the world. Other parts of Southeast Asia (and much of the rest of Indonesia) actually have relatively low population densities.

Shatterbelt Diversity

Situated between South Asia (dominated by India), East Asia (dominated by China) and Oceania, Southeast Asia is a cultural shatterbelt where many different cultural influences have merged and/or influenced each other through history. (This is similar to the Balkan Peninsula and the Transcaucasia area between Europe and Asia.) Today, the dominant cultures in the region include Burmese (Myanmar), Thai, Vietnamese, Cambodian (Khmer), Malay, Chinese, Indian and European. Southeast Asia's diversity extends to many other aspects of the region as well. Large numbers of people are followers of Buddhism, Islam and Catholicism, all of which may be combined with elements of each other, and older animist traditions. Ruling governments adhere to political ideologies that include communism, capitalism, democracy, dictatorships, monarchies and military rule. Large areas were formerly under the influence or colonial control of imperial China, the Netherlands, Great Britain, Spain, France, Japan and the U.S. The diversity of economies in Southeast Asia today includes some of the wealthiest countries in the world (Singapore and Brunei) and some of the poorest (Timor Leste, Laos, Myanmar and Cambodia), as well as some of the most important emerging economies (Malaysia).

Physical Geography of Southeast Asia

The major physical geographical distinction in Southeast Asia is between peninsular Southeast Asia and insular Southeast Asia. Peninsular Southeast Asia consists of most of the countries directly attached to the Asian continent: Myanmar, Thailand, Laos, Vietnam and Cambodia. Insular Southeast Asia consists of the mostly island countries of Malaysia, Singapore, Brunei, Indonesia, Timor Leste (East Timor) and the Philippines.

Peninsular Southeast Asia

Peninsular Southeast Asia is very mountainous in its northern areas that border China, though these areas are also sparsely settled. The mountain ranges extend southward from southern China and the Tibetan Plateau and are rich in minerals, including tin, iron, bauxite (aluminum) and oil. The highest peak in Southeast Asia is Hkakabo Razi (19,240 ft, 5,881 m), which is actually on the edge of the Himalayas in northern Myanmar. Between these mostly north to south trending mountain ranges run the major rivers of Southeast Asia, which have their origin on the Tibetan Plateau and in Southwest China. Most of the population of peninsular Southeast Asia resides along these rivers, especially on the coast where they enter the sea. The major rivers include:

- Mekong River—touching all of the countries in peninsular Southeast Asia and entering the South China Sea near Ho Chi Minh City, Vietnam;
- Irrawaddy River—in Myanmar and entering the Andaman Sea near Yangon (Rangoon);
- Salween River—in Myanmar, touching on the border with Thailand and entering the Andaman Sea in eastern Myanmar;
- Chao Phraya River—in Thailand and entering the Gulf of Thailand near Bangkok (Krung Thep);
- Red River—in Vietnam and entering the Gulf of Tonkin near Hanoi.

The Mekong River is the most significant of these because it connects to all of the countries of peninsular Southeast Asia and the ethnically diverse province of Yunnan in Southwest China. Regional development programs have focused on the Mekong River Basin to encourage a more integrated transportation system among the countries of peninsular Southeast Asia (and China), and to promote international tourism.

The Golden Triangle

A popular tourist point along the Mekong River is where the countries of Myanmar, Thailand and Laos meet. Known as the *Golden Triangle*, this region was once infamous for growing poppies for heroin production. Today, however, former heroin traders run teashops and souvenir stands for tourists who flock to the Golden Triangle. There Myanmar has built a large casino and tour boats to take visitors over the border to Laos (without a visa) so they can send postcards with relatively rare Laotian stamps. Trekking trips to visit the minority hill tribes in the mountains of northern Thailand, out of the cities of Chiang Mai and Chiang Rai, is a long popular activity that has also spread to Laos and northern Vietnam since the mid-1990s.

Although the river systems of Southeast Asia are extremely important for trade and human settlement the low-lying river estuaries are also threatened by climate change, especially from storm events and tidal surges. There is therefore growing attention being given to the ways in which cities such as Bangkok (Thailand) and Ho Chi Minh City (Vietnam) can become more resilient to climate change. This may involve both changes to urban design as well as trying to ensure that natural protection against storms and floods, such as mangrove swamps, receive better protection.

Insular Southeast Asia

Insular Southeast Asia consists of a large arc of volcanic islands formed by the collision of the Eurasian continental plate with the Pacific Ocean floor (Pacific plate and Philippine plate) and the Indo-Australia continental plate. In the middle of this arc is the island of Borneo, which is politically divided between three countries: Malaysia, Indonesia and Brunei (Brunei Darussalam). Borneo is not geologically volcanic, but it is instead part of the Asian land mass. It is the second-largest island in the world (after Greenland), and its highest peak is Mount Kinabalu (4,095 m, 13,435 ft), which is a popular hiking trip out of the town of Kota Kinabalu in the state of Sabah, Malaysia.

Scuba Diving in the Coral Triangle

Sabah also has one of the world's leading scuba diving destinations on the island of Pulau Sipadan, which is the tip of an ancient volcano covered with coral, and a popular breeding ground for sea turtles. Sipadan was contested between Malaysia, Indonesia and the Philippines until 2002 when the International Court of Justice granted the island to Malaysia based on historical grounds. The Philippines also argues that it has historical claims to the entire state of Sabah, Malaysia.

Similar spectacular diving is found throughout insular Southeast Asia, especially in the *Coral Triangle* area that extends from Luzon (the largest island in the Philippines) to Borneo and over to New Guinea and the Solomon Islands. At the center of the Coral Triangle is Raja Ampat, a small island group in Indonesia at the western tip of New Guinea. The Raja Ampat islands are believed to have the highest recorded diversity of marine life on the planet. In 2006, international biologists exploring the waters in this area discovered 50 new species of fish, shrimp and coral in an area that contains more than 1300 fish species and some 600 species of coral, whales, sea turtles, crocodiles, giant clams, manta rays and dugongs (related to the manatee).

Volcanic Islands

While few of the internal islands of insular Southeast Asia are volcanic, almost all of the outer islands are, from Indonesia's Sumatra to Java, to Timor and New Guinea on the west and south of the region, to the Philippines in the northeast. All of these islands have extremely active seismic systems that produce rich agricultural lands and lush green landscapes, punctuated by steep volcanic peaks that occasionally wreak havoc on the surrounding countryside and affect air travel. In addition, the plate boundaries are also noted for their high levels of seismic activity meaning that the volcanic areas are also often subject to large earthquakes as well.

Climate and Ecology

Insular Southeast Asia sits directly astride the equator and receives year-round rain fall from the Intertropical Convergence Zone (ITCZ). As discussed in Chapter 3, the ITCZ is an area where air from the north and south collides and rises into the upper atmosphere, forming clouds and rain. This occurs throughout the year, though the ITCZ does follow the sun, moving northward in summer (June–August) and southward in winter (December–February). Along the equator, winds and ocean waters mostly move from the east toward the west.

To the east of Southeast Asia is the vast Pacific Ocean, which gets quite warm in the summer months and causes a lot of moisture to evaporate into the atmosphere. This rising air forms tropical storms that occasionally become typhoons (in Asia), cyclones (in Australia) and hurricanes (in North America, including Hawaii). In the Pacific northern hemisphere summer, the typhoons tend to hit the Philippines more than any other place, causing widespread flooding and landslides somewhere in the islands every year. From the Philippines, the typhoons tend to move northward toward China. (In the southern hemisphere summer, the Pacific cyclones mostly hit northern Australia and the islands of the South Pacific.) Peninsular Southeast Asia and the southernmost portions of insular Southeast Asia (the southern portions of Indonesia) experience a hot summer dry season because they are further away from the ITCZ and its cooling rains.

Southeast Asia Monsoons

Like South Asia, all of Southeast Asia has a *monsoonal climate* pattern. In the northern hemisphere winter, the Siberian High Pressure system circulates air over the Asian continent in a clockwise direction, bringing moisture to Southeast Asia from the South China Sea, which is located northeast of Southeast Asia. In the summer, the Siberian Low Pressure system circulates air over Asia in a counterclockwise direction, pulling moisture into Southeast Asia from the Indian Ocean. For peninsular Southeast Asia, the winter monsoon usually brings more moisture than the summer monsoon. The Korat Plateau in northeast Thailand is the driest area in peninsular Southeast Asia, as well as the poorest region in Thailand. Because it is in the interior of peninsular Southeast Asia, it is in the rain shadow of mountains that keep out both the summer and winter monsoon rains.

The Wallace Line

The tropical rainforests of Indonesia are home to the second greatest terrestrial biodiversity in the world, after Brazil's Amazon rainforest. Its flora and fauna are a mix of Asian and Australian species. The line between these species, known as the Wallace Line after British naturalist Alfred Wallace who first described it, roughly runs between the islands of Borneo and Sulawesi, and between Bali and Lombok in southern Indonesia.

The Wallace Line follows a deep-water trench between the Asian and Australian continental landmasses that many animals could not cross when sea levels were lower during the ice ages. The area to the west of this line is clearly part of the Asian continent, and much of it was exposed land in the past, allowing it to be populated by tigers, rhinoceros, orangutans, elephants and leopards. While once common, these large animals are rare today, especially with the annual practice of burning rainforests in the spring dry season to expand land for agriculture. These mostly illegal burnings reduce the habitat for large and small animals and cause widespread air pollution through large areas of Southeast Asia, threatening the survival of over 140 species of mammals, as identified by the World Conservation Union (lUCN).

West Papua and Timor Leste

East of the Wallace Line the flora and fauna are more closely related to those of Australasia. The island of New Guinea was part of the Australian continent several times during the ice ages. This means that New Guinea has species such as tree kangaroos and eucalypt trees. This southeast corner of Southeast Asia is less populated and less visited than most of the rest of the region. War and civil unrest in Timor Leste and on the Indonesian side of New Guinea (formerly known as Irian Jaya, but changed to West Papua in 2006) have largely kept outsiders to a minimum in this part of the world. In 2005, an international group of scientists visiting the Foja Mountains (also known as the Pegunungan Van Rees) on the north coast of West Papua discovered a virgin tropical rainforest that is one of the few places on the planet not directly impacted by humans. Although the situation has improved in recent years, a separatist movement continues to struggle against Indonesian control of West Papua and the international timber and mining companies that operate there. Travel to this emergent ecotourism destination, however, is slowly growing, making it one of the more challenging, and potentially rewarding, destinations in Southeast Asia.

Orangutan

One of the most critically endangered mammals is the orangutan (for *orang hutan*, meaning old man of the forest). Orangutans once roamed throughout Southeast Asia but today are only found in the rainforests of Borneo and Sumatra—and in the homes of people throughout Asia who adopt infant orangutans as pets. They are known for their reddish-brown hair and incredible high level of intelligence, and they are becoming increasingly rare in the wild. While cute and cuddly as infants, when they become adults they can easily toss a human across a room. Several centers have been established on Borneo and Sumatra to rescue orangutans displaced by fires (mostly started to clear forests for palm oil plantations) or abandoned as pets, and to reintroduce them into the rainforests. These centers have also become popular ecotourism destinations. These support their operations, although their biodiversity goals usually take precedence. However, in the longer term the security of the orangutan will only be gained by limits to *deforestation*.

Human Geography of Southeast Asia

Negritos

Humans have lived in Southeast Asia almost as long as they have resided in Africa. The Negrito people are recognized as having lived the longest in Southeast Asia among the region's current population. Negrito groups include the Orang Asli ("original/aboriginal people") in peninsular Malaysia; the Aeta, Ati and at least 25 other tribes in the Philippines; the Mani of Thailand; and twelve Andamanese tribes of the Andaman Islands in the Indian Ocean. They may have migrated to Southeast Asia from Africa as early 70,000 yrs ago, though they are not closely related to any modern-day African ethnic group. They have one of the purest DNA of any contemporary genetic group, indicating a very long period of isolation. They are short in stature (about 4 to 5 ft) and black skinned. Many still practice hunting, gathering, and slash-and-burn agriculture in the mountains of Southeast Asia.

The Negrito populations comprise a very small percentage of the population of Southeast Asia today, though they were once the dominant group. They were widespread during the European colonial period, but the remaining 50,000 or so today are mostly found in the interior mountains of Borneo and the Philippines, and on the Andaman Islands (part of India). Negritos are generally considered members of the Australoid race, which includes the Aborigines of Australia, the Papuans on the island of New Guinea, the Melanesians in the South Pacific, the Veddahs of Sri Lanka, and possibly the Ainus of northern Japan.

◀ *Rice paddy fields on a hillside in Bali, Indonesia. The year-round tropical rains in insular Southeast Asia allow intensive and diverse agricultural activities even on steep slopes in Indonesia, Malaysia and the Philippines.*

Photo: Alan A. Lew

Malays

Insular Southeast Asia is dominated by the Malay race. The Proto-Malay (early Malay) probably originated in Southwest China (Yunnan Province) and migrated into Southeast Asia about 5000 BCE. There are only a few pockets of Proto-Malay groups remaining in Southeast Asia, mostly in southern Vietnam and southern Thailand and on some of the islands of Indonesia. The rest of the Malay population (including most of the people of the Philippines, Indonesia and Malaysia) are a mix of Proto-Malay with other migrants from South Asia, China and the Tibetan highlands. The Malay language is the national language of Malaysia (Bahasa Malaysia) and Indonesia (Bahasa Indonesia).

Early Malay society was matriarchal. Women were responsible for household chores and agriculture, and they held the rights to property. This gave women considerable economic influence, and in traditional Malay culture the man moved to the woman's household after marriage. Even today, Malay wedding rituals are based on a ceremony in which the man is welcomed into the woman's household. Most of the Malay race today practices Islam, although there are some major exceptions, including Filipinos and East Timorese, who adopted Catholicism under Spanish and Portuguese rule, respectively, and the Balinese who still practice a form of Hinduism that was common in Southeast Asia prior to the arrival of Islam.

The Khmer

The Mon-Khmer people migrated into peninsular Southeast Asia about 3,500 yrs ago, after the Malays. It is possible that Tibetans pushed them into the region. They became a dominant civilization in peninsular Southeast Asia, displacing the early Negrito and Proto-Malay populations. Today, the Mon are situated in Myanmar on the Thailand border, while the Khmers comprise about 90 percent of the population of Cambodia. One of the peak periods of Khmer political power was during the Angkorean Empire in the ninth to 13th century CE. At this time, the Khmer controlled most of present-day Cambodia, Laos and Thailand, and they built the capital city of Angkor and its famous Angkor Wat (Angkor Temple).

With over one million people at its peak, Angkor was once the largest pre-industrial city in the world. The rise of Tai kingdoms in the 13th and 14th centuries, along with the over-exploitation of Cambodia's natural resources in support of the city, eventually resulted in the almost total destruction of Angkor in 1431 by Tai armies, and the enslavement of some 90,000 of the Khmer elites. Today, large populations of Khmer-speaking people reside in areas of Thailand and Vietnam adjacent to Cambodia.

Hill Tribes

The mountains of peninsular Southeast Asia also contain many smaller ethnic groups, most of whom reside in more than one country. These groups are generally referred to as Hill Tribes, and visits and stays in their villages are a major part of the trekking economy in their mountain homelands. Some of these (and their approximate populations) include;

- Akha (120,000 people) in Myanmar, Laos, Thailand and Vietnam
- Cham (20,000) in Cambodia (where they are Muslims) and Vietnam (where they are Hindus)
- Kachin (or Jingpo) (1,500,000) in Myanmar and China
- Shan (6 million) and Hmong or Meo (1.6 million) in Myanmar, Laos, Thailand, Vietnam and China
- Jarai (220,000) in southern Vietnam
- Karen (2 million) in Myanmar and Thailand
- Lisu (500,000) in China, Myanmar and Tibet
- Montagnards (800,000) in Vietnam's highlands
- Wa (400,000) in Myanmar and China
- Yao (1 million) in China, Vietnam and Laos

Indochina

Over 2000 years ago, during the Han Dynasty in China, the Empire of China made a major expansion out of North China and into South China. That move initiated a major migration of non-Chinese people of modern day South China into Southeast Asia. The Vietnamese, also known as Annamese, migrated from Southeast China down the east coast of Southeast Asia. Vietnam has historically had the closest cultural ties to China of any country in Southeast Asia. It was administratively part of China until the tenth century CE, and the Vietnamese practice the same form of Mahayana Buddhism mixed with Confucianism that is common in China. They also practice a very labor-intensive form of rice cultivation similar to that of southern China. Because of these similarities, older geography textbooks often included Vietnam in East Asia instead of in Southeast Asia.

Vietnam, Cambodia and Laos comprise the former French colony of Indochina, although Indo-China was also sometimes used to describe all of Peninsular Southeast Asia as well. France controlled this region from the 1880s, taking some of it in a short war with Thailand in 1893. Japan took control of Indochina in World War II, and France tried to take it back after the Japanese surrendered in 1945. However, the communist Viet Minh, under the leadership of Ho Chi Minh (whom the United States supported against the Japanese), fought to keep the French out in the first Vietnam War (1945-1954). After that war, Vietnam was divided into North Vietnam (that was communist ruled and supported by China) and South Vietnam (that was capitalist and supported by the United States). With China's support, North Vietnam sought to reunite the country by force. This resulted in the Second Vietnam War, which lasted from 1959 until the North Vietnamese victory in 1975. The Communist Party of Vietnam still rules the country, though today it is widely open to tourism and international investment.

Early Indianization of Southeast Asia

About 2000 years ago, traders from China to the north and India to the west entered Southeast Asia. For most of Southeast Asia, it was the Indian influence that was most significant. The Indians referred to Southeast Asia as Suvardvira (or Suvard Bumi), the Golden Land. Most of the Indian traders were Sudras, from the bottom of the caste system, and were generally uneducated adventurers. Around the first century CE, the higher Kshatriya caste entered Southeast Asia as merchants. This was the second-highest caste group in India, and they were primarily looking for goods to trade with the Roman Empire. They were a more educated caste and they introduced Indian political structures into Southeast Asia.

At about the same time, the village chiefs in Southeast Asia were gradually expanding their areas of rule to larger territories. Some of these local rulers traveled to India where they were impressed by the large Indian armies and cities and the priestly elites. The Kshatriya merchants brought Brahmin priests to Southeast Asia to perform ceremonies that helped legitimize the emerging local rulers. Buddhist monks also performed these types of ceremonies in Southeast Asia. As late as the 1930s, Hindu priests from India were still given the role of inaugurating the kings of Southeast Asia.

God Kings

The Brahmin priests and Buddhist monks from India brought new ideas to Southeast Asia, including the idea of a god king. This view promotes the king as a reincarnation of the Buddha or of a Hindu god. This was the origin of Cambodia's Angkor Wat, which was originally built for the Hindu god Vishnu. The Indian priests and monks also brought new architecture, a bureaucratic and court system, and a military framework. Mahayana Buddhism was also flourishing in India, especially in the fourth to sixth centuries CE. Buddhist monks brought it to Southeast Asia. By 800 CE, Hindu and Buddhist-based kingdoms were common throughout Southeast Asia. This is when construction is believed to have started in Java,

Indonesia, on the Mahayana Buddhist monument of Borobudur, the largest Buddhist structure in the world. Southeast Asia's first major empires emerged at this time on the island of Java, on the island of Sumatra, and in Cambodia, Thailand and Myanmar. The largest of these was the Majapahit Empire centered on Java in the 13th century.

Hinduism and Buddhism

Hinduism and Mahayana Buddhism in Southeast Asia were largely restricted to the ruling elites and, as such, are often referred to as court cuts. Their ceremonies were used to grant legitimacy to the rulers and conquerors. Their religions did not filter down to the common peasants, who continued to practice animist traditions. It was not until Theravada Buddhism was introduced to peninsular Southeast Asia, and Islam and Catholicism to insular Southeast Asia that the common people moved somewhat away from their traditional belief systems. In the twelfth century CE, refugees from internal wars in Myanmar fled to Sri Lanka where they learned Theravada Buddhist teachings, which they then brought back to Myanmar. From there, Theravada Buddhism spread rather quickly throughout peninsular Southeast Asia. Theravada Buddhism is closer to the original teachings of the Buddha, which focuses more on self-development, and eschews praying to saints and other beings. In practice, however, it is mixed with an older Hinduism, and with Mahayana Buddhism and animist beliefs.

Borobudur

Borobudur, as mentioned above, is the largest Buddhist monument ever built. It is believed to have been constructed from the mid-eighth century CE and completed about 825 CE. It is a Mahayana Buddhism structure, which is the form of Buddhism practiced today in Tibet and East Asia, and located in Central Java, Indonesia. The monument comprises six square levels topped by three circular levels adorned with 2,672 relief panels and 504 Buddha statues. Pilgrims follow a path from the base to the top of the stupa that takes them through the three levels of Buddhist cosmology: Kamadhatu (the world of desire), Rupadhatu (the world of forms), and Arupadhatu (the world of formlessness). Borobudur was abandoned in the 14th century after Islam became the dominant religion in insular Southeast Asia. Today the restored Borobudur is the single most visited site in Indonesia and is a pilgrimage destination for Mahayana Buddhists from Japan, China and Korea.

Angkor Wat

Construction of Angkor Wat first began in the 1100s as a Hindu temple at the center of the Khmer capital city. The original temple, however, was largely destroyed by the Cham Kingdom in Vietnam in 1177. It was later rebuilt and changed to a Theravada Buddhist temple around 1400, which it continues to be today. The temple includes the temple mountain (the highest point, representing Mount Meru, the home of the Hindu gods) surrounded by temple galleries with bas-relief carvings and a moat. It is the epitome of classical Khmer architecture style. It has also become a symbol of modern-day Cambodia, appearing on its national flag, and it is one of the leading tourist attractions in Southeast Asia. Many other temples and areas of former human settlement are scattered in the Angkor National Park area, each built by a different Khmer ruler. Indeed, the area of the Angkor complex is now realized to be so large that appropriate conservation and protection measures are becoming increasingly problematic.

Angkor Wat underwent major renovation in the 20th century, mostly by foreign aid organizations. Renovations were interrupted while the communist Khmer Rouge controlled Cambodia in the 1970s and 1980s, but they began again in the 1990s. A rapid growth in tourism to Cambodia in recent years has seen a correspondingly large increase in visitors to Angkor Wat. The temple complex has been modified

▶ *The main entrance to the Angkor Wat Temple compound in Cambodia. Angkor Wat, which is surrounded by a massive moat, is the largest of a collection of adjacent temples built by different Khmer rulers. The entire area is today part of the Angkor Wat National Park, and it is the single most visited attraction in Cambodia.*

Photo: Alan A. Lew

only slightly to accommodate the increase in visitors, with ropes and wooden steps used to protect the bas-reliefs and floors.

Islam and Christianity

Muslims were in power in South Asia in the 14th century, and large numbers of Arab traders were visiting Southeast Asia. During this time Marco Polo visited China and returned to Europe through Southeast Asia. The Arabs introduced Islam to the trading ports of insular Southeast Asia, where many of them had settled. Local leaders gradually adopted Islam, mostly to strengthen their trading ties with the Muslim merchants. Through this process, Islam became the dominant religion in insular Southeast Asia by the 16th century.

The result of this history is that peninsular Southeast Asia has a mix of Hindu and Buddhist cultural practices, while insular Southeast Asia has a mix of Hindu and Muslim traditions. The major exceptions to this include Vietnam, which has a strong Chinese influence, including Mahayana Buddhism (originating from Tibet), Confucianism and Taoism; and the Philippines and Timor Leste, which are mostly Roman Catholic due to European colonial influences. Some of the minority hill tribe populations, especially in Myanmar, have also adopted Christianity through the work of missionaries. While most of the country of the Philippines is Catholic, its southern island of Mindanao has a large Muslim minority that has been fighting for greater autonomy and independence since Spanish colonial times. Although dangerous to visit in the past, peace talks in recent years have opened Mindanao to tourism, with the old sea gypsy and pirate port of Zamboanga being the island's top tourist destination. Piracy is still practiced on an occasional basis in the Sulu Sea near Zamboanga.

Early European Colonies

The Portuguese were the first of the European powers to establish a colony in Southeast Asia. They conquered the port city of Melaka (also spelled Malacca) in modern-day West Malaysia in 1511, which was a year after they had established their colony of Goa in India. In 1556, they established control of the eastern half of the island of Timor. By controlling Melaka, the Portuguese controlled the Strait of Malacca, the most heavily used trading route between China and India. They could easily stop each ship traveling through the narrow passage between modern day Malaysia and the island of Sumatra, and force it to pay a tax to continue.

Spain in the Philippines

Magellan arrived in the Philippines in 1521, during his voyage around the world. Starting in the 1560s, Spain sought to conquer the sparsely settled and remote islands of the Philippines. In 1571, they captured Manila, the largest town in the Philippines (which at that time had a population of about 15,000). Under Spanish rule, large numbers of people, especially from China, migrated to the Philippines, and by the late 17th century over half of the population of the Philippines was Chinese. The Spanish language and culture formed the basis of the Filipino elite. Following the Spanish American War in 1898, the Philippines became a colony of the United States, which lasted until 1946—much longer than Filipinos had expected. During that time, much of the Spanish influence was eroded and replaced by American cultural influences, although most place names and family surnames still reflect the Spanish period.

The Netherlands in Indonesia and Melaka

In 1596, Dutch ships entered Indonesia for the first time after they had been banned from Portuguese ports. In 1610, they signed a treaty with the King of Jakarta allowing them to sail to the Spice Islands (the Moluccas). The most valuable product from insular Southeast Asia for Europeans was the spices, especially black pepper, which they used to preserve and enhance meat. From their initial base in Java, the Dutch were soon able to dominate almost the entire territory that is today the country of Indonesia. In 1641, the Dutch completely levelled the Portuguese colony of Melaka, which then became a Dutch colony. They were helped by local Malay sultans who had come to hate the ruthlessness of the Portuguese.

The British in Singapore and Malaysia

The British were latecomers to European colonization in Southeast Asia. In 1685, they established themselves in Bencoolen, on the south side of the island of Sumatra. In 1796, they signed a treaty with local Malay rulers to establish a colony on the island of Penang, located at the north end of the Strait of Malacca. This was an effort to break the Dutch monopoly over trade in Southeast Asia. In 1819, they signed a treaty that allowed them to establish the city of Singapore, which would eventually become the dominant economic center for all of Southeast Asia. The British were much more diplomatic than the Portuguese, and even the Dutch.

To establish Singapore, they utilized a conflict between two brothers over which of them would become the sultan of the territory where the island of Singapore was located. Sir Stamford Raffles, who spoke fluent Malay and was a scholar of the history of the region, signed the treaty for Singapore with the older brother who had fled into hiding on the island of Sumatra. The younger brother could not stop the power of the British Navy, which supported British claims that it had signed a treaty with the rightful ruler of the territory. (The descendants of the ruler continued to reside in the Sultan's compound in Singapore until the early 2000s when the site was renovated into the Malay Heritage Centre.) The British came to control Myanmar as a further extension of their colonization of India, and eventually they expanded their control from Singapore to all of present-day Malaysia.

Peranakan Chinese Culture

The former British colonies of Penang, Melaka and Singapore together are known as the Straits Settlements. All three of them have predominantly Chinese populations. Some of these Chinese are descendants of early Chinese merchants who married Malay women and settled in the Strait of Malacca 700 years before the arrival of the Europeans. In Malay, they are known as Peranakan Chinese, and they exhibit a distinct culture that is a fusion of Chinese and Malay. Peranakan culture is seen in the architecture, cuisine and traditional clothing of the cities of the three Straits Settlements. The old

▶ *Four large Christian churches, each representing a different denomination, can be seen in this remote village on the Lembeh Strait near Manado on the island of Sulawesi, Indonesia. The influence of European colonialism in Southeast Asia is visible in areas where Christian missionaries successfully converted local populations. Manado is a world-renowned scuba diving destination, located near the center of the biodiversity rich Coral Triangle region of Southeast Asia.*

Photo: Alan A. Lew

central town of Melaka probably demonstrates Peranakan culture best, as well as the ruins of the Portuguese era and administrative buildings dating back to the Dutch period. The Malaysian government is increasingly promoting the colonial heritage as a tourist attraction.

Countries and Tourism of Southeast Asia

The political boundaries of the countries of Southeast Asia today directly reflect the European colonial boundaries of the 19th century. Vietnam, Laos and Cambodia (Indochina) were French colonies; the Philippines was a Spanish and U.S. colony; Indonesia was a Dutch colony; Timor Leste (East Timor) was a Portuguese colony; and Myanmar, Malaysia, Singapore and Brunei were British colonies. Thailand was the only country of Southeast Asia that was never colonized.

Thailand and the Tai

The Tai people migrated from Southwest China and formed the basis of a language group that today is spoken in Thailand (Thai), Laos (Laotian), Myanmar (Shan minority group) and the southern China provinces of Guangxi (Zhuang minority group) and Yunnan (Dai minority group). Outside of Thailand, some 35 million people speak versions of the Tai language and its dialects. This does not mean that all of the different Tai speakers can speak to one another. Laotian, for example, is part of the Tai language group; however, Laotian itself has four major dialects, each of which is further subdivided, and most of which are mutually unintelligible.

Thailand's population is almost 64 million (2010 estimate), with about 75 percent Thai, 15 percent Chinese and 3 percent Malay. The Chinese are mostly in the larger urban areas. Thailand is a constitutional monarchy and democracy, and the King commands great respect in the country. In fact, speaking poorly about the Thai king can result in a prison sentence.

Siam

In the 19th century, Siam (the former name for Thailand) was able to play off rivalries between the British and French colonizers of Southeast Asia, which enabled it to be the only territory in Southeast Asia that was never made a colony. Concessions to the British, however, resulted in three mostly Malay provinces in the south becoming part of British Malaya. However, one other Malay-dominated province remained in Thailand, and Muslim separatists in that province (which borders Malaysia) have caused civil unrest and terrorist attacks on police and Buddhist Thais in recent years.

Buddhist Kingdoms

The Thai are the predominant people in the country of Thailand (Thai here means "freedom" in the Thai language). The Buddhist kingdom of Sukhothai (1238–1368 CE) is considered the first of the Siamese kingdoms. The second Siamese kingdom of Ayutthaya (1350–1767 CE) eventually conquered it. Following the destruction of Angkor in 1431, most of the Hindu Khmer courtiers were brought to Ayutthaya, and the Siamese court adopted many Khmer customs and rituals. Ayutthaya was sacked by Buddhist Burmese in 1767, and in 1782 the capital of Siam was moved further south to Bangkok. The ruins of Ayutthaya are today a major cultural heritage site and tourist attraction in Thailand, although massive monsoonal floods in 2011 did significant damage to the ancient Thai capital, affecting not only the built heritage there but tourists' visits to the place.

Bangkok

Thailand today is one of the most visited countries in Asia, and its capital, Bangkok, is a major hub for air travel into and out of Southeast Asia. Tourism to Thailand rose rapidly in the 1970s after the Vietnam War, when Thailand offered a popular break for United States servicemen. Much of Thailand's tourist reputation has been influenced by the sex industry, which grew to serve visiting soldiers. That still exists today, though it is a much smaller part of the formal tourism economy, which now includes temples, other cultural sites and shopping in the Bangkok area. The market for sex tourism in Thailand primarily comes from male visitors from Australia, China, Japan, North America and Europe who travel, sometimes exclusively, for sexual encounters. This form of tourism has received a great deal of negative press, and the Tourism Authority of Thailand has made efforts to downplay its pervasiveness and replace it with new forms of tourism.

Thailand received 26.5 million visitors in 2013. Of these, about 65 percent of the visitors to Thailand come from the Asia-Pacific region, with Malaysia, Japan and China being the main sources. Western tourists to Thailand also favor the country's beach resorts on the Gulf of Thailand and on many of Thailand's scenic islands (especially Phuket and Koh Samui). Asian tourists tend to focus on Bangkok and

◀ *The Damnoensaduak floating market, west of Bangkok, Thailand. At one time, canals bisected much of Bangkok due to its low elevation. Most, however, have been covered to build roadways for urban growth. This market is the most visited by international tourists of the remaining floating markets in Thailand.*

Photo: Alan A. Lew

such historic sites as Ayutthaya. *Medical tourism* is a growing segment of Thailand's economy (and also popular in Singapore, Malaysia and India). Medical tourists tend either to come from highly developed countries (where medical care is very expensive), or from developing countries where the more wealthy have difficulty finding high-quality medical care.

However, both medical tourism and leisure tourism have been affected by the country's periodic political instability. For example, since 2010 rural political parties have been winning in national elections, with the results have been rejected by more urban political parties. Political protests have affected airport access at times, while street protests in Bangkok create a negative media image, even if tourists remain well looked after. In May 2014 a *military junta* took power with the apparent support of the Thai monarchy. Nevertheless, significant divides between rural and urban Thais remain with long-term political stability continuing to be a problem in the absence of armed force and royal encouragement of a full democratic tradition.

Vietnam

Vietnam, with over 90 million people (2010 estimate), has the largest population in peninsular Southeast Asia. It is 87 percent Vietnamese, with the remainder being ethnic Chinese and various hill tribe populations. In Vietnam, tourism is expected to generate 5.6 million direct and indirect jobs by 2020, which is about 10 percent of the total jobs in the country. One of Vietnam's most popular tourist attractions is the World Heritage Site of Ha Long Bay, located on the coast near the border between China and Vietnam. It consists of a large cluster of over 1,900 limestone islands that rise steeply from the sea. Many of the islands have limestone caves with stalactites and stalagmites, and some have floating villages attached to them.

French and American Impacts

Both the French and American legacies provide foundations for tourism to the country today. The French supported the establishment of hill resorts in the early 20th century, which were connected to major cities by train. Many of these, such as Sapa, remain tourism destination to the present day. Hanoi, the capital of colonial Indochina, is the most French city in Asia and renowned for its European-influenced cuisine and architecture. American soldiers who fought in Vietnam have been taking tours to the country since the late 1990s to revisit sites that they knew in the war and to reacquaint themselves with the people of Vietnam under more favorable circumstances.

Like many Southeast Asian countries, Vietnam has embraced tourism as a means of local economic development. The International Labor Organization (ILO) has reported that in some provinces in Vietnam, communities have increased their earnings by developing products such as traditional weaving and handicrafts, spices, tea gift baskets and homestay services for tourists. New tourism developments include the development of cruise ship travel on the Mekong River between Vietnam and Cambodia. In 2014, luxury travel company Aqua Expeditions commenced cruising on the Saigon offering a product similar to its Amazon cruises. Such products further reinforce Vietnam's positioning as an international ecotourism destination.

Cambodia

Cambodia is one of the smallest and poorest countries of Southeast Asia. At the end of the Vietnam War (1975), the Khmer Rouge (Red Khmer) communist government came to power and adopted policies that resulted in the death of 1.7 million of its population of 8 million at that time. This event is popularly known as the Killing Fields after the movie made about it in 1984. In 1978, Vietnam's army invaded Kampuchea (the country's name at the time) and removed the Khmer Rouge from power.

Cambodia today is a constitutional monarchy, with a strong one-party rule. Tourism has begun to flourish there, with Angkor Wat being the country's premier attraction. The "*dark tourism*" attraction of the Killing Fields also attracts curious visitors, and the country has had the unfortunate designation of being one of the most popular child prostitution destinations in Asia. In 2013 the tourism sector in Cambodia generated more than US$2.5 billion in national revenue last year and created more than 500,000 direct jobs and many other indirect jobs.

Tonle Sap Lake

Most of Cambodia is very flat, situated at the southern end of the Annamite Mountains. The Tonle Sap is a unique geographic feature located in this flatness near the center of Cambodia. It is a "river lake" that changes dramatically in size between the dry and rainy seasons. During the summer monsoon rainy season, the swollen Mekong River backs up into the Tonle Sap, causing it to expand from 1,000 sq miles (2,700 sq km) to 6,200 sq miles (16,000 sq km), flooding adjacent fields and forests. The currents in the lake also change direction in the rainy season, and the lake is possibly the most productive inland fishery in the world. Tourists can visit some of the many floating villages, created by refugees from the Vietnam War who were not able to settle on Cambodian soil.

Laos

Laos is the only landlocked country in Southeast Asia and, like Cambodia, it is also among the poorest. It is located between the Mekong River, separating it from Thailand to the west and the Annamite Mountain Range, which separates it from Vietnam to the east. About half of its population is ethnic Lao, who live in the lower valleys, while the other half is a mix of hill tribes that live in isolated and remote mountainous areas. These hill tribes make Laos one of the most ethnically diverse countries in Asia, with about 100 different language dialects spoken. The villages are also popular destinations for international trekking tourists.

Vientiane is the capital of Laos, although Luang Prabang is the more popular tourist destination. Luang Prabang was the pre-communist capital of the kingdom of Laos and has been designated a World Heritage Site by UNESCO for its many temples. Due in part to the low level of economic development, both Cambodia and Laos are perceived as among the most corrupt countries in Southeast Asia. Only Myanmar, which is ranked the second worst for corruption in the world (after Somalia), ranks lower on Transparency International's listings.

Myanmar

Myanmar (also known as Burma) has a population of 55.7 million (2014 estimate), about 68 percent of whom are Burmese (Bamar), with the remainder being various hill tribe groups (Shan and Karen together comprise 16 percent), as well as small populations of Chinese and South Asians. The South Asians, mostly from India, reside in the lowland cities, such as the former capital of Yangon (Rangoon), where most of them are merchants. The Chinese are found in cities along the border with China. Myanmar is also among the poorest countries in the world and has long been ruled by a military junta, which took power during a coup in 1988 and also changed the country's name from Burma to Myanmar. In protest over the dictatorial and oppressive policies of the military regime, some countries (including the United States, the United Kingdom and Canada) have officially rejected the name change, though that is slowly changing with recent political reforms.

Elections in 2011 began the creation of a parliamentary political system. Many improvements in the political life of Myanmar have occurred in recent years. In 2012, for example, many political prisoners were released, peace treaties were negotiated with most of the warring ethnic factions in the country, and the way has been paved for a multi-party electoral system in the near future.

Isolation and Boycotts

Politically and socially, Myanmar has been the most isolated and least westernized of all the countries in Southeast Asia. Calls for tourism boycotts of the country have been common in the past because money spent there supports the military, which most observers and foreign governments deem brutal in its use of villagers as forced labor and lacking in human rights. At the same time, the untouched nature of the country makes it a very special and memorable place to visit for those who have traveled there, and the country's cultures and built heritage are among the most spectacular in the region. Since the improvement in the domestic political situation tourism has become a focal point of government economic policy as it provides foreign exchange and employment. However, the tourism infrastructure remains well behind that of neighboring countries in the region such as Thailand.

Malaysia

Malaysia's 30 million people are also ethnically diverse, comprised of 50.4 percent Malay, 23.7 percent Chinese, 7.1 percent from South Asia, 11 percent from indigenous tribes and other minority groups, and 7 percent non-citizens. Large numbers of ethnic Chinese and South Asians are in Malaysia because of its importance along the trade route between India and China. They were also brought to Malaysia by the British to work on rubber and palm oil plantations. Malaysia is the world's largest producer of natural rubber and palm oil, as well as tin.

Malaysia is divided into West Malaysia, which is predominantly Malay, and East Malaysia, where non-Malays are the majority population. The two parts are separated by the South China Sea. West Malaysia is attached to peninsular Southeast Asia, whereas East Malaysia is comprised of two states, Sarawak and Sabah, on the island of Borneo. Malaysia gained its independence from the United Kingdom in 1963,

▶ *Roadside fruit stand on the slopes of Mount Kinabalu, in Malaysia's Sabah Province on the island of Borneo. Among the fruits are rambutans (the hairy, red golf ball-sized fruits on the right), mangosteens (the purple balls with green tops on the bottom left), known as the Queen of Fruits in Southeast Asia, and durians (the large green and spiky fruit below the hanging bananas). The durian is the King of Fruits and the most difficult for visitors to appreciate due to its pungent smell. Some of the fruits at this stand can only be found on the island of Borneo.*

Photo: Alan A. Lew

bringing together the British colonies on the Malay Peninsula (Malaya and Singapore) and the two British colonies on Borneo. Ethnic tensions between Malays and Chinese were a major source of friction within Malaysia in the early years, and they resulted in the island of Singapore becoming a separate country from Malaysia in 1965.

Malaysian tourism is based largely on urban heritage (e.g., Melaka and Penang), beach resorts along the country's coastlines, diving and water activities, and rainforest-based ecotourism mostly on Borneo. There is a steady flow of cross-border travel between Malaysia and its southern neighbor, Singapore, particularly for shopping by Singaporeans in the border town of Johor Bahru.

Singapore

Singapore is an island *city-state* at the southern tip of the Malay Peninsula. It is 237 sq miles (697 sq km) in size, with 5.6 million people. Ethnic Chinese are predominant at 77 percent of the population, although they come from several different areas of China and traditionally spoke a variety of Chinese dialects. Malays comprise 15 percent of Singapore's population, and Indians make up 6.5 percent. Singapore is one of the wealthiest countries in the world and enjoys a very high standard of living. The country consistently ranks as having one of the least corrupt governments in the world, although it is often criticized as having a lack of press and political freedom.

Singapore is a very important gateway and air hub for all of Southeast Asia as well as being a significant stopover between Europe and Oceania (Australia and New Zealand in particular). Shopping tends to dominate tourist activities in the downtown area, and it is the major attraction for Asian visitors. Casino tourism and major events, such as the hosting of a night time Formula 1 grand prix, have also become a significant part of Singapore's tourism product in recent years.

Because Singapore is a small country with a large and urbanized population, Singaporeans often seek opportunities to travel abroad. Given its small size and close proximity to neighbors, this is quite easy to achieve, with frequent day and weekend trips across the borders into Malaysia and Indonesia. Quick ferry services to Indonesia's nearby Bintan and Batam islands allow easy access for Singaporeans to golf courses and beach resorts as well as second home developments.

Although some tensions continue to underlie politics and race relations in Malaysia and Singapore, the different ethnic groups live harmoniously together and provide these countries with a colorful display of festivals, food and ethnic shopping districts, such as Chinatowns and Little Indias. Most of the street food hawkers are now in government-regulated food centers where they serve Malay, Chinese and Indian specialties, as well as fusion cuisine unique to the Malay Archipelago.

Brunei

Brunei Darussalam (Brunei) is one of three countries sharing the island of Borneo. It is an oil-exporting Islamic sultanate that intentionally did not achieve independence from the United Kingdom until 1984. When Malaysia gained its independence from Britain in 1963, Brunei decided to remain a protectorate under British control. Brunei is about the size of the U.S. state of Delaware, making it one of the smallest of the world's countries, though it is also one of the largest oil producers east of the Middle East. Because of its vast oil deposits, Brunei's ruling sultan is one of the wealthiest men in the world. Although tourism is not a high priority for Brunei, there is some, particularly sport tourism and business-related travel.

Indonesia

Indonesia, with a population of 253.6 million (2014 estimate), is demographically the fourth largest country in the world, after China, India and the United States. It has more than 18,000 islands (6,000 of which

▶ *The Petronas Towers in central Kuala Lumpur. While they are no longer the world's tallest building, the Petronas Towers (named for Malaysia's state oil company) remain the world's tallest twin towers and possibly the best-known icon of Malaysia. Combined with the classic, colonial-era hotel in the lower right, Kuala Lumpur (known to many simply as KL) represents the vibrancy and dynamism that the Southeast Asian economy has experienced since the end of the Vietnam War in 1975.*

Photo: Alan A. Lew

are inhabited), more than 300 distinct ethnic groups and more than 250 different languages. As a lingua franca that could be under stood by inhabitants of the Indonesian archipelago, the Indonesian language (Bahasa Indonesia, which is closely related to Bahasa Malaysia) was invented in the 1920s and became the country's official language at independence after World War II. The Javanese people (on central and eastern Java) comprise 42 percent of the total population, followed by the Sundanese at 13 percent (on western Java) and ethnic Chinese at 3 to 4 percent.

World's Largest Muslim Country

Islam is the religion of 90 percent of the population in Indonesia, making it the world's largest Muslim-populated country. However, the vast majority mixes Islamic faith with pre-Islamic animist and Hindu influences. Indonesia's economy is based primarily on tropical agriculture, including coffee, palm oil, tea, tobacco and tapioca. Indonesia claimed independence from Dutch colonial rule in 1945, although the Dutch did not relinquish full control until 1949. Over 136 million people live on the island of Java, where the country's capital of Jakarta is located. The Hindu island of Bali is also part of Indonesia and is a major beach resort destination, especially for Southeast Asians and Australians.

Ecotourism

Despite the impacts on Indonesia's rainforests of logging, much of which is illegal, and palm oil production, trekking and touring the jungles and islands of Indonesia is very popular, making it one of the leading ecotourism destinations in Asia. In the 1990s, most of the tourists who traveled to Indonesia for trekking, bird and animal watching, scuba diving and visiting the diverse ethnic communities scattered throughout the islands were from the more developed countries of the world. This has changed in recent years, as Asians, and especially Indonesians, have become wealthier and more interested in the natural wonders the country has to offer. The rise of intra-regional travel within Southeast Asia has also been prompted by the growth of low-cost airline carriers, making travel to nearby destinations more attractive than long-haul trips to Europe or North America.

As noted earlier in the chapter, Borobudur is one of Indonesia's most notable attractions. It is located near the city of Yogyakarta on Java, which also contains the spectacular Prambanan Hindu temple complex (another UNESCO World Heritage Site). Gamelan music, religiously-based dances, living folkways and cultural performances provide most of the backdrop for Indonesia's cultural tourism.

Islands

While Indonesia's tourism has historically been centered on Java and Bali, many other islands have begun to appear on the tourist circuit. Lombok, just east of Bali, is growing as a beach resort destination. The Komodo Islands are home to the famous and endemic Komodo Dragons (the world's largest land lizard). Sulawesi, Sumatra, Kalimantan (on Borneo) and many of the other outlying island areas have grown into important tourist destinations, visited for their cultural traditions, village life and natural environments. Orangutan, as noted earlier, and the highly endangered Sumatra rhinoceros and tigers are examples of famous fauna species found in Indonesia that add to its ecotourist appeal.

Tourism Challenges

Several events have influenced the growth of tourism in Indonesia in recent years, including the December 2004 tsunami, which affected a large part of Sumatra and islands in the Indian Ocean. Indonesia's Aceh Province was hit particularly hard, with hundreds of villagers losing their lives, and tourism to Indonesia nearly collapsed in the aftermath of the event. Terrorist bombings in 2002 and 2005 in the popular tourist destination of Bali killed many Indonesians and foreign tourists, and especially hit the Australian and New Zealand markets. These events, following upon ethnic unrest after the Asian Economic Crisis in 1997 that toppled the country's long-time leader, General Suharto, together with the outbreak of the SARS and the Bird Flu epidemics early in the 2000s, have kept tourism in Indonesia from reaching its full potential. Indonesia today is a parliamentary democracy with a healthy and growing economy, though poverty and corruption are still problematic.

Timor Leste

Timor Leste (also known as East Timor) comprises the eastern half of the island of Timor, the western half of which is part of Indonesia. When Portugal withdrew from Timor Leste in 1975, the Indonesian government forcibly took control of the former colony, an illegal action not recognized by the United Nations and the international community. Following a bloody civil war, Timor Leste gained its independence from Indonesia in 2002. In 2004, Timor Leste had the lowest per capita GDP (roughly equates to per capita income) in the world, at about US$500. Living conditions for its approximately one million citizens have been very difficult as it has attempted to recover (with considerable Australian aid) from the ravages of war. Political instability plagued the new nation, and today there are widespread worries of government corruption, violence and other criminal activity—all of which have slowed economic development in this fragile country.

The Philippines

The Philippines has more than 7,000 islands and like the rest of Southeast Asia is also a very diverse country. The largest ethnic group is Cebuano (around the island of Cebu), comprising 24 percent of the country's population of 107.7 million (2014 estimate). Tagalog is the national language, although the Tagalog ethnic group makes up only 14 percent of the population. More than 90 different Malay languages are spoken in the Philippines, though at least 50 percent of the population uses English as their lingua franca—a legacy of a half century of American rule (1898–1946). The Philippines is 85 percent Roman Catholic, 8 percent Protestant and 5 percent Muslim.

Under Spanish rule, land was used to reward those who supported the Spanish. The result was an elite class that controlled the best land in the country, while the impoverished peasants had the worst or no land. Land reform today is still the number one political issue in the Philippines. Muslim separatists on the southern island of Mindanao have been attempting to break off from the Catholic Philippines since the country's independence and have resorted to terror tactics and kidnappings (including tourists) over the years. These conditions have limited tourism to the country although relations between the Muslim areas and the central government have improved.

The Philippines and the United States

The Philippines has had a long history of close relations with the U.S. military and, like Thailand, it has a reputation for *sex tourism* that was at least partially related to the foreign military presence. This declined, though it did not completely disappear, after the withdrawal of the last U.S. troops in 1992.

▶ *Ha Long Bay on the coast of Vietnam near its border with China is a dramatic landscape of almost 2000 limestone islands. Numerous fishing settlements are nestled among the islands. This World Heritage Site is one of the more popular nature-based tourist destinations in Vietnam, as well as being a stop for Asian cruise ships.*

Photo: Alan A. Lew

Filipinos comprise one of the largest and fastest growing immigrant populations in the United States and the Middle East. The mailing of *remittances* (money) back to relatives in the Philippines is important to the survival of many Filipino families. Filipinos can regularly be seen at major international gateway airports in the U.S. with large packages of goods to take back to their friends and relatives in the Philippines.

The U.S. military personnel that were formerly based in the Philippines also comprise a sizeable portion of the country's international visitors. From a tourism standpoint, the country today is best known for tropical resorts and rainforests where people undertake wildlife watching, hiking and mountain trekking, while the capital, Manila, has been trying to position itself as a major conference and event center.

4.3 | EAST ASIA

Current East Asia Issues and Resources (http://wrgeography.com/eastasia.html)

East Asia is racially one of the most homogeneous regions of the world. China is 90.5 percent ethnic Han Chinese, Japan is 98.4 percent Japanese, and Korea is about 99 percent Korean. By comparison, the United States is about 60 percent non-Hispanic Caucasian. Even though it covers an area larger than most of the other regions discussed so far, there are only five countries in East Asia: China, Japan, North Korea, South Korea and Mongolia. In addition, there are the Chinese territories of Taiwan, Hong Kong and Macau, which are largely autonomous parts of Mainland China, and often treated as separate countries in international statistics. Culturally, East Asia has been dominated by China, and all of the region's countries look to classical China as part of their historical influences. For example, both Mahayana Buddhism (which came to China from Tibet) and Confucian values predominate throughout East Asia.

Physical Geography of East Asia

The western reaches of East Asia extend into the center of the Eurasian continent. Western China and Mongolia share many similar physical and human geographical characteristics with the countries of Central Asia. It is an area of high mountains and vast basins, most of which are very arid. From the Pamir Knot in Tajikistan, the Tian Shan Mountains extend northwest to form China's western border with Kyrgyzstan. The Altun Shan and Kunlun Shan mountain ranges form the northern boundary of the Tibetan Plateau, while the Himalayas comprise its southern boundary. The Tibetan plateau (known as the Qinghai-Xizang Plateau in China) has a base elevation of about 15,000 ft (4500 m), from which higher mountains rise. All of the world's highest peaks (above 26,000 ft, 8,000 m) are located on the edges of the Tibetan Plateau.

North of the Tibetan Plateau are the deserts and grasslands leading to Central Asia and the border between China and Mongolia. These include the Tarim Basin (home of the Taklamakan Desert) and the Gobi (or Shamo) Desert. Southeast of the Tibetan Plateau is a rugged mountain region with steep valleys through which many of the great rivers of East and Southeast Asia flow. In one area, the Salween (Myanmar), Mekong (Indochina) and Yangtze (China) rivers flow parallel to one another within a 50 mile (80 km) distance. These mountains become lower further away from the Tibetan Plateau, though most of southern China is still mountainous all the way to the South China Sea.

The North China Plain

North of the Yangtze River lies the North China Plain, and north of that, across the Bo Hai Sea, is the Manchurian (or Heilongjiang) Plain. The North China Plain was once covered with water. It was gradually

Photo: Alan A. Lew

▲ *The Limestone hills of Guilin, China have long been a popular element in Chinese ink painting. This scene is near the town of Yangshuo, which was made a popular international tourist destination primarily through the Lonely Planet guidebooks.*

filled in by sediment carried from the west by China's two major rivers. In the north, the Yellow River (or Huang He) flows from the Tibetan Plateau through the Gobi Desert and on to the North China Plain. The Yellow River is known as China's Sorrow because it is "undrinkable, unnavigable and unforgiving" in its flooding of the North China Plain. It is 'Yellow' because of the large amounts of mud that it carries from the *loess* areas of the Gobi Desert. Loess is a type of loosely-compacted soil made up of wind-blown dirt. Further to the south is the Yangtze River (Changjiang, or 'Long River', in Chinese), which is the most important river in China. It flows from the Tibetan Plateau, through the Red Basin of Sichuan Province (the most densely populated region in China), through the Three Gorges of the Yangtze, through the Twin Lakes area near Wuhan, and then out to the East China Sea at Shanghai.

The Pacific Ring of Fire

Coastal East Asia is part of the Pacific Ring of Fire, which refers to a ring of volcanic and earthquake areas that encircles much of the Pacific Ocean. In Asia, this ring of active seismic activity extends from Alaska's Aleutian Islands, down the Kamchatka Peninsula of Russia, to Japan, Taiwan and the Philippines, then to New Guinea and New Zealand. Between these islands and peninsulas and the East Asian mainland are major seas, including the Sea of Okhotsk (off Russia) and the Sea of Japan (between Japan and Korea; Koreans refer to it as the East Sea); the Yellow Sea (between the Korean peninsula and China, and north of Taiwan); and the South China Sea (between China and the Philippines, south of Taiwan).

Japan's Mount Fuji is part of the Ring of Fire. It is a large volcano cone and the highest peak in Japan (12,389 ft, 3,776 m), which last erupted in 1707 CE. It is a popular hiking destination, with many overnight huts and panoramic views of Tokyo and Honshu Island. However, such volcanic attractions are not without risk. In September 2014, Mount Ontake, a popular hiking destination, about 200km (125 miles) west of Tokyo, suddenly erupted killing 47 people, mostly hikers. Volcanic activity is most frequent in Kamchatka to the north and the Philippines in the south. Earthquakes are frequent throughout the Ring of Fire, and Japan is one of the world's leading countries in adapting its architecture to withstand earthquake shaking. This, however, was not enough to prevent the massive tsunami waves (up to 33 ft, 10 m high) that engulfed Sendai, Japan, following a 9.0 magnitude quake on March 11, 2011.

Climate and Agriculture

East Asia lies in the northern latitudes. It extends from the tropics in the southernmost areas of China to cold northern latitudes in Mongolia, Manchuria (China) and Japan's island of Hokkaido. Mongolia and western China experience a harsh continental climate that is very cold in the winter and hot in the summer. The eastern coastline of China, along with Japan, Korea and Taiwan, has some maritime (ocean) influences that help to moderate temperature extremes. However, because they are on the east coast of the Eurasian land mass, they are downwind from the cold winter winds and the hot summer air masses that originate in Siberia and Central Asia. This makes the difference between summer and winter weather conditions much greater than one would find in coastal areas on the opposite side of the continent, such as in Europe. This is especially true for the more northerly latitudes of China, Korea and Japan.

Along the Pacific coast and the seas of East Asia, the climate varies with changes in latitude (from south to north). The southernmost areas are tropical and subtropical, meaning they are located below the Tropic of Cancer and have the sun directly overhead at the peak of their summer months. The southernmost island of Hainan in China has a tropical monsoon rainforest climate. Further north from Hainan, the summers are also hot and humid, but cooler winter temperatures create a subtropical climate. The warm and humid climate of southernmost China allows for two, and sometimes three, crops of rice to be grown in a single year. Tea and upland dry rice are often grown in the mountainous areas of southern China.

Typhoons

East Asia is also subject to *typhoons* that form in the Pacific Ocean east of the Philippines in the summer months. From the Philippines, the typhoons turn northward, as they follow the counterclockwise wind flow of the summer Siberian High Pressure system. They usually make landfall along the coast of southern China (from Hainan to Taiwan). Occasionally they will even make their way as far north as southern Japan. In the winter, the Siberian High Pressure system, which churns in a clockwise direction, brings occasional blasts of frigid cold out of the interior of Asia to much of East Asia. In an extreme year, it will even bring snow to southernmost China, where it borders Southeast Asia. While this sounds like a monsoonal climate, with wind directions changing from summer to winter, the winters are not usually described as monsoonal because they are relatively dry.

A humid subtropical climate exists in most of southern Japan, South Korea and eastern China (south of the Yellow River). However, because they are further to the north, the winter months are longer and colder. This is similar to the climate in the southeastern United States, which is characterized by relatively warm temperatures and rainfall throughout the year, though especially in the summer months. The growing season for rice, however, is shorter, and only one crop a year is common. Even further north, wheat is grown instead of rice.

Continentality

The densely populated North China plain near Beijing experiences more intense winter conditions due to the large Eurasian land mass, which due to its continentality, gets extremely cold in the winter months. In this area, wheat and soybeans are the dominant agricultural crops. The more sparsely settled area of Manchuria is a major breadbasket for East Asia. The Korean Peninsula and the islands of Japan have a similar transition from rice growing in the south to soybeans and other crops in their northern areas. East Asia consists of more than just these coastal areas, though this is where most of the region's population resides.

There is also a major east to west climate transition. The further west it is, the drier the climate becomes. This is true of all continental areas that are not situated on the equator. While the coast's humid and warm conditions allow for intensive agricultural activity, further to the west agriculture becomes extensive, with a focus on the production of wheat and other grains and the grazing of animals. In the far western reaches of China, oasis agriculture is common due to the harsh arid conditions. This is true in both the Gobi Desert and on the Tibetan Plateau.

Human Geography, Countries and Tourism of East Asia

Since the end of World War II, East Asia has undergone a remarkable transformation in social and economic development. Japan led the way, quickly recovering from the ravages of war to become one of the wealthiest countries in the world, with one of the most successful economies. The Japanese enjoy one of the longest life expectancies in the world, and their approach to export-based economic development has served as a model for South Korea, Taiwan, Hong Kong and Singapore, which were among the first newly industrializing countries (NICs) in the developing world. These countries were known as Asia's little dragons, with Japan being the big dragon. All of these countries, including Japan, have a strong Chinese cultural influence.

The country that has moved the most from poverty to wealth has been South Korea, which was one of the poorest in Asia after the Korea War (1952–1955) and is today ranked among the top 15 wealthiest countries in the world. More recently, China, too, has moved from being one of the poorest countries in the world to becoming a leading world economic power in only two decades. China, Japan and Korea also share the challenge of being among the most rapidly aging countries in the world. In East Asia, only the countries of North Korea and Mongolia have been struggling in their economic development efforts.

China's Dominance

With a population of about 1.3 billion people, China is the most populated country on Earth. It is also one of the world's oldest, continuous civilizations and nations. Although at times it was fragmented into opposing feudal kingdoms, it has had an identity as a single cultural realm since at least 2000 BCE, and probably earlier. Versions of Chinese writing date back to 3000 BCE, and the earliest Chinese historical records make reference to the Xia (or Hsia) Dynasty 1994 BCE. The Xia Dynasty was located on the Yellow River, just west of the North China Plain near the city of Xi'an, which was the cultural hearth of Chinese civilization. The earliest dynasty that has significant historical records is the Shang (or Yin) Dynasty, which existed from 1523 to 1027 BCE. This dynasty had a complex agricultural society and government bureaucracy, along with a well-established social class system. It was during this time that contemporary Chinese writing and the Chinese lunar calendar were established.

Confucius and Unification

The Chou Dynasty, 1027 to 256 BCE, was concurrent with the early Greeks in Europe and is considered the classical period of Chinese civilization. This was the time that Confucius (Kung Fu-zi, or Master Kong) and Lao Tze, the founder of Taoism, lived. Early Chinese laws were codified at this time, and the world's first paper money was introduced. It was also a time of feudal divisions and warfare on the North China Plain (known as the Warring States Period).

During the Qin (or Chin) Dynasty (247-221 BCE) all of the smaller kingdoms of China were unified for the first time by Qin Shihuang (the Yellow Emperor of the Qin Dynasty). He was an extremely ruthless emperor, known for massive book burnings and the murder of scholars, resulting in the loss of much of the earlier historical records of China. He was also responsible for creating the first Great Wall of China by connecting a series of smaller defensive walls built in earlier times. His tomb in Xi'an, with its over 8000 life-size terracotta soldiers, is one of China's most significant archaeological finds. However, the Qin Emperor's dynasty did not last beyond his death. The Han Dynasty followed in 202 BCE and lasted up to 220 CE. It was concurrent with the height of the Roman Empire, with which the Chinese traded along the Silk Route. The Han Dynasty continued the unification and expansion of China, moving into what is modern-day southeast and southwest China.

The Great Wall

The Great Wall of China is the country's most recognized cultural icon, and it has much symbolic significance for how China relates to the outside world. On the one hand, it was built to protect China from periodic invasions by nomadic northern tribes. The Huns, for example, invaded China in the fourth century BCE. (These were the same Huns that invaded Europe and sacked Rome in the fourth century CE.) The Great Wall also played a role in keeping China "Chinese" by maintaining cultural uniformity to the south of the wall. In some periods of Chinese history, ethnic Chinese were not allowed to leave China, and they were considered no longer Chinese if they did.

Chinese civilization has historically been based on settled agriculture. This made China very different from the nomadic cultures that existed in the grasslands and deserts to the north and west of the North China Plain. The Great Wall marks the change in vegetation between the drier region to the north that is better suited to animal grazing and the more humid region to the south of the wall. Throughout the world a cultural division exists between settled agriculturalists and nomadic people. This was seen previously in Sub-Saharan Africa (section 3.3), and to a lesser degree in North Africa (section 3.2). Nomadic people see themselves as free and self-reliant. They view settled agriculturalists as weak and dependent. Settled agriculturalists, on the other hand, see themselves as civilized and educated, and they view nomadic peoples as vulgar and ruthless. Historically, nomadic peoples have demonstrated military prowess and have often invaded settled agriculturalist societies. This was also seen in the Indo-Aryan invasions of South Asia (2000 BCE).

The Mongols

Nomadic peoples tend to be easily absorbed into the cultures of the settled populations they conquer. This has happened in China, though only two nomadic groups ever completely controlled all of China. The Mongols under Genghis Khan did so from 1260 to 1368 CE (Yuan Dynasty), when China was part of the largest empire that the world has ever known. This was also true when Marco Polo visited China. After the fall of the Mongolian Yuan Dynasty, the Chinese Ming Dynasty (1368–1644) revived the idea of the Great Wall and built the current structure that millions of tourists visit today. It marked the northern border of China, separating it from Mongolia.

There is a famous and often repeated claim that the early American astronauts said the Great Wall was the only human-made object they could see from space. Actually, the Great Wall cannot be seen from space because it is much narrower than any American freeway or German autobahn. What the astronauts probably saw was the change in vegetation that occurs along the approximate location of the Great Wall.

The Ming Dynasty also built the Imperial Palace (known as the Forbidden City) in Beijing. The Ming Dynasty fell to non-Chinese nomads from Manchuria that established the Qing (Ching) Dynasty from 1644 to 1911 CE. It was the last dynasty of China before the establishment of the Republic of China. Under the Qing Dynasty, China became the wealthiest empire in the world up through the 18th century. However, China was in considerable decline by the time European traders started arriving by sea in the 19th century.

Post-Dynastic China

For most of the 19th and 20th centuries, China was in a state of political disarray and internal turmoil. In the final years of the Qing Dynasty, China had lost large amounts of territory to neighboring countries, and China's 21 provinces and five autonomous regions were battling amongst each other over who should rule the country. When Dr. Sun Yat-sen established the Republic of China in 1911, this period of internal turmoil had not yet been settled. The situation was further complicated by the rise of the Chinese Communist Party, first in the major cities, and later in the more rural parts of the country. After World War II, Chinese Communist soldiers, under Mao Zedong, pushed the government of the Republic of China out of the mainland and onto the island of Taiwan.

People's Republic of China

In 1949, the People's Republic of China was established, and for the following three decades China was essentially closed off to the rest of the world. For China, this was a period of intense political infighting centered on the ideology of communism. With some exceptions, mass starvation was mostly eliminated under the communists, though standards of living stagnated. In 1976, China's revolutionary leader, Mao Zedong, died and Deng Xiaoping came to power. His reforms gradually, but successfully, brought free market economic principles to China, although the Chinese Communist Party still maintains strict political control. The result has been a rapidly rising standard of living for most of the people of China. However, those in the coastal provinces have benefited more than those in the interior of the country, and the gap between the rich and poor has increased dramatically from Mao Zedong's China.

China's Tourism

One indicator of how China has changed since the death of Mao Zedong is seen in its tourism economy. In 1978, China received about 230,000 international foreign tourists, mostly because of the severe limitations the government placed on who was allowed to visit the country and who was not. In 2006, China received over 22 million foreign visitors. These numbers are calculated by China's National Tourism Administration (CNTA) and include day visitors who do not stay overnight in China, but they exclude visitors from the so-called compatriot territories of Hong Kong, Macao and Taiwan. The latter would add another 100 million visitors to China's total.

Using the data from the World Tourism Organization (UNWTO), which only includes overnight visitors and also includes the Chinese compatriots, China received 56 million visitors in 2010, making it the third most visited country in the world (behind France and the United States). China received a similar number of arrivals in 2013 making it fourth in the world in arrivals as well as fourth in receipts (US\$ 52 billion). If its current growth rates continue, which some researchers feel is a questionable assumption, China will be the most visited country in the world in 2017. In addition to the large numbers of international tourists to China, the country recorded over 1.6 billion domestic trips taken by Chinese in 2009.

Chinese Outbound Tourism

As an export industry (one that brings money into a country or destination), the large numbers of international tourists are good for China. However, the most significant impact upon the rest of Asia and the world is the rapidly growing numbers of Chinese tourists who are traveling to other countries. In 2014, over 100 million Chinese trips were taken outside of China. In 2013 China had become the number one tourism source market in the world, spending US\$ 129 billion on international tourism. China has more outbound tourists than any other country in Asia, though their travels are mostly concentrated in Hong Kong and Macao (71 percent).

The impact of Chinese outbound tourism is starting to be felt in Asia (17 percent), with far fewer going to Europe (5 percent) and elsewhere. Arrivals to Australia though have grown tremendously in the past few years since it was placed on China's approved destination list. The U.S. was given this status in 2008 and Chinese tour groups are now a common sight at major U.S. tourist destination. In addition, members of the rapidly growing middle class in China are among the biggest spenders when they travel overseas, making them a highly desired market.

By 2020 China is projected to produce 100 million outbound trips going to every corner of the globe. This will make it by far the largest producer of tourists in the world (surpassing the current world leaders of Germany and the United States). Destinations should prepare for this onslaught of Chinese tourists, as the population of that country is increasing in affluence and interested in seeking out more and more remote corners of the planet. For example, in 2014, Chinese tourists became the third largest market for Antarctic tourism, after the United States and Australia.

Landscape and Philosophy

How people experience and interact with the natural environment is influenced by the culture they grew up in. The people of China have a distinct perspective, shared with most other East Asian cultures, on the relationship between humans and the environment, and how that relationship is expressed. Mountain peaks have been seen as places of respite and contemplation, with monasteries being one of the few proper forms of development allowed.

This is best seen in the traditional landscape paintings made with simple black ink on an off-white paper. The classical painting would show a series of steep mountains, similar to the limestone (karst) mountains of Guilin, with one or more very small representations of people and buildings scattered in the valleys and on mountain trails. The scenery represents the *Taoist philosophy* of harmony and balance, values that are actually common in many traditional societies.

Landscape Stories

The close cultural relationship to mountain areas means that most mountains show some signs of human impact, typically in the form of a pagoda, a rock carving or a monastic building situated in particularly attractive locations. Pristine wilderness is rare, though it does exist. Many foreign visitors are surprised when they visit a popular mountain and see a large Chinese character carved into a rock face; they are even more surprised by the Chinese interpretive material, which typically makes reference to ancient Chinese history and legends that all Chinese know, but that are totally unknown to foreign visitors. (These stories are taught through the textbooks used for learning to read and write Chinese.)

The more fundamental difference is that for the Chinese, virtually every mountain and rock feature is associated with an element of Chinese history and legend, and it has a cultural significance. The scientific significance is also there, but for the masses the cultural meanings are more accessible and immediate. For many foreign visitors from the West, the opposite is true.

China's Minority Areas

Most of the ethnic Han Chinese in China live in the eastern, coastal parts of the country. The more remote northern, western and southwestern regions have large non-Chinese populations, who together number over 120 million (about 9.5 percent of China's population). China's *one-child policy* does not apply to its minority groups, which are growing at rates many times faster than Han Chinese. The eleven largest non-Chinese ethnic groups, among the 55 officially recognized by China, are:

- West Southwest China (bordering Southeast Asia)
  - Zhuang (16 million)—Tai-speaking people mostly in the Guangxi Zhuang Autonomous Region
  - Miao (8 million)—several different hill tribe ethnic groups, including the Hmong, who live along China's border with Southeast Asia
  - Yi (7 million)—Tibeto-Burmese speaking minority who live along China's border with Southeast Asia
  - Tujia (7 million)—Tibeto-Burmese speaking minority who mostly live near the borders of Guizhou, Hunan and Hubei Provinces
  - Buyei (3 million)—Tai-speaking people who live along China's border with Vietnam
- Northern China
  - Manchu (10 million)—in Northeast China, which was formerly known as Manchuria
  - Mongols (5 million)—in the Inner Mongolia Autonomous Region of northern China
  - Koreans (2 million)—along the border with North Korea
- Western China
  - Hui (9 million)—a Muslim population in Xinjiang and other western provinces
  - Uyghur (7 million)—a Central Asian Muslim people mostly in the Xinjiang Uyghur Autonomous Region
  - Tibetans (5 million)—a Buddhist people in the Tibet Autonomous Region (recognized by some people as an occupied territory)

Each of these ethnic groups, and the many others that are officially and unofficially recognized, have a distinct material culture and traditions around which they have built tourism economies. Visiting ethnic villages that illustrate one or more of China's 55 minority groups has become one of the most sought-after experiences in the country. Dozens of minority cultural centers have been established in recent years throughout China. These are especially popular among domestic travelers and of increasing interest to foreign visitors as well.

Southwest China and Tibet

Southwest China is known for its many different ethnic groups, most of which are also found in neighboring countries in Southeast Asia. The old town of Lijiang in Yunnan Province has been identified as one candidate for the legendary Shangri-La, and it is also in a popular mountain trekking region. Western China is best known for the ancient Silk Road trading route, which includes oasis cities and high mountains in China and neighboring Central Asia. Tibet is a very special place where traditional Mahayana Buddhist practices are seen in their purist forms, and which has also been suggested as the inspiration for the legendary Shangri-La. The region has also developed rapidly since the Qinghai-Tibet Railroad (the highest in the world) was opened in 2006, although ethnic tension between Tibetans and Chinese remains challenging in some areas.

Compatriot China: Taiwan, Macau and Hong Kong

Taiwan, Hong Kong and Macau are all considered Chinese territories. However, they were not under the control of the Chinese Communist Party when the People's Republic of China (PRC or Mainland China) was established in 1949. The government of the Republic of China (ROC) fled to Taiwan and reestablished itself after the communist takeover of the mainland. Hong Kong had been a British colony

◄ *A popular view of Hong Kong Harbor from The Peak on Hong Kong Island. Hong Kong was a British colony until 1997 when it became a Special Administrative Region of China. In many ways, however, it continues to be a separate entity, requiring international visitors to have a visa to cross from Hong Kong to Mainland China. The mountainous terrain and high population require the maximum utilization of every bit of flat land, which has resulted in the tall residential towers in the upper class neighborhood below the Peak.*

Photo: Alan A. Lew

since 1842, while Macau had been a Portuguese colony since 1557. In 1997, Hong Kong was returned to Mainland Chinese control, while Macau reverted back to Mainland control in 1999. The mainland refers to the Chinese of these areas as "compatriots", and gives them special status to visit China and invest in the mainland's economy. The mainland and compatriot areas are also referred to as "Greater China", though that term can sometimes include diaspora ethnic Chinese living in Southeast Asia.

Taiwan

Although China considers Taiwan part of its sovereign territory, Taiwan (which still refers to itself as the Republic of China) officially considers itself the original China and an independent government, separate from the PRC. Although politically separate, the two countries are closely linked economically.

Historically, the island of Taiwan was not an effective part of China until quite recently. The indigenous people of Taiwan are Malayo-Polynesians, who probably migrated to the island from the Philippines some 12,000 years ago, and from their extended their cultural legacy into Southeast Asia and the Pacific. Genetically, the twelve official aboriginal nations (tribal groups) on the island are among the most homogeneous people on Earth, indicating a very long period of isolation.

The first Chinese to settle on the island arrived in the 1400s, although some historians dispute this. A wave of Chinese settlers also arrived after the Dutch colonized Taiwan in 1624. The Dutch were ousted from Taiwan by the Chinese pirate, Koxinga, in 1662 when he established the Kingdom of Tungning. His kingdom lasted until 1683 when Mainland China took over the island as part of its efforts to stop piracy in the South China Sea. In 1895, following China's defeat in the Sino-Japanese War, Japan took control of Taiwan, which it ruled until 1945, when it was returned to the Republic of China at the close of World War II.

Taiwanese and Mainland Chinese

Some two million refugees fled from Mainland China to Taiwan in the period between 1945 and 1949, while the mainland was in a state of civil war. Today, the population of Taiwan is almost 23.3 million (2014), of which about 85 percent are native Taiwan Chinese (not related to the post-World War II refugee migration). Thirteen percent are Mainland Chinese, and less than 2 percent are indigenous peoples.

The original indigenous Taiwanese number less than 400,000 and live as minorities in the island's high mountains.

Taiwan has one of the highest per capita incomes in Asia (behind only the very small states of Singapore and Brunei, but ahead of Japan and South Korea, as well as the United Kingdom and France), and its capital, Taipei, had the world's tallest building until 2007. An important attraction in Taipei is the National Palace Museum, which is one of the finest museums of Chinese art in the world. Most of the items on the rotating display were brought from Mainland China by refugees fleeing the communists.

Although Taiwan does not have large volcanoes, it is part of the Ring of Fire and has dramatic mountains that reach over 13,000 ft (3960 m), most of which are protected with national park status. The Taroko Gorge, which is a national park, is a deep and narrow marble canyon that cuts through the center of the mountains with a dramatic road that was carved into its granite walls during the Japanese occupation.

Macau

Macau was given to Portugal in 1557 by the Emperor of China as a reward for Portuguese assistance in defeating a notorious pirate in the South China Sea. It is a very small area, consisting of a peninsula and two small islands. In 1974, following a leftist revolution in Portugal, the Portuguese offered to give Macau back to China. China was not ready to accept the return of Macau at that time, but it officially designated it as Chinese territory under Portuguese administration. Historically, Macau has been a backwater to nearby Hong Kong, and it has been much more politically dependent on Mainland China than was its British counterpart. The small population of Macanese culture is a mix of Chinese and Portuguese, similar to other places where the Portuguese have intermixed with local populations, such as in Melaka (Malaysia) and Goa (India).

Gaming in Macau

Macau is mostly known for its gaming industry, and it serves both nearby Hong Kong and Mainland China as a gambling center. In 2004, the first international casino companies were permitted to open in Macau, bringing new life (and Las Vegas-style casinos) to the city's former single-owner monopoly gaming industry. The rise in Mainland Chinese tourists to Macau has transformed it into the largest gaming city in the world, in terms of profits earned. The Macau Grand Prix has also been a major event for several decades. Besides gaming and tourism, major areas of employment in Macau are textiles, garments, fireworks, toys and electronics. Many people visit Macau on side-trips from Hong Kong, as the two territories are well connected by fast boats and hydrofoils.

The Opium Wars and the Founding of Hong Kong

In the early 1800s, European trade with China was basically a one-way arrangement. The Europeans purchased silk, tea and fine Chinese pottery from China. There was little, however, that the rather sophisticated Chinese economy was interested in purchasing from the Europeans in exchange. So the European traders had to use silver to purchase Chinese goods, which put a strain on the central banks of European countries.

The Portuguese were the first to introduce opium into China as an alternative product for trade. The opium was grown in South and Southeast Asia. Opium was illegal in China but was traded openly in the southern port city of Guangzhou. Because of the opium trade, silver began leaving China, and the Chinese government decided to confiscate the opium from the foreign settlements to stop its trade. They also forced most of the foreigners out of the port cities in southern China.

Most went to Macau, but the Portuguese did not allow the British to go there, so instead the British anchored at a nearby fishing village called Hong Kong. The British traders called on the British Navy to retaliate against China, and the British Navy subsequently bombed cities along the Chinese coast all the way to Beijing. In 1842, they forced the Qing Emperor in Beijing to grant the island and nearby peninsula of Hong Kong to the British. Other Europeans also fought opium wars with China to force China to allow them to establish trading communities in the major port cities. But only the British actually established a permanent colony.

Hong Kong Today

Under British law, Hong Kong flourished as a trading center. It offered a much better port facility than Macau, and the British had a reputation for treating most people fairly in a largely free-market economy. Hong Kong today, even under Mainland Chinese rule, is consistently rated as having one of the most free and open economic systems in the world. Less can be said for democratic rule, which even the British suppressed until their last days in Hong Kong, and in October 2014 it became the subject of significant public protests.

As much as 75 percent of the land in Hong Kong is mountainous and covered with lush subtropical vegetation. This steep terrain makes the harbor and city view from Victoria Peak on Hong Kong Island one of the most scenic urban landscape views in the world. The steep terrain also exasperates the density of crowding in the city. Hong Kong could not survive without considerable support from China in providing basic water, food, petroleum and electricity needs. Hong Kong today continues to be a major global financial center, though it has lost much of its manufacturing base to less expensive areas in neighboring Mainland China.

Hong Kong is the most visited international tourist destination in Asia, receiving well over 20 million visitors a year. The majority come from Mainland China and are considered international tourists, despite their both also being considered the same country. Mainland Chinese come to Hong Kong to shop for high-end goods that are sometimes more difficult to find in China. Hong Kong has also been a major gateway to China, especially in the period before the 1990s when direct international flights into China were very limited. Taiwanese business people and tourists needed to enter China through Hong Kong until 2008 when the first scheduled direct flights between Taiwan and Mainland China were introduced.

◀ *Typical commercial street in Tokyo, Japan. With a population of 35.5 million people, the Tokyo metropolitan area is, by far, the largest urban region on the planet. Despite this massive population agglomeration, Tokyo is also one of the most crime-free cities on Earth. Japanese culture, perhaps along with its tenuous position on the Pacific Ring of Fire, helps to maintain civil behavior in a potentially stressful environment.*

Japan

Japan has a population of more than 127 million people (2013), making it the tenth largest country in the world. The land area of Japan is slightly smaller than the U.S. state of California, but with 3.4 times as many people. It is an insular country made up of four large islands and about 3000 smaller islands that extend almost to Taiwan. Like much of East Asia, the Japanese have a tendency to occupy lowland areas only, leaving the interior mountains uninhabited, except for temples and resorts.

World's Largest Urban Area

Most Japanese live in very high-density urban areas on the southern coast of the island of Honshu, which is the largest island in Japan. This includes the Tokyo-Yokohama metropolitan area—the world's largest urban agglomeration at 35.7 million people (2007). Kyushu and Shikoku are the two southernmost islands and are warmer and more subtropical than the rest of Japan. Hokkaido is the northern frontier island and is famous for its winter Snow Festival and large ice sculptures in Sapporo. Hokkaido is relatively sparsely settled and connected to Honshu by the 33.4-mile Seikan Tunnel (53.9 km), an underwater railway that is slightly longer than the Channel Tunnel between England and France.

Ethnic Homogeneity

Japan is ethnically the most homogeneous country on Earth. As mentioned above, 98.4 percent of the population is Japanese. Koreans and Chinese each make up 0.4 percent of the population, and the Ainu of Hokkaido (the earliest settlers of Japan) comprise some 0.1 percent. The Japanese take great pride in their racial purity. A former prime minister, Yasuhiro Nakasone, commented in 1986 that the United States' education system was falling behind Japan because of its ethnic impurities. (The United States' view is that the multiplicity of ideas and perspectives are what has made the United States the success that it is today.)

The homogeneous culture of Japan is due, in part, to its geographic isolation. Its islands are physically separated from the main areas of the Asian continent, with the Korean Strait and the Sea of Japan providing protection from invasions. The Mongols, for example, were not able to incorporate Japan into their massive empire despite several attempts in the 13th century CE. In addition to geographic isolation, the Japanese have been isolated culturally, taking considerable measures historically to protect themselves from outside influences.

Safety and Uniformity

Japan has some of the largest and most crowded cities in the world, yet it is also one of the safest countries in the world. This is a benefit of the powerful influence of Japanese culture, which includes a strong adherence to cultural and social norms of behavior. Deviants are treated harshly. Improper behavior results in a loss of face and social ostracism, not just for the individual, but also for the groups the individual belongs to, including family, community, school and workgroup. All these groups lose face and share in the blame of an individual's improper behavior. While such characteristics are common in many Asian cultures, it is especially pronounced in Japan and in neighboring Korea.

This strict adherence to social norms has its cost. The Japanese education and work system tends to wear people out by their early thirties because of its strenuous social demands. In addition, Japan has one of the highest child suicide rates in the world, and it also has the highest suicide rate among elderly women, who find that they have only a limited role to play in society after their children leave home. The result is a disciplined society that is adaptable to large-scale organization and mass production assembly manufacturing that produces highly skilled labor. This has placed Japan at the cutting edge of industrial and high technology development and contributed to the country's economic recovery after World War II.

Otaku Culture

Two of the most popular tourist destinations in Tokyo are Ginza, famous for its department stores and luxury brands, and Akihabara, considered by many the electronics capital of the world, as well as for *Otaku culture*. Otaku culture refers to the behavioral traits of the obsessive hobbyist, which in Japan typically means those who follow anime (Japanese animation) and manga (Japanese graphic novels), computer nerds and game players, military geeks, pop culture fans and followers of other fads and fetishes.

Japan and China

Japan has a history of openness to incorporating and adapting foreign ideas. During China's Sui Dynasty (581–619 CE), Tang Dynasty (618–907 CE) and early Song Dynasty (960-1279 CE), Japan sent scholars and government officials to China to learn everything they could about Chinese arts, culture and government. (This was roughly concurrent with the Arab Empire and the Dark Ages of Europe.) Most went for a two to three year period and upon their return were placed in positions of authority to implement what they had learned. Buddhist teachings and Chinese art, architecture and city planning were adopted from China at this time.

One example of this is Japan's former capital city, Kyoto, which was designed after the Tang Chinese capital city of Chang'an (modern day Xi'an), using principles of Chinese geomancy (feng-shui). Kyoto was the imperial capital of Japan for eleven centuries, although Japan was not fully united for all that time. Spared from the fire-bombings of World War II, Kyoto is one of the best-preserved cities in Japan. Today it has some 2000 Buddhist temples and Shinto (Japanese traditional religion) shrines, plus many palaces and gardens. It provides a welcome contrast from the ultra-modern Tokyo landscape, and tourism is a major part of Kyoto's economy.

Feudal Japan

The tradition of sending Japanese to China lasted some 600 years, after which Japan entered a feudal Dark Ages in the twelfth century. During this time powerful clans and regional families (daimyo), as well as the military rule of warlords (shogun), fragmented Japan. Throughout the European colonial period (starting in the 15th and 16th centuries), Japan benefited from its great distance from Europe. It was essentially closed to the rest of the world, until 1850 when the U.S. Navy, under Admiral Perry, forced Japan to open its doors to trade with the West. Concerned about encroaching Western powers, the Japanese came together in 1868 and overthrew the Shogun of Japan, the most powerful feudal lord. They restored the Emperor Meiji to power in an effort to re-centralize control over the entire country. This is known as the Meiji Restoration.

Meiji Restoration

Following the Meiji Restoration, Japanese students and entrepreneurs roamed the world to learn all they could about why and how the West had become so successful economically and militarily. They quickly learned and adopted Western technology and management practices, and in the 1890s they fought successful wars against Russia and China. As Japan's economy grew, it sought to expand its colonial control over most of East Asia to secure access to raw materials for its industries. This was similar to the colonial motivations of the European powers in their industrial revolution in the 1800s. These moves ultimately resulted in World War II, and although Japan lost that war, today it is a major global economic power, with considerable influence throughout the realm of East and Southeast Asia.

Three major tourist attractions reflect this history of Japan. The Nijo Castle in Kyoto was built in 1603 by Tokugawa Ieyasu, one of Japan's most powerful shoguns and the founder of the Tokugawa Shogunate.

The almost entirely wood castle was a symbol of the power of the Tokugawa and is today filled with fine art. The Meiji Restoration moved the capital from Kyoto to Tokyo, where the Meiji Shrine is among the most visited temples in Japan. It is a Shinto temple and museum and was built after the death of the Emperor Meiji (died 1912) and his wife, the Empress Shoken (died 1914), though the original building was destroyed in World War II.

Hiroshima

The third major historical attraction is the Hiroshima Peace Memorial Park and the Peace Memorial Museum, which commemorate the destruction of Hiroshima by the world's first atomic bomb used in military conflict on August 6, 1945. The park, which was built in 1949, also includes the Children's Peace Monument and the ruins of the A-Bomb Dome, and it is a symbol of international peace. In addition to these historic sites, Mt. Fuji is a symbol of Japan known throughout the world that has become something of an iconic image and tourism brand for the country. The powerful March 11, 2011, earthquake and tsunami in Japan (in conjunction with the powerful earthquakes in Haiti in January 2010 and Christchurch, New Zealand, in September 2010 and February 2011) brought additional light to a form of tourism that has long existed but that neither scholars nor the media understand well.

Disaster-based tourism manifests in several forms and includes elements of volunteer tourism and dark tourism. In the Japanese situation, the disaster immediately resulted in tens of thousands of people travelling within and to Japan to offer relief, to assist in rescuing survivors and to reconstruct destroyed buildings and infrastructure. In the weeks and months that followed, international tourists began arriving to witness the damage that the worldwide media had so abundantly described. The Japanese example illustrates some interesting trends regarding how people react to mega-disasters, in terms of their desires to assist and their curiosity about the loss of life and physical destruction.

The Korean Peninsula

Korea has long served as a buffer between China and Japan. Chinese culture entered Japan mostly through the Korean Peninsula. The Go-Joseon Empire (also called Tan- gun Joseon, which means land of the morning calm) rose in 2333 BCE in Korea and at one time controlled Manchuria to the north and coastal areas of the North China Plain. This empire lasted until its defeat by the Han Dynasty of China in 108 BCE. Korea has been considered a tributary state of China for most of the time since then, which meant that it maintained a degree of autonomy but paid regular tribute to the Emperor of China. Korea was under Japan's control from 1905 to 1945.

Unlike other areas of East Asia that Japan colonized, the Japanese fully annexed Korea in 1910 because they saw the Koreans as more like themselves than other Asians. The Japanese then tried to erase Korean culture forcibly and replace it with Japanese culture. This led to great devastation in Korea and resentments that are still strong today.

North and South Korea

At the end of World War II, Korea was divided into a Soviet Union sphere of influence in the north and a United States sphere of influence in the south. (Germany and Austria were similarly divided in Europe.) The Korean War was fought from 1950 to 1953 after North Korea attacked South Korea, resulting in an estimated 1.5 million civilian deaths. United Nations troops (that included soldiers from the United States, the United Kingdom, Turkey, Canada, Australia, France and the Philippines) pushed the North Korean soldiers to the border with the People's Republic of China. China, however, decided to enter the conflict and pushed the UN armies southward to where the present border lies between North and South

Korea. The border today is known as the *Demilitarized Zone* (DMZ), which remains an uninhabited area 155 miles (248 km) long and about 2.5 miles (4 km) wide.

The DMZ

The DMZ today is a significant tourist attraction on both sides of the border. Tours from Pyongyang (the capital of North Korea—the Democratic People's Republic of Korea) visit, as do tours from Seoul (the capital of South Korea—the Republic of Korea). One interesting facet of these tours on the south side is that Korean and U.S. military personnel serve not only as security officers on these trips but also as the guides who explain their versions of the Korean conflict. The abandoned village of Panmunjom is in the center of the DMZ and is where North and South Korean representatives meet to discuss border issues. Other than brief visits to Panmunjom, few humans have entered the DMZ since 1953, making it one of the best-preserved temperate wildlife areas in the world. However, the large numbers of land mines that both sides have placed there over the years complicate the situation with respect to tourism development.

North Korea

Although ethnically one people. North Korea and South Korea are vastly different countries. North Korea is ruled by a dictatorial state communist government that controls all aspects of daily life, including recreation and tourism. It is very mountainous and has an abundance of hydroelectric capacity. The mountains also contain considerable mineral wealth, including coal and iron, which are key for heavy manufacturing. North Korea's climate is much more continental than South Korea's, and it has a much smaller population. North Korea is very dependent on industrial factories because it has limited agricultural land. It must sell manufactured products to pay for food imports to feed its people. Malnutrition and starvation have been major problems in North Korea since the collapse of its primary supporter, the Soviet Union, with China now taking the Soviet Union's place as its main ally.

◄ *South Korean tourists at Mount Kumgang (Geumgang) in North Korea, which is a major historical and cultural site for all Koreans. This is one of only a few destinations in the North that South Korean tourists can visit, although even this site is sometimes closed for lengthy periods when tensions increase between the two Koreas.*

Photo: Dallen J. Timothy

Mount Geumgang / Kumgangsan

North Korea is one of the most closed countries in the world today, though it has very gradually opened to tourism since about 2000. Until recently, there were no communications (e.g., mail, Internet, transportation, telephone) connections between North and South Korea, and citizens from either country were not allowed to visit the other. Today, however, South Koreans can visit a few selected locations in the North, most notably the Mount Geumgang (or Kumgangsan) area, which is considered sacred for all Koreans. In the late 1990s, organized tours began taking southerners to the north by boat around the DMZ and, more recently, across the DMZ in motor coaches. Access to this and other sites in North Korea varies depending on the often changing state of relations between country and the rest of the world. Even U.S. citizens, who traditionally have not been allowed to enter the North, were permitted to visit on these pre-arranged tours, and U.S. citizens can now visit the country easily as participants in pre-organized package tours.

With the 2008 shooting death of a South Korean tourist at the Geumgang area at the hands of North Korean soldiers, the South boycotted the program and tours were cancelled. In retaliation for lost revenue from the boycott, North Korea confiscated South Korean-owned hotel properties. In late 2014 some signs of improved relations between North and South Korea appeared following visits of senior North Korean officials to the Asian Games in Seoul. However, without considerable improvements in diplomatic and border relations, tourism between the Koreas and to the North from international visitors will remain limited.

Most tourist activities in North Korea are geared toward extolling the virtues of the Great Leader Kim Il Sung and his son, Kim Jong-Il, and Kim Jong-un, who ascended to the party leadership following the death of his father, Kim Jong-Il, in 2011. Visits to schools, monuments to the leaders, factories that are set up to show the country's industrial success, circuses and festivals comprise the bulk of tour itineraries in North Korea. Independent travel is not permitted, as one must always be with a North Korean guide.

Most North Koreans are not permitted to leave their country, with the exceptions of government officials, ship workers, airline attendants and a handful of others. Domestic travel is virtually non-existent, as the government rarely allows people to leave the villages and cities where they are assigned to live. Special passports and permits are required even for domestic travel. There is a steady stream of *defectors* who manage with great difficulty to get to China and South Korea, and the Republic of Korea welcomes them with open arms.

South Korea

South Korea also has mountains and a number of ski resort areas, but it has much more lowland area that is suitable for agriculture. It has a more humid temperate and subtropical climate, which even allows double cropping in its southernmost areas. South Korea is a food exporter, despite having a population that is over twice the size of North Korea. It has the largest cities on the Korean peninsula including Seoul, with 25.6 million (2012) in its metropolitan area, and Pusan, the main industrial city on the peninsula. South Korea is a close ally of the United States, and it has had successful democratic elections in recent years. South Korea is a newly developed country with a strong and globalized economy and per capita incomes just below those of Japan. Tourism in the south centers primarily on cultural festivals and religious historic buildings. The southernmost area of the peninsula, particularly subtropical Jeju Island, is considered a very desirable sun and beach resort destination among domestic Korean tourists and a favorite honeymoon getaway for young Korean couples.

Mongolia

The popular view of Mongolia is of a broad expanse of monotonous landscape with little differentiation. While true for a large part of the country, the eastern portion does contain the high, snow-capped Altai Mountains and the Mongolian Plateau. Some of the major rivers of Central Asia originate in this area, which has many lakes and streams. These rivers have created rich pasturelands that have been central to Mongolia's economy throughout its history. Most of Mongolia, however, is arid with steppe grassland vegetation. This is because it is too far from the ocean to benefit from the prevailing onshore summer monsoon winds.

The Gobi Desert

The Gobi Desert area has a hard surface and is known for its strong winds caused by the heating of the land. The climate of the Mongolian steppe has made the development of agriculture challenging. Mining, however, is the country's biggest foreign exchange earner. Industrial activities are centered in the capital, Ulaanbaatar (or Ulaan Baatar), and are primarily related to mining and agriculture, with some handicraft production. Mongolia's population of just under three million (2014) is comprised of 82 percent Kahlkha, followed by Kazakhs (4.4 percent) mostly near the Kazakh border, and over 15 other ethnic groups. In addition, there are four million more Mongols in China's neighboring Inner Mongolia Autonomous Region (Nei Mongol), although they only comprise 17 percent of the population there. Historically, Mongolia has had a major impact on its larger neighbors, China and Russia.

The Mongol Empire

The Mongol Empire in the 1300s was the largest empire the world has ever known, stretching from China to Eastern Europe and the Middle East. Today the Mongolian language (Khalkha Mongol) is spoken by people extending to the north and south beyond the country's borders. Mongolian is an Altaic language, related to the languages of Central Asia, Turkey and Manchuria. Culturally, Mongolia is very homogeneous, with 96 percent of the population practicing the Lamaism form of Buddhism (closely related to Tibetan Buddhism). Herding and hunting form the bases of traditional life among the nomads of Mongolia.

Yurts

One of the most popular international tourist activities in Mongolia is a stay in a yurt (known locally as a *gir*) tent village on the Mongolian steppe. It includes eating traditional food (e.g., mutton, camel meat and fermented horse milk) and riding horses and camels. The country has several national parks and is home to some of the earliest paleontological (prehistoric life) sites in Asia. The annual Naadam Festival, which takes place in Ulaan Baatar, is a salient part of the tourism product of Mongolia. The festival focuses on the three national sports–archery, horseracing and wrestling—and draws thousands of urban dwellers and nomads from all over the country to compete and observe. Naadam has also become an important part of international itineraries in Mongolia, and the event has been inscribed on UNESCO's List of Intangible Cultural Heritage.

Modern Mongolia

In 1921, Mongolia gained independence from China and was quickly backed by the Soviet Union, which established a puppet communist regime in Mongolia in 1924. Concurrent with the collapse of communism in Central and Eastern Europe, Mongolia ceased to be a totalitarian communist state in 1991 and has since seen considerable economic, social and political reforms. Because the Mongolians are a nomadic people who did not live in urban settlements, the Soviets (who essentially controlled the internal affairs of Mongolia until 1991) insisted on the establishment of cities and towns, and they erected many buildings similar to the bland Soviet style in Russia.

Today, approximately 69 percent of the Mongolia's population is urbanized, with the rest continuing to practice traditional herding and migratory lifestyles. Ulaanbaatar (population 1.2 million) has little to offer visitors other than its role as the nation's gateway and government center, although the city has a high population of German and other European expatriates. One of its most important market segments is Koreans, who, it is believed, have racial and genetic connections to the Mongolians.

4.4 | OCEANIA: AUSTRALIA AND THE PACIFIC ISLANDS

Current Oceania Issues and Resources (http://wrgeography.com/oceania.html)

This region encompasses a vast area of relatively low population density, a lot of open water, and inhospitable deserts, tropical rainforests and vast fields of ice. It includes the continents of Australia and the adjacent islands of New Zealand, as well as the islands of the South Pacific. These lands are the most southerly on the planet, with New Zealand (and the southern tip of South America) being closest to Antarctica and the South Pole. New Zealand and Australia (along with South America) have been isolated from much of the rest of the world in recent geologic history. This has produced unique flora and fauna in these lands. In addition, Australia and New Zealand (like North and South America) have been major colonial settlement areas for Europeans. The South Pacific islands have also had a long history of European colonization, though actual European settlement has been much more limited.

The Continent of Australia

Australia is regarded as either the smallest continent or the largest island in the world. It is three-fourths the size of Europe, the second smallest continent. However, it is three times the size of Greenland, the world's largest island. Australia is considered a continent rather than an island in part because it is culturally different from surrounding regions. Culturally, Australia is primarily European with a small population of Asian and Aboriginal minorities. If it had been settled by Malayo-Polynesian peoples, such as Indonesia to the north and Polynesia to the east, Australia may have been considered an extension of Asia and the world's largest island, instead of a separate continent.

A significant geological reason for regarding them separately is that the island of New Guinea and the continent of Australia share a distinct continental plate. In fact, during the Ice Ages, when sea levels were significantly lower, the two islands were part of a single land mass. Even now the islands off Cape York in the far north of Queensland in the Torres Strait are so close together they can be traveled by canoe all the way to New Guinea on the northern edge of the strait.

Australia's population is over 23 million people, which is about the same population as the islands of Sri Lanka and Taiwan, but it has considerably less than the annual increase in population of India or China. A major limitation on Australia's population is the continent's low carrying capacity. Only 8 percent of Australia's land is arable (suitable for agriculture), and only 1 percent is suited to intensive agriculture, such as that found in the central United States and in Ukraine. Climate change is having a profound impact on Australia's agricultural production as the country is getting warmer, and the increased likelihood of low-frequency, high-magnitude weather events, such as floods and droughts, has increased.

To the north of Australia is the island of New Guinea (comprised of Papua New Guinea and Indonesia's West Papua Province). At 320,000 sq miles (828,796 sq km), it is the world's second-largest island (after Greenland). New Guinea is part of the Australian continental plate and was pushed up by Australia's movement to the north and collision with plates that make up the Pacific Ocean floor. To the southeast of Australia is New Zealand, which is about a third the size of New Guinea.

The Islands of New Zealand

New Zealand is about the same size as the British Isles or the Philippines and, like New Guinea, was formed by the collision of the Australian plate with the Pacific plate. New Zealand has a population of

◀ *The Southern Alps on New Zealand's South Island. New Zealand, unlike its larger neighbor, Australia, is on the southwestern-most edge of the Pacific Ring of Fire, giving it active volcanoes, dangerous earthquakes and dramatic mountain scenery. Because it is so far south, glaciers and fjords are also part of the New Zealand landscape.*

Photo: C. Michael Hall

over 4 million people, 75 percent of whom are of European origin. The Maori, the Polynesians who were the first human settlers of New Zealand approximately 1000 years ago, comprise 15 percent of the country's population, while other South Pacific peoples make up 6.5 percent, and Asians constitute 6.6 percent. (The total is more than 100 percent because people are permitted to designate up to three ethnicities in the national census.). Both Australia and New Zealand are experiencing an increase in the proportion of the population with an Asian background as a result of migration, while both countries have moved much closer to Asia in political and economic terms.

Physical Geography of Australia and New Zealand

The supercontinent Pangaea split over 200 million years ago, forming the continents of Laurasia and Gondwanaland. Some 135 million years ago, Australia, New Zealand and Antarctica broke away from Gondwanaland to form their own separate continent. This has allowed the development of unique plant and animal species. Approximately 80 million years ago Australia and New Zealand detached from Antarctica, and from each other, and moved northward, while Antarctica migrated south to the South Pole. This breakup was relatively late in geologic time and occurred at about the same time the Rocky Mountains, the Alps and the Andes Mountain Range rose, during the period the dinosaurs came to an end.

Australia's Physical Geography

Australia has the lowest average elevation and relief (change in elevation) of all the continents. Much of Australia is ancient, exposed rock from the old Gondwanaland shield. The continent's highest mountains, the Great Dividing Range, extend along much of the east coast, with its highest peaks reaching over 7,000 ft (2,135 m) in the Snowy Mountains between Sydney and Melbourne. However, much of the Great Dividing Range is below 5,000 ft (1,525 m) in elevation in southeastern Australia, and below 2,000 ft (610 m) in the northeast of the continent. The northern and northeastern parts of Australia have a rainforest climate with occasional typhoons coming from the ITCZ along the equator. Along the ITCZ, air moves from east to west, so most of the tropical storms that hit Australia come from the Coral Sea and across the Great Barrier Reef.

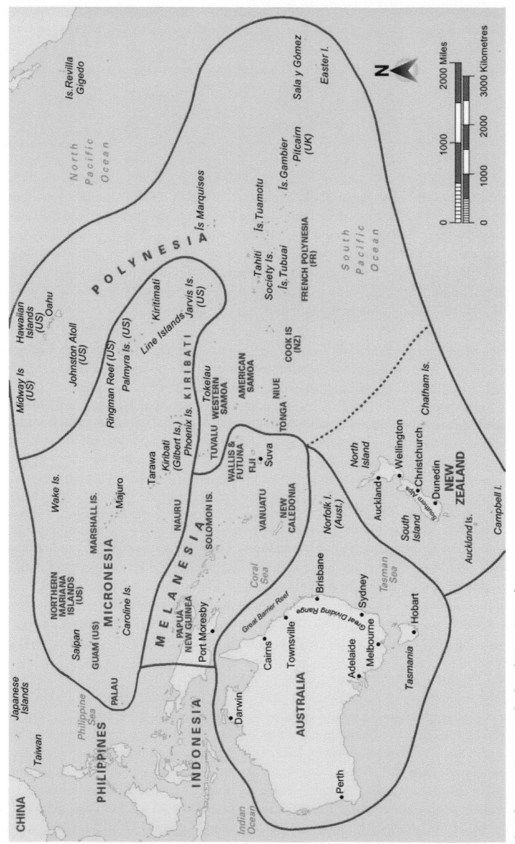

▲ Subregions, Countries, Islands and Major Cities of Oceania

There are a few other highland areas in Australia, all with their highest peaks below 5,000 ft (1525 m) in elevation. These older mountain systems have been heavily eroded, exposing considerable mineral wealth. Coal deposits, for example, are widespread and able to meet most of Australia's energy needs and are a significant, although controversial, export because of their contribution to carbon emissions. Australia also has major deposits of iron, lead, zinc, copper and uranium, and it is a major supplier of raw materials to Japan and China.

The Outback

To the east of Australia's Great Dividing Range is a large grassland area, which is home to some of the largest sheep flocks in the world and, in its more humid areas, millions of cattle. Further westward the grasslands turn into tropical savannas in the north and deserts in central Australia. This is the famous Australian Outback, which contains the Simpson Desert, the Great Sandy Desert and the Great Victoria Desert, among others. Nearly in the middle of the continent is Uluru, also known as Ayers Rock, the world's most famous monolith (single rock outcrop), and an Aboriginal sacred site. The Australian Outback is situated in the band of high air pressure that resides north and south of the ITCZ. This means that dry air descending from the upper atmosphere is the dominant climate pattern for much of the interior of Australia, which causes desert-like conditions.

Population densities throughout the grasslands, savannas and deserts of Australia are very low, and people are widely dispersed, making transportation and communication a challenge. At the far south-western tip of Australia is another grassland region where the city of Perth is located.

Australia's Climate

The southernmost regions of Australia are part of the southern upper latitude low-pressure storm track. Low air pressure storms of hurricane intensity (called cyclones) move from west to east, approaching southernmost Australia from the Indian Ocean. Perth has a Mediterranean climate, with a dry summer (in January) and wet winter (in July). Most of the population of Australia lives on the southeast coast of the continent where it is relatively warm and humid throughout the year. The island of Tasmania has a Marine West Coast climate (similar to the British Isles and North America's Pacific Northwest) because it is much closer to the South Pole than to the rest of Australia and is the first landfall for many of the weather systems that have travelled along the Southern Ocean picking up considerable amounts of moisture on the way.

New Zealand Physical Geography

New Zealand consists of two major islands: the North Island and the South Island. They both have mountains, with those on the South Island referred to as the Southern Alps. The mountains in the north are smaller and less dramatic, with most of the flat farmland located on the North Island. Human settlement is primarily in coastal areas due to the mountainous interiors of the islands. New Zealand is further south than most of Australia and has a more humid Marine West Coast climate with year-round rain fall. The highest mountains of New Zealand reach over 12,000 ft (3,660 m) on the South Island and are home to permanent glaciers. New Zealand is an active volcanic and earthquake region, which is especially evident in the volcanoes and thermal areas on the North Island. These geothermal phenomena provide much of the North Island's tourist appeal in the form of geysers, volcanic calderas, mineral springs and hot mud baths.

Australia's Biogeography

In Australia, 95 percent of the fauna and 85 percent of the flora diversity are endemic, which means they are found only in Australia. Among Australia's native plants and animals, non-endemic species are mostly found in the northernmost part of the continent, having migrated there from Southeast Asia. The deserts and steps of central Australia kept these migrants to the north of the continent and away from the endemic species further to the south. Among Australia's endemic animal species, 58 percent are placental mammals and 43 percent are marsupial mammals. The placental animals probably migrated to Australia from Southeast Asia long ago. Nearly a third of the placental are rats and mice, 25 percent are bats, and the largest is the dingo dog (wild dog) at 0.3 percent.

Marsupials

Marsupials are a very old type of mammal, which originated in North America during the dinosaur era (80 million years ago) and arrived in Australia after traveling through South America and Antarctica. Two-thirds of all marsupial species today are found in Australia and New Guinea, with most of the rest in South America. Marsupials carry their growing fetuses in pouches, as opposed to placental mammals, which carry their growing fetuses in the mothers' wombs. This added protection has allowed placental animals to replace marsupials throughout the world, except in Australia, which has a large variety of marsupial species, and South America, which has a few. In North America the only native marsupial is the possum, which was probably reintroduced from South America.

Marsupials come in many different forms to fill a wide variety of ecological niches. In Australia there are marsupial mice, cats, anteaters, moles, wolves (the extinct Tasmanian tiger), marmots (wombats), dogs (Tasmanian devil) and koalas, in addition to the more distinctive wallabies and kangaroos (of which there are 50 species). Large kangaroos are grazing animals, similar to cows and sheep in other parts of the world. Marsupial cats and wolves are meat-eating predators, whereas koalas are tree-dwelling vegetarians. Australia is also home to the most ancient type of mammal, the monotreme, which nurses its young with milk–that all mammals share in common–but lays eggs like reptiles. The only existing species are the platypus and the echidna (resembles a hedgehog).

New Zealand's Biogeography

New Zealand has been isolated from Antarctica for an even longer period than Australia. There are no marsupials in New Zealand apart from introduced species such as possums and wallabies. The possums are regarded as a major pest because of the damage they cause to native forests. In fact, prior to human settlement there were only three native species of mammals, all bats, one of which is now extinct.

Like other Gondwanic island systems, such as Madagascar, New Zealand was home to a spectacular array of birds. Many are now extinct or endangered, including some endemic flightless birds (e.g., kiwis and the extinct moas). The first wave of introduced species, such as rats and dogs, probably arrived by island hopping through the South Pacific, along with the Maori Polynesians who settled in New Zealand. Today, numerous placental animals have been introduced to Australia and New Zealand from other continents, and they compete for the ecological niches with native animals. Sheep are the major grazing animal in both Australia and New Zealand, which are leading exporters of wool and lamb. Domestic sheep, wild goats, wild pigs and placental rats and mice have led to the loss of numerous animal and plant species in both Australia and New Zealand.

Bushfires and Cyclones

A major challenge for environmental interpretation for tourists in Australia is explaining the impacts of natural changes in the environment. These can be such things as bushfires or the effects of cyclones. Bushfires are an essential element of the environmental health of most of Australia's ecological communities, which are highly fire-adapted. Many species, such as acacias and banksias, actually are fire dependent

---

ISSUES AND INSIGHTS | **World's Largest Sand Island**

Fraser Island in southeast Queensland, Australia, is the world's largest sand island. It was named after Eliza Fraser, the wife of the captain of a ship that was wrecked on the island in 1836. Europeans had extinguished the island's Aboriginal population by 1900. For Europeans, the island provided an important source of hardwood timber that was used all around the world for pier supports in harbors. During the 1960s and 1970s, the island was the subject of disputes over the mining of mineral sand leases, which were issued by the Queensland government despite the island's environmental and recreational significance. However, for the first time in Australian environmental law, the Commonwealth (national) government used its constitutional powers to prevent the export of mineral sands, thereby making mining uneconomical.

The island was listed as a UNESCO World Heritage Site in 1992 because of its natural heritage value with respect to being an outstanding example of ongoing ecological and biological processes and containing superlative natural phenomena. In addition to the significance of the sand dunes, the island also has important forest ecosystems and over half of the world's perched dune lakes, including the world's largest, Lake Boomanjin. The island's natural history and its high profile in Australian environmentalism have assisted it in becoming a significant international and domestic nature-based tourism destination. However, the growth of tourism, coupled with continued demand for recreational access for beach fishing and off-road vehicles (ORVs) from a growing regional population, is placing increased pressure on the island's environment.

---

for the dispersal of their seeds. In addition, bushfires are an excellent method of eliminating weeds and other undesirable exotic plant and animal species. Indeed, the biodiversity and high degree of endemism associated with many Australian natural heritage sites, such as national parks or even World Heritage areas like Kakadu, are critically dependent on the maintenance of fire regimes. However, for many people bushfires are perceived as a negative because of the blackened landscapes they produce in the short-term.

Cyclones

Cyclones and storm events also can have a major impact on the environment, but they should also be regarded as part of the natural cycle of environmental change. In the case of Cyclone Larry, which hit the tropical rainforest of northern Queensland in 2006, the immediate impact was the full or part closure of 30 Queensland parks and state forests, as well as physical damage to rainforest habitat and several communities. However, although unsightly, occasional disturbance may actually contribute to biodiversity by helping maintain mosaics of different stages of forest growth and canopy cover in which different species occupy different habitats. Yet, this situation occurs in healthy rainforests. The problem in much of Queensland, as elsewhere in Australia, is that the forest has become fragmented and isolated as a result of human clearing, so that their capacity to regenerate has been substantially reduced. Just as with bushfires in Australia's pyrogenous forests, the impacts of cyclones in rainforests also need to be explained to visitors as a natural part of the ecosystem.

Human Geography of Australia and New Zealand

Aborigines

The Aborigines of Australia were the continent's first people, although they never grew into a very large population. Today they comprise approximately 1 percent (200,000) of the country's over 20 million people, which is about half their number at their maximum population size (about 250,000 to 500,000 in the late 1700s). Only about a third of them are pure Aboriginal; they are usually regarded as descendants of the Negrito peoples of South and Southeast Asia, although evidence suggests an earlier wave of human settlement. The traditional lifestyles and religious beliefs of Australia's indigenous peoples are closely related to their environment.

▶ *On Michaelmas Cay (Island) in the Great Barrier Reef, off the coast of Australia near Cairns. The Great Barrier Reef is the world's largest reef area, and it is also among the most endangered. Coral, which is a living animal colony, is highly susceptible to increases in water temperature, which can kill the coral and turn it into a colorless white debris (known as coral bleaching).*

Photo: Alan A. Lew

Prior to European settlement Australia was culturally part of Melanesia, including the island of New Guinea. They arrived in Australia at least 60,000 yrs ago, and 40,000 yrs later they had settled the entire continent, including Tasmania. Under white European rule, they have been victims of considerable discrimination and poverty.

European Discovery Australia

Arab and Chinese legends told of the presence of Australia, but without any degree of accuracy. The Portuguese were probably the first Europeans to see the continent, which appeared on their maps as early as 1528. The Spanish probably sailed the northern coast of Australia in the early 1600s in an expedition that originated in Peru, and the Dutch drew the first navigational chart of Australia's northern coastline. Tasmania was named after a Dutch sea captain (Abel Tasman), who sailed there from Batavia (modern-day Jakarta, Indonesia) in 1642. It is likely he was also the first European to visit New Zealand.

In 1770, Captain James Cook reached Australia and made landfall in several places, accompanied by Joseph Banks, a botanist who later became director of Kew Gardens in London and a leading natural scientist of the day. Cook and Banks also recognized the potential for colonization and settlement. The French were also active in exploring and attempting to colonize Australia in the 1770s.

Penal Colonies

The American Revolution of 1776 had a major impact on Australian settlement. Prior to the Revolution, British prisoners were sent to the American colonies, but after 1776 they were sent to new penal colonies in Australia, with the first settlers arriving in 1788. It is therefore not surprising that Australian historian Robert Hughes described Australia as the world's first gulag. Later these penal colonies became known as free colonies after the prisoners had served their sentences.

Gold Rush

Other periods of major European population growth in Australia occurred in 1851 when gold was discovered in the southeast, and in 1891 when gold was discovered in Western Australia. The sheep industry

Photo: C. Michael Hall

▲ *The Sydney, Australia, skyline, with the famous Sydney Opera House on the right. Sydney is the most popular urban destination for visitors to Australia, especially after it hosted the Summer Olympic Games in 2000.*

also grew to importance in the late 1800s in southeastern Australia. For many years Australians with a convict heritage often tried to hide their embarrassing ancestry, but from the 1970s it became a source of pride as Australia increasingly began to express its own cultural identity.

The Commonwealth of Australia

Beginning with New South Wales in 1788, several colonies were established under British rule on the continent. Each was based on a settlement that later became a major city: Sydney (New South Wales), Hobart (Tasmania), Melbourne (Victoria), Brisbane (Queensland), Adelaide (South Australia) and Perth (Western Australia). Later, Darwin became the capital city of the Northern Territory, although the Territory is still denied full statehood and the federal government retains the right to overrule Territory law. These coastal cities and their colonies often competed with one another for resources and immigrant populations, but in 1901 they joined together to form the Commonwealth of Australia as a federation of states and territories.

The Commonwealth of Australia consists of six states and two territories joined in a single federal state or government. In the federal system, each state retains certain rights, with the other specific rights granted to the central government. The system is similar to that established in the United States, but Westminster parliamentary traditions in Great Britain also greatly influenced it. In Australia the central government retains less power than in the United States, but the areas it has authority over have increased over time. For example, aviation did not exist in 1900 when the Australian Constitution was enacted, but under the constitution new areas became the responsibility of the Commonwealth (federal) government. It should be noted though that some areas under the United States Constitution, such as the right to bear arms and the freedom of expression, do not exist under the Australian constitution.

Australia is therefore somewhat similar to the European Union, although the central government is much weaker in the EU than in Australia or the United States. Also, people primarily identified with their state rather than with Australia for many years. However, as communication and transport linkages grew, and wars were fought, a sense of nationhood developed, although state rivalries still exist. A single-gauge rail linking all the capital cities was not completed until 2004 with the line between Adelaide and Darwin.

Canberra

Competition between Sydney and Melbourne also led to the creation of a new capital, Canberra, and the Australian Capital Territory, which is approximately midway between these two principal cities of Australia. The American architect Walter Burley-Griffin designed Canberra as well as the planned capital Brasilia, Brazil. Today, Canberra is also known as the bush city because of the extent to which native vegetation has been conserved.

Australia's Race Policies

Historically, Australia had a selective immigration policy aimed at making and keeping the country racially white. Some 80 percent of Australia's population descends from the British Isles, with the rest originating in other parts of Europe. Since World War II, Australia received considerable criticism of its white-preferred population policies, which have been a major topic of debate in the country. Australian Aborigines only became citizens of their own country in 1966 following a national referendum.

Today the Australian government has established reparation programs for the Aboriginal people, who lost much of their traditional lands, and the country has become more accepting of its absolute location next to Southeast Asia. Nevertheless, significant issues still remained with the Federal Liberal Government under Prime Minister John Howard during the late 1990s and the early part of the 20th century. That government was often regarded as racist because of its treatment of Aborigines and its policies toward illegal migrants or refugees.

Such was the extent of perceived racism that it was even regarded as being at odds with national tourism promotion and the attempt by the country to portray itself as a welcoming destination for visitors. The Australian government is increasingly active in Southeast Asian politics and recognizes common security interests with its neighbors to the north. An increasing number of Southeast Asians have immigrated to Australia both legally and illegally, which is changing the complexion of the country.

New Zealand Maori

As mentioned above, New Zealand has a sizeable minority population made up of Maori and other Polynesian peoples. The Maori are known for their traditionally fierce hunting skills. Maori has cognates in other Polynesian languages, such as the Hawaiian Maori, the Tahitian Maori and the Cook Islands Maori, which all share similar meanings. The word as used in contemporary English means native, indigenous or aboriginal (with a small a). Maori people themselves often use the term Tangata whenua (people of the land) to describe their relationship with a particular area of land. A tribe (iwi) or sub-group (hapu) is therefore tangata whenua in one area, but not others. Archaeological and linguistic evidence indicates that there were probably several waves of migration from Eastern Polynesia to New Zealand between 800 and 1300 CE.

Maori oral history describes the arrival of the ancestors from Hawaiki (a mythical homeland in Polynesia) by large ocean-going canoes (waka). There is possible evidence of earlier Polynesian settlement, although the numbers of people, if any, were likely very small. The settlement patterns of Maori were primarily determined by the climate and the capacity to grow kumara (a type of sweet potato) of which the southern limit is about the area of Christchurch in the South Island; below that latitude Maori were primarily dependent on fishing and marine gathering, hunting and trade.

Therefore, the vast majority of Maori were settled in the North Island and in the far north of the South Island, with the capacity to grow sweet potato being a major limit on settlement patterns. As population pressures built up, intertribal warfare by Maori for access to resources became increasingly common, particularly as the Maori killed off native species such as the moa (a flightless bird endemic to New Zealand), which was a major food source.

Christian Missionaries

In 1769, Captain Cook claimed New Zealand for England, and Christian missionaries from various countries and faiths soon followed. Christianity and the eventual acceptance of British law eventually brought an end to Maori intertribal warfare and cannibalism, although different coalitions of Maori tribes fought against each other in cooperation with the British. The New Zealand Maori Wars or Land Wars were not over until the 1870s, with passive resistance remaining until 1881. The outcome was a massive loss of Maori land to the British, reparations for which are still being sought and decided to the present day.

Today there is considerable interracial harmony between Maori and the European populations of New Zealand. In recent decades, there has been a major revival of Maori culture in New Zealand. Auckland is the largest city in New Zealand, and the largest Polynesian city in the world, with over 195,000, Samoans, Tongans, Cook Islanders and other Pacific Islanders, along with a significant Maori population, and the city often uses this feature in its tourism promotion.

Ethnicity in New Zealand

The presence of a large indigenous population makes ethnicity a topic of considerable debate in New Zealand. Multiple ethnicity is increasing, and people's ethnicities may change over time. In the 2001 Census, 9 percent of New Zealand's population identified with more than one ethnic group; this increased to 10.4 percent in 2006 and 11.4 percent in 2013. In 2013, 22.8 percent of people under 15 years of age identified with more than one ethnicity. Indigenous Maori people are the largest non-European ethnic group, accounting for 14.9 percent of the population in the 2013 census. In 2013, 11.8 percent of the population identified themselves as Asian, up from 9.2 percent in 2006, while 7.4 percent are of Pacific Island origin. Although people could select more than one ethnic group, slightly more than half (53 percent) of all Maori identified themselves solely as Maori. According to Statistics New Zealand , the population projection in 2021 will comprise:

- Maori—17 percent;
- Other Pacific ethnicity—9 percent
- Asian ethnicity—15 percent.

Population projections suggest that 70 percent of the population will identify with a European ethnicity in 2021, making European the largest ethnic group. Although it is often noted that the country has a strong British influence in its political and cultural institutions, it should be emphasized that there are significant regional differences in settlement patterns that have lingering effects to the present day.

Dunedin

For example, the Otago and Southland areas had a substantial Scottish Protestant influence to such an extent that the small city of Dunedin has much of its old streetscape laid out in a manner similar to the city of Edinburgh from which it derives its name. However, as a rule, the larger the urban settlement is, the more diverse the population. Therefore cities such as Auckland, Wellington and Christchurch have the most diverse populations, as well as the most diverse economic bases.

Geographic Similarities between Australia and New Zealand

Both Australia and New Zealand have a strong British heritage and are major agricultural countries where sheep are the most common agricultural animal and cattle are second. Both countries are major exporters of agricultural products, with Australia being a major cereals exporter and New Zealand exporting dairy products, such as milk and cheese. Both countries also export significant amounts of horticultural

ISSUES AND INSIGHTS   Australia's The Murray-Darling Basin: Environmental Management on an Arid Continent

The Murray River is Australia's second-longest river (1,470 miles; 2,530 km). Together with the Darling River (1,700 miles; 2,740 km) and Murrumbidgee River (1050 miles; 1,690 km), the Murray comprises the Murray-Darling Basin. The Murray-Darling Basin covers 409,835 sq miles (1,061,469 sq km), which is equivalent to 14 percent of Australia's total area, extending over the states of New South Wales, Victoria, Queensland, South Australia and the Australian Capital Territory (Canberra). The importance of this basin is that it accounts for over 40 percent of Australia's gross value of agricultural production and approximately 70 percent of all water used for agriculture in Australia is used by irrigation in the basin.

The Murray River is of immense cultural, economic and environmental significance. Gravesites at Roonka and Big Bend demonstrate that Aboriginal communities lived there continuously for at least 35,000 yrs. Hamilton Hume and William Hovell discovered the river in 1824. It was originally named the Hume River, but Captain Charles Sturt renamed it the Murray in 1830 after Sir George Murray, the British Secretary of State for the Colonies. The publication in London, in 1833, of Sturt's account of his river explorations indirectly led to the establishment of the state of South Australia.

For many years during the 19th century, the Murray was Australia's main inland transport route. In 1852, the first shallow draft paddle steamer from South Australia reached Echuca in Victoria. Such was the economic importance of the Murray that in 1863, almost 40 years prior to federation, a conference was held between New South Wales, Victoria and South Australia regarding improving the navigability of the river. Although the river is still used for river boats, including tourist craft, the loss of natural water flow, as a result of upstream vegetation clearance, dams, irrigation and silting up, has created some major environmental issues with respect to salination, species loss and water quality, which the state, territory and federal governments are seeking to solve.

Since the early 2000s, sand dredging machines have operated at the mouth of the Murray River to maintain a minimal flow from the sea. Without the dredging, the river mouth would silt up and close, cutting the supply of fresh seawater into the Coorong lagoon system, which is part of the Murray Estuary. Unfortunately, experts expect the environmental problems of the Murray to only get worse given the climate change scenarios for Australia.

► *Chinatown in Melbourne, Australia. The large numbers of Asian immigrants to Australia over the past several decades have created many colorful ethnic neighborhoods and restaurants, as well as helping to increase the country's political and trade connections to Asia.*

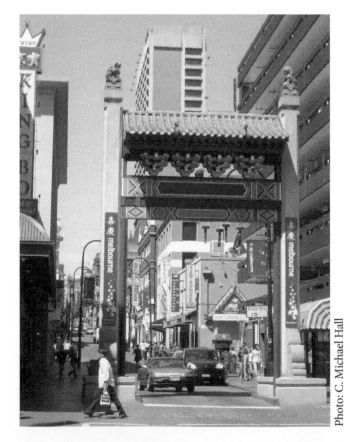

Photo: C. Michael Hall

produce to Asia and North America. In recent years specialist production has also started to develop, such as truffles and boutique cheese and wine in Tasmania, Australia, and almonds, saffron and walnuts in the Canterbury region of New Zealand.

Australia and New Zealand share the problem of having small internal markets that are not large enough to support the local production of most consumer goods without tariff protection, which they therefore import from abroad. They are both geographically isolated from their main markets, although they have become much more oriented toward Asia in recent years. Despite having strong traditions associated with the bush or the countryside and natural areas–images that are often part of national tourism promotion—they are both highly urbanized, with a large proportion of their population living in cities: 85 percent in New Zealand and 86 percent in Australia.

They both have social welfare systems similar to Northern Europe, with medical care, housing assistance, education and old-age care supported by government, although these systems are undergoing increasing change. In both countries the gap between rich and poor is also rapidly increasing following years of market deregulation and the privatization of many government services and agencies. This is now becoming a source of significant national debate as both countries also have strong traditions of equality of access and social mobility as part of their *national identity*. This is referred to colloquially as a 'fair go'.

Geographic Differences between Australia and New Zealand

Physical geography is a major difference between Australia and New Zealand. Australia is many times larger than New Zealand and is known for its vast desert outback, while New Zealand is known for its green lowlands and Southern Alps. However, the southern state and island of Tasmania (Australia) is much more comparable to New Zealand in terms of its geography.

ISSUES AND INSIGHTS ◖ Tourism and Urbanization

Tourism is regarded as a major contributor to urbanization, especially in coastal areas. It does this in two main ways. First, it is a very specific form of development in which urban settings are produced to meet the demand for tourism by the temporary resident or mobile visitor. These can be described as urban tourism destinations. Second, tourism is a part of broader processes of urbanization in which certain locations within the urban area are seen as having a tourism or leisure enabling function even though urban development is not being undertaken just for tourists. There are a number of features that can be used to distinguish places associated with tourism urbanization. They are:

- Spatially and functionally different from other urban places. The amenity-driven nature of much tourism means that along many coastal areas tourism urbanization is highly linear and runs along the coastline where permitted. For example, in Cyprus 95 percent of the tourism industry is located within 1.25 miles (2 km) of the coast;
- Symbolically different, with various brands, images and symbols used to promote the tourist function;
- Characterized by rapid population and labor force growth in the early stages of development. Even as urban growth

slows, the population and labor force tend to have a relatively high degree of transience, with considerable underemployment and unemployment as a function of construction cycles and tourism seasonality;
- Distinguished by flexible forms of production, particularly in terms of a highly flexible labor force organized to meet daily, weekly and seasonal change in demand (usually through high rates of short-term, part-time and casual employment and low rates of unionization);
- Dominated by government intervention and public-private partnerships, whereby government indirectly invests in facilities infrastructure with a view to encouraging further inward investment;
- Associated with large-scale tourism service production that simplifies the local economy and requires the substantial importation of goods, services, water and energy from outside the resort region in order to meet the demands of a highly mobile population;
- Sites of substantial transformation of natural or rural landscapes for tourism production and consumption; and foci of transport networks because of the need to import and export not only goods and services but also the tourists themselves.

▶ *Uluru, or Ayers Rock, in Central Australia. Aboriginal people discourage climbing the rock, but it is still legal to do so, as the hikers high on the ridgeline show in this photo. The situation reflects a tension that exists in Australia over the relationship between the dominant society and the Aboriginal people.*

Photo: Alan A. Lew

Australia's indigenous heritage is Negrito-Melanesian, whose modern-day Aboriginal descendants have had very poor relations with the European settlers. In Australia, unlike New Zealand (and the United States), no treaties were ever signed between the indigenous peoples and the invading colonial powers. In contrast, New Zealand is part of the Polynesian world and modern-day Maori have a relatively good relationship with the dominant European (Caucasian) settlers.

Both countries have become much more cosmopolitan and economically diversified in recent years, with Australia arguably being more conservative, particularly with respect to politics at the national level. Nevertheless, the two countries are very closely related. Under the Australian Constitution, for example, New Zealand is entitled to become a member of the Australian Commonwealth. There is an economic union between the two countries known as the Closer Economic Relationship (CER) and something approaching a common labor market. The extent of the closeness is also judged by the fact that tentative discussions have begun to develop about having a common currency.

There are also a number of governmental organizations shared between Australia and New Zealand, especially where common regulations exist. The two countries have also fought beside each other in a number of wars, which means they share the same remembrance day (ANZAC Day, April 25), so perhaps it is a matter of time before economic union starts becoming a political one.

Indigenous Tourism

The indigenous people of both countries form a considerable part of the tourism product. Tourism in New Zealand focuses overwhelmingly on nature (mountains and thermal areas) and Maori culture, and the Maori people are heavily involved in the production of tourism experiences. Likewise, by law they are active participants in tourism legislation, and the Maoris themselves must approve all use of their cultural emblems for tourism purposes.

This is not the case in Australia, where the Aboriginals and Torres Strait Islanders have historically had little say in tourism development and the use of their heritage for tourism purposes. Nonetheless, their sacred places, musical and dance traditions, and bush culture have become popularized aspects of Australian tourism. *Film-induced tourism* plays a major role in both countries. Movies associated with Australia, such as *Crocodile Dundee, Kangaroo Jack* and *Wolf Creek*, have bought international visibility to the Australian Outback and draw many tourists to the country. Similarly, in New Zealand, the *Lord of the Rings* and *The Hobbit* movie series has also presented new opportunities for tourism promotion.

Geography and Tourism of the Pacific Islands

Australia and New Zealand are part of a much larger human and physical geographical realm that is generally known as Oceania. Oceania extends from the islands of Australia, New Guinea and New Zealand, across the Pacific Ocean as far north as Hawaii and as far east as Easter Island off the coast of Chile. Antarctica is sometimes included in this realm as well. What this vast region shares in common are its ethnicity and its close relationship with water and ocean travel. With some major exceptions, the biogeography of Oceania is also very similar. Australia and New Zealand are the major international tourist destinations in the region accounting together for about 1.5 percent of global international tourism arrivals.

Island Diversity

Islands dominate most of Oceania. If it were not a continent, Australia, at 3,000,000 sq miles (7,769,964 sq km), could be considered the world's largest island. New Guinea, at 320,000 sq miles (828,796 sq km), is the world's second-largest island (after Greenland). Other large landmasses include the islands of New Zealand and

◀ *The World Centered on Australia. This map of the globe shows what one would see of the planet from a satellite directly over the continent of Australia. What is prominent in this map is the great extent of water that covers most of the planet, as well as the large gap between Australia and Antarctica. That gap comprises a large part of the Southern Ocean.*

© Volina, 2011. Used under license from Shutterstock, Inc.

New Caledonia (a French overseas territory). Most of the rest of the islands of this vast realm are much smaller. Many are volcanic in origin, and some are very active, including Hawaii and Fiji, while others are coral atolls that have developed over many millennia.

Microstates

The term South Pacific is generally applied to the island microstates of the southern Pacific Ocean that lie south of Hawaii, although the political and economic organizations of the region often include Hawaii (USA), Papua New Guinea, Australia and New Zealand. The island microstates, which are dominated by the Polynesian, Melanesian and Micronesian cultures, have few natural resources except for large ocean territories (i.e., fishing) and access to sun, sea and sand ("the three Ss"). The islands' isolation has resulted in substantial species endemism, dependence on the natural environment, diversity of cultures and language, and vulnerability to natural disasters. The region's colonial heritage means that states remain economically linked to their former colonial occupiers, while the region's image as an island paradise is also a colonial legacy rather than an actual reality, given histories of occasional political unrest, poverty and natural disasters (e.g., cyclones and tsunamis).

Pacific Island Tourism

The South Pacific region accounts for only 0.15 percent of global international tourist arrivals, and most of the travelers come from Australia, Japan, New Zealand and the United States. Nevertheless, tourism is a vital component of the regional economy and a major, though seasonal, employer. Cooperative agreements between South Pacific nations have been established with respect to economic development, fishing, conservation and tourism.

The South Pacific Tourism Organization (SPTO) is the lead intergovernmental tourism agency for the region. SPTO members include the following Pacific Island nations and territories: the Cook Islands, Fiji, Kiribati, New Caledonia, Niue, Samoa, the Solomon Islands, Tahiti (French Polynesia), Tonga, Tuvalu, Vanuatu and Papua New Guinea, as well as China, New Zealand and Australia, and over 180 private tourism operators. The SPTO was initially established as a development and marketing organization for tourism in the region with EU support, but the agency has increasingly narrowed its focus to marketing and promotional activities.

High and Low Islands in the South Pacific

There are basically two physiographic types of islands within the area known as the South Pacific (east of Australia, New Zealand and the islands of the Pacific Ocean), and the northern Pacific islands; these are high and low. High islands typically have a tall volcanic mountain in their center. This mountain causes the humid ocean winds to rise, forming clouds and precipitation. High islands tend to have year-round rainfall (an important source of fresh water) and tropical forests. The populations tend to be much larger on high islands because the rainfall and rich volcanic soils are able to support agriculture.

Low islands are usually coral atolls. They were once high islands but were eroded away by wind, rain and waves, and the original islands are now often well below sea level. Coral has grown on top of the original island, and the hard outer shelled material has eroded to form sandy stretches of land above the sea. Some of these are quite large, and they sometimes have a freshwater lagoon in their center. Because their elevations are not high, they do not capture much precipitation from the passing winds. They tend to be much hotter and drier, with some exhibiting almost desert-like conditions. Fishing is the main source of food on low islands, and their populations tend to be smaller than those on high islands. Both high and

◀ *A horseback rider in the interior hills of Viti Levu, the main Island of Fiji in the South Pacific. This is an example of a large, high island, with a relatively extensive inland interior that offers a very different experience, including trekking and camping, in comparison to the low coral islands of the Pacific.*

Photo: Alan A. Lew

low islands in the Pacific were formed primarily by collisions of several large oceanic plates that caused earthquakes and volcanic activity. Many of the low islands are regarded as some of the most at-risk countries to climate change and sea level rise in the world today.

Human Geography of the South Pacific
Melanesia

The Negrito people (first introduced in the Southeast Asia section) were the earliest to settle Oceania. As mentioned previously, it is uncertain how long they have lived in Southeast Asia and Australia, though they arrived at least 40,000 BCE and possibly as early as 70,000 BCE and they clearly predate the arrival of the Malayo-Polynesian peoples. Today, their descendants survive in small numbers on the islands and interior hills of South and Southeast Asia, but in much larger numbers in Australia and the Melanesia region of the South Pacific. "Mela" means black, and Melanesia is one of the three major regions of the South Pacific. Included in Melanesia are New Guinea, which is home to the diverse Papuan peoples, the Solomon Islands and New Caledonia, all of which are among the largest islands in the South Pacific, as well as Vanuatu and Fiji.

Malayo-Polynesian People

The Malayo-Polynesian people, also referred to as Austronesian people, most likely migrated out of Taiwan about 3000 BCE, moving down through the Philippines and into insular Southeast Asia (about 2000 BCE). From there they spread out into the Pacific Ocean (about 1500 BCE). By 1000 BCE, they had settled most of the South Pacific. There last major migrations took them to Madagascar (about 300 CE) off the coast of Africa, the Hawaiian Islands (about 500 CE) and New Zealand (about 1100 CE). The Polynesians used stars (their height above the horizon) and ocean currents (by which they could also detect nearby islands) to navigate the vast waters of the Pacific Ocean.

Micronesia

The Malayo-Polynesians have three major groupings. The first is located in insular Southeast Asia, including Malaysia, Indonesia, the Philippines and aboriginal groups in Taiwan and possibly Japan. This grouping is not considered part of Oceania or the South Pacific. The second major group is in a region known as "Micronesia", which consists of many small islands located east and northeast of peninsular Southeast Asia as far as the international dateline. The major racial distinction of Micronesians is their straight hair and sometimes smaller stature. Legends indicate that they may have been the earliest inhabitants of the Hawaiian Islands. The islands of Micronesia are almost all low islands.

Polynesia

The third major settlement area of the Malayo-Polynesian peoples is the area known as "Polynesia". This consists of the islands located in a triangle from New Zealand to Hawaii, all the way to Easter Island off the coast of South America. Racially, the people of Polynesia tend to have curly hair and lighter skins than Melanesians. The word poly means many, referring to the large number and variety of Islands found in Polynesia. Both low and high island lifestyles and environments exist in Polynesia. The major high islands are the Hawaiian chain and the archipelagos of Samoa, Tonga and Tahiti. The indigenous people of New Zealand and Madagascar are also Polynesians.

South Pacific Subregions

The three subregions of the South Pacific, therefore, are Melanesia, Micronesia and Polynesia.

1. Melanesia includes the countries of West Papua (Indonesia), Papua New Guinea, the Solomon Islands, Vanuatu, New Caledonia and Fiji.
2. Micronesia includes Palau, the Federated States of Micronesia (FSM), Nauru, Kiribati, the Marshall Islands, the Northern Marianas (USA), Wake Island (USA) and Guam (USA); Guam is the largest of the Micronesian islands.
3. Polynesia encompasses all of the island groups east of Melanesia and Micronesia, from New Zealand in the southwest, to Hawaii in the north, and to Easter Island in the southeast Pacific Ocean. Prominent among these are Midway Islands, Samoa and American Samoa, Tonga, Tuvalu, Cook Islands and French Polynesia (which includes Tahiti). Mutineers from Captain Bligh's ship, the Bounty, settled Norfolk Island and Pitcairn Island. They married Polynesian women, creating a mixed-blood contemporary population.

Modern Pacific Migrations

Migration and mobility are inherent to the peoples of the Pacific Ocean. Mobility remains a central component of the lives of many Pacific islanders, from Polynesian migration throughout the islands of the South Pacific (including New Zealand) during the pre-European period to the labor migrations in comparatively recent decades from the islands to Australia, New Zealand and the United States. Yet although island populations are highly mobile in terms of employment and education, substantial bonds of kinship and relationships to village and land remain. The magnitude and importance of migration from the Pacific Islands has strong historical roots.

The interface created between expanding colonial powers and indigenous peoples in the 19th century helped broaden the base of movement in the region. Between the 1950s and 1970s, the rate of migration from the Pacific Islands, particularly Samoa, to New Zealand was substantial. The size of the Samoan population in New Zealand rose from 6,481 in 1961 to 27,950 in 1976. The primary reasons for migration were employment and family connections. Labor shortages in the New Zealand economy, which existed until the mid-1970s, provided both short- and long-term migration opportunities. In the 1950s, the main point of migration was to seek financial returns for the home community (remittances). Communities often selected single females to migrate because they regarded them as more likely to remit their wages.

By the 1960s, migration began to have a greater family focus, with reunification becoming an important migration impetus.

Pacific Island Economies

While post-war economic growth within the countries bordering the Pacific Islands has, overall, been strong, many island nations have experienced considerable fluctuations in their own economies. Many Pacific nation-states, for example, struggled during the recession in the early 1980s, which was largely a result of the reliance on relatively few economic activities (i.e., agriculture and fishing). As a consequence, while the region was on the extreme economic periphery, it facilitated movement to nations that were more developed, such as Australia, New Zealand and the United States, where some family connections also lay (such as for American Samoans).

As more migrants elected to emigrate, family members often joined their kin in their new homelands. Policies in the receiving countries with respect to migration, particularly the restriction of numbers, have also played an important role in influencing migration patterns. Many restrictions operate within the larger sphere of international economic conditions and thus dictate the rate and flow of migrants. However, Australia and New Zealand have developed special working visa schemes for residents of several Pacific Island countries to enable them to come to work for short-term periods, usually for three months, in agricultural and horticultural activities such as harvesting, before returning to their own countries. These schemes together with the geographical spread of Pacific Islanders, the importance of family connections, and long traditions of travel also mean that much of the tourism to Pacific Island countries, as well as the travel by Pacific Islanders themselves, is based on VFR (Visiting Friends and Relatives) tourism.

Political Ties

Some of the islands also have direct political connections to more developed countries, such as American Samoa and the United States. Nationals from the Cook Islands, Niue and Tokelau are also citizens of New Zealand, so their movement is legally and politically unrestricted. In addition, political unrest has influenced some migration flows in the region, particularly the Fijian Indian community, in the case of military coups that have occurred several times in the island nation. Unfortunately, in the case of Fiji, this has often meant that many of the best educated of the Fijian Indian community (e.g., physicians, teachers, academics) have emigrated, especially to Australia and New Zealand.

Tourism in the South Pacific

In terms of tourism and broader development, the major handicaps of the South Pacific islands are their small size and isolated/remote location. Because the islands are small, they tend to have high population densities and land use pressures relative to their carrying capacities. Because they are remote, shipping goods and transporting visitors to and between the islands is very expensive. Likewise, several of the islands are so remote that access can be gained only once or twice a week by air or monthly by ship. Although some have had useful mineral resources in the past, for the majority the advantages are natural and cultural amenities, such as natural landscapes, warm climates, unique cultures and recreational opportunities.

Early European Arrivals

Europeans arrived in the South Pacific in the latter half of the 1700s. Captain Cook made several sailings between 1768 and 1779. Captain Bligh (made famous by the book, *Mutiny on the Bounty*) charted the South Pacific from 1787 to 1789. Cook's and Bligh's charts were used by European navigators well into the 20th century. In the early 19th century, whalers and missionaries entered the Pacific, settling on islands that had been claimed by various European powers. In the late 19th century, steamships and the first

Western tourists arrived in the Pacific; the steamships reduced the time it took to reach the South Pacific from the coast of North America to only a few weeks.

Pan Am's China Clipper

Air travel has played a major role in overcoming the isolation of the South Pacific, though it is still quite expensive to reach many of the islands. In 1935, Pan Am's China Clipper made the first trans-Pacific flight from the United States' west coast to Australia. The trip took six days with 60 hours of flying time. The flight followed the charts drawn by Captain Cook and Captain Bligh and was funded by the United States Post Office, which initially used the China Clipper for postal services. Later, commercial and government/diplomatic travel were the primary users of the China Clipper, although the first tourists to fly across the Pacific did so as early as 1936.

Following World War II, there was a major expansion of air travel across the Pacific. The war made Americans more aware of the Pacific (through the soldiers who fought there and media coverage) and generated more interest in traveling there. In 1947, it still took six days to travel by ship from California to Hawaii, 14 days from California to Asia, and 20 days from California to Australia. In contrast, by airplane it only took 13 hours to fly from California to Hawaii. In 1957, Boeing introduced the first successful passenger jet aircraft (the 707), which brought flying times closer to those of today (about six hours from California to Hawaii). The development of airline technology after World War II led to airports being built and upgraded throughout the Pacific region, along with new and expanded accommodations and services for tourism development.

Fiji and Tourism Dependence

Today, many of the islands are heavily dependent upon tourism for their financial wellbeing. North Americans, Australians and New Zealanders especially favor Fiji, the most popular destination in terms of annual arrivals. The country has been the focus of much resort development in recent decades, with new resorts currently being built. Fijian culture, ecotourism and beaches are the main products in Fiji, which include visits to cultural centers. Many of its smaller, outlying islands, have become resorts in their own right.

As alluded to earlier, military coups have plagued Fiji for the past few decades. A non-violent coup in December 2007 effectively halted tourism to the islands for a few months, although it has since recovered well. In September 2014, Fiji again became a parliamentary democracy following the first general election since the 2007 coup. Most of the islands (e.g., Cook Islands, Samoa and Tonga) have the similar products of famous sun, sea and sand destinations. Political unrest, interethnic violence (between the Solomon Islanders and ethnic Chinese), and natural disasters in recent years have all but completely curtailed tourism in the Solomon Islands, and the country has appeared on warning lists throughout the world.

Nauru

Nauru is a unique case, wherein the island desires to grow tourism, but it has little by way of tourist offerings. It has a few beaches, but most of the island was destroyed in the 20th century by phosphate mining, which has since ceased production. There is little industry on the island, although there is a minor fishing sector. Most citizens rely on monthly remittances from overseas, and the island's budget is based largely on the interest from a lawsuit won against the British and Australians for depletion of the island's only natural resource (phosphate). Travel to Nauru is expensive and infrequent. The island has its own airline, but it operates at a financial loss. The government of Nauru subsidizes it heavily, and the airline provides only periodic flights to a few other islands in the region.

▲ *The island of Bora Bora, in the Society Islands of French Polynesia. The highest peak of Bora Bora, Mount Otemanu (2835ft (727 m)) can be seen here, along with the ring of coral that surrounds the old volcanic core of the island. Also seen are some of the other Society Islands in the distance. French Polynesia is often referred to as Tahiti, although that is only the name of the largest of the Society Islands.*

© wilar, 2011. Used under license from Shutterstock, Inc.

Scuba in Micronesia

The Micronesian islands, particularly Palau, the FSM, Guam, and the Marshall Islands, are well known for scuba diving, depending on the countries' white sand, blue waters and vast array of coral reefs. Some of the Micronesian island countries are also trying to establish heritage tourism based on their living cultures and World War II battle sites and memorials. Guam is still an integral part of the United States, but FSM, the Marshall Islands and Palau only recently (1980s and 1990s) gained their independence from the United States-administered UN Trusteeship, whose sovereignty over the islands resulted from the defeat of Japan in World War II. As a result, relations with the United States are still very strong, and most of the islands' tourists come from the United States and Japan.

4.5 | ANTARCTICA AND THE OCEANS

Current Antarctica and Ocean Issues and Resources (http://wrgeography. com/oceania.html)

The landmass surrounding the South Pole is the continent of Antarctica. It is larger than Europe, but it has no permanent human population and only minimal forms of biological life. Massive sheets of ice and almost constant stormy seas kept Antarctica isolated from much of the rest of the world until the late 1700s and early 1800s. Antarctica remains a frontier region hostile to human habitation and seldom visited except by scientists and a small number of cruise passengers and adventure tourists.

▶ *Antarctica*

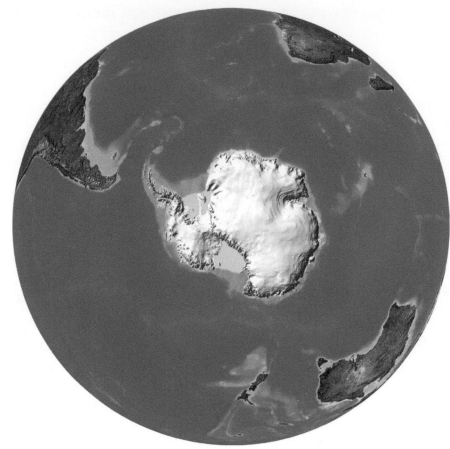

Physical Geography of Antarctica

The Antarctic was part of the supercontinent Gondwana around 200 million years ago, along with South America, Africa, Madagascar, India, Australia and New Zealand. At that time, Antarctica was a land of lush vegetation and many animals. Fossils and thick coal beds are what remain of that time today. However, a number of *Gondwanic* species can still be found in the Americas, Australia and New Zealand, including Nothofagus (beech), rattites (flightless birds such as Darwin's Rhea, emu and ostrich) and marsupials (in Australia and the Americas). Beech forests existed in Antarctica as recently as three million years ago.

Ice

Ice covers nearly all of Antarctica; its average thickness is 7,000 to 8,000 ft (2,134 to 2,440 m), although at its deepest it is almost 3 miles (5 km) thick. The ice sheets are as old as 14 million years and contain 90 percent of all the ice and 70 percent of all the fresh water in the world. Should global warming melt the thinnest portions of this ice sheet, which is possible, world sea levels would rise about 20 ft (6 m), entirely endangering the physical existence of countries such as Tuvalu and the Maldives. Some 2 percent of Antarctica is ice free; these small areas are known as oases and are located on the continent's edge.

Antarctic ice flows very slowly into the oceans that surround it, and it extends over the oceans in ice shelves. The largest of these, the Ross Ice Shelf, is larger than France and is an average of 1,000 ft (300 m) thick. In winter, which is July in the southern hemisphere, the outer edge of the ice shelves can extend up to 1000 miles (1,600 km) from the continent. The edges of these shelves periodically break off and float as icebergs northward toward warmer waters.

The Antarctic Convergence

The *Antarctic Convergence* is a zone located approximately 1000 miles (1,600 km) off the coast and is the transition area between the icy cold waters that surround Antarctica and the warmer ocean waters further to the north. Also surrounding the continent is the Antarctic Circumpolar Current, which moves clockwise, or eastward, around the continent and is the longest ocean current in the world. Antarctica has the world's lowest recorded temperature, -128.6°F (-89.2°C). The average temperature in the interior is -70°F (-57°C), while the warmest temperatures are on the Antarctic Peninsula and can reach as high as 59°F (15°C) in the Antarctic summer (January). The interior of Antarctica is in fact a desert, owing to the high-pressure air system that resides over the South Pole. The average precipitation is only 1 to 2 in (2.5–5 cm) of rainfall equivalent a year.

Endemic Species

The limited plant and animal life in Antarctica includes many species that are only found there (endemic). Some bacteria, lichens and algae grow close to the South Pole. Insects are the most abundant animal life; there are 76 species, nearly all of which are only found in Antarctica. Forty-five species of birds live south of the Antarctic Convergence, including several penguin species. Four species of seal live only in the Antarctic region. Fish are quite abundant and have evolved blood that enables them to live in waters as cold as 28°F (-2°C). Krill are a small shrimp that exist in huge numbers in the waters surrounding Antarctica and comprise the basic food for birds, seals and whales. Increased pressures on Antarctic krill and fish stocks is coming from fishing in Antarctic and Southern Ocean waters where, even here so far from permanent human populations, overfishing is a growing concern.

Human Geography and Tourism of Antarctica
European Exploration

Although there may be been earlier visitors, James Cook was the first known to have seen Antarctica between the years 1772 and 1775, though he never landed on the continent. In 1820, British and American seal hunters joined a Russian expedition and made the first landfall on the Antarctic Peninsula, but the first recorded landing on the main continent of Antarctica did not occur until 1895.

On December 14, 1912, the expedition led by Roald Amundsen of Norway became the first people to reach the geographical South Pole, arriving their before the British expedition led by Sir Robert Falcon Scott, which perished on their return from the Pole. This period of geographical exploration is often referred to as the 'heroic age' of Antarctic exploration. But in many ways this terminology hides the fact that polar exploration was grounded in imperial and mercantile aims as much as scientific exploration, with many expeditions financially supported because of their potential to locate new whaling, sealing and fishing grounds.

Modern Research Claims

Antarctica entered the modern age in the International Geophysical Year (IGY) of 1957–1958. That year some 50 permanent research stations were established, including one at the geographical South Pole (the center of the earth's, rotational axis), and one at the geomagnetic South Pole (the center of the earth's geomagnetic field and of the southern hemisphere auroras). (In addition, there is the magnetic South Pole, which is the point on which compasses align.) Seven countries have claimed pie wedge-shaped sectors extending out from the South Pole. Three of these sectors overlap and one region remains unclaimed.

While the claimant states insist that their Antarctic claims are part of their national territory, most other countries of the world do not recognize these claims, and these areas, according to international law under the Antarctic Treaty, are not considered part of the sovereign territories of the claimants. Some claimant countries maintain and treat their Antarctic areas as integral parts of their national territory with 'permanent' settlements, postal services, issuing of birth certificates, and other government administrative duties (e.g., Chile and Argentina).

Sovereignty

The legal criteria for a country establishing sovereignty over an area is the ability of the area to support human habitation, a demonstrated a history of active occupation, and the effective exercise of state functions in the asserted region. Despite the fact that the Antarctic Treaty disallows the universal recognition of the Antarctic claims as integral territories of the administering states, extant government services, together with tourism, are used as a means of trying to satisfy the otherwise legal definition of sovereign control.

Environmental Protection

Twelve countries were part of the IGY agreements to protect Antarctica's resources and preserve them for scientific use, which was codified in the 1959 Antarctic Treaty. These included Argentina, Australia, Belgium, Chile, France, Japan, New Zealand, Norway, South Africa, the United Kingdom, the United States, and Russia (the Soviet Union at the time). The Antarctic Treaty does not allow military activities except for scientific and peaceful purposes. Other countries have since joined the Antarctic Treaty.

Antarctic Tourism

Tourism to Antarctica occurs in the summer months, between the end of November and early April. According to the International Association of Antarctica Tour Operators (lAATO), approximately 40,000 tourists, along with staff and crew, visit Antarctica each year, with Australia, the United States and China are largest markets for tourists to Antarctica (as of 2014). Norway has become a major player in Antarctic tourism, with nearly 20 percent of Antarctic tourists visiting on Norwegian cruise ships. Many other cruises depart regularly from the Chilean and Argentinian portions of Tierra del Fuego.

A typical 15-day cruise, including airfare from North America and twelve nights on the cruise ship, costs approximately US$5,700. Cruise tourism is primarily self-regulated under IAATO and the regulations of Antarctic Treaty countries, and its impacts have so far been relatively minimal. The greatest concerns are the introduction of exotic plant and animal species that could endanger the unique biogeography of Antarctica and major shipping accidents that lead to oil spills.

The Ocean World

An Unknown Geography

As a terrestrial animal, humankind tends to emphasize the diversity of life and landscapes on the 29 percent of the earth comprised of continents and islands. However, most of the planet (71 percent) is actually covered by water, and the oceans comprise 97 percent of that. The oceans are a major source of food for humans, accounting for approximately 10 percent of human protein consumption. They are also a very important medium for human migration, commercial transportation, and recreation and tourism. The more shallow continental shelves have also served as important sources of fossil fuel and mineral extraction. And recent research has demonstrated the major role that the oceans play in both regulating climate change and directing its impact in some locations over others.

Despite its substantial importance, much of the geography of the oceans is still unknown to this day. For example, new animal species and ways that the oceans impact the global climate are continually being discovered. Our lack of adequate knowledge of ocean ecosystems often results in human-caused environmental disasters, including oil spills, overfishing and large-scale pollution.

The Pacific Ocean

There are five oceans that most geographers agree on today. The Pacific Ocean is the largest, covering 28 percent of the planet (about the same as all of the land areas combined). It has an average depth of 13,215 ft (5,028 m), and also contains the deepest spot on the planet in the Challenger Deep in the Mariana Trench (near Guam, south of Japan). That point is 35,840 ft (10,924 m) below sea level. Most of the Pacific Ocean is surrounded by the Pacific Ring of Fire, as discussed in the East Asia section of this chapter. This is where various oceanic and terrestrial tectonic plates are colliding, causing frequent earthquakes and volcanic and other mountain-building and island-creating activities.

Underwater earthquakes can result in large tsunami events (multiple large waves), caused by a sudden shift in ocean floor elevation. These can flood island and coastal areas around the entire Pacific Ocean basin. A Pacific-wide tsunami warning system has been created which includes floating wave detectors (buoys) that help to confirm if an earthquake-generated tsunami might threaten life and property. The Atlantic Ocean is the second largest ocean at about half the size of the Pacific Ocean.

The Atlantic Ocean

Unlike the Pacific Ocean, the Atlantic is an area where the major tectonic plates are moving away from one other, with North and South America moving westward and away from Europe and Africa. The equator divides the Atlantic and Pacific oceans into northern and southern portions.

The Indian Ocean

The Indian Ocean is the third largest ocean and is only a little smaller than the Atlantic. Because most of the water area of the Indian Ocean lies right on the equator, it is also the warmest of the oceans overall and a major source of moisture for the South Asian summer monsoons.

The Arctic Ocean

Both of the polar regions have their own distinct oceans. The Arctic Ocean, which covers the North Pole, is the smallest of the world's oceans, and also the shallowest. Historically, it has been covered with a layer of ice about 30 ft (10 m) thick through most of the year, though the extent of summer melting has been increasing since the early 2000s. The melting of ice has been accompanied by an increase in shipping and of cruise ships around the fringes of the Arctic Ocean. (See Chapter 6 for more on the Arctic Ocean and climate change.) The countries bounding the Arctic Ocean have also become more assertive in claiming territorial rights to potential ocean floor mineral resources.

The Southern Ocean

The Southern Ocean is the newest ocean in terms of recognition, being designated by the International Hydrographic Organization in 2000. Although it appears to be the southern portions of the Atlantic, Indian and Pacific Oceans, it is functionally very distinct from its adjacent siblings to the north. Officially,

▲ *The North and South Poles, showing the recently designated Southern Ocean that surrounds Antarctica, and the Arctic Ocean, which is likely to become ice-free for the first time in human history during most of our lifetimes.*

it includes all of the ocean waters south of 60° south latitude. Its dominant feature is the Antarctic Circumpolar Current, which is a continuous current of cold water and upper latitude cold fronts (cyclonic storms) that encircle the continent of Antarctica. The eastward (clockwise) flowing current is the longest ocean current in the world, and the storms in the air above it have the strongest average winds in the world, making travel to Antarctic a challenge. A large part of the Southern Ocean freezes in the southern hemisphere winter months when most of the ocean is without sunlight.

Ocean Threats

Global warming, over fishing and litter pollution are among the major threats to the oceans today. As the planet has increased in temperature since the 1960s, the oceans have also been getting warmer. This has resulted in massive ice melting in the polar regions and the death of vast areas of coral (a living animal) in many coastal areas. Ocean water levels, which vary around the world due to differences in the shape of the earth's crust, are also gradually increasing and causing more frequent flooding of coastal lands and settlements. These disasters will impact up to a billion people in the coming century.

Climate Change and the Oceans

Ocean temperatures have a huge, and poorly-understood, impact on climate and weather patterns. As large bodies of water, the oceans are very slow both to gain and lose heat (as opposed to the continents, which gain and lose heat from the sun very rapidly). Even if the earth were to stop warming now, it would take several decades for the oceans to cool again to the temperatures they were a couple of decades ago. In addition, ocean currents circulate warm and cold water across the surface of the planet, which also helps to stabilize the much more dramatic temperature changes that occur on the earth's continents. This may be why the overall temperature of the planet has not increased significantly in the 2000s, even though land temperatures have increased significantly.

Overfishing

It is estimated that in 1900 there were six times as many fish in the oceans than there are today. The greatest loss of fish has been of the larger, predatory species, including tuna and cod. One of the reasons for this is because no one owns the *open seas* (beyond 200 miles off the coastlines), which allows anyone with a large enough net to catch an unlimited number of fish. Overfishing in waters close to coastlines can be managed to some degree, though this depends on poorly developed science and effective (and costly) government actions.

The biggest threat to fishing comes from bottom trawling (also known as bethnic trawling), in which massive nets are dragged along the bottom of the oceans, capturing or killing everything in the way. It is estimated that the amount of sea floor that is scraped by bottom trawling is 150 times the amount of forests that are clear cut worldwide each year.

Industrial scale fishing of this kind is a major focus on government regulation and environmental groups. However, about 90% of the world's fishing is undertaken by small scale operators who depend on fish resources for the family's livelihood. Awareness by consumers and better ocean and coastal management are essential to address the loss of our underwater marine populations and ecosystems. Some success in the recovery of bluefin tuna in the Atlantic and Mediterranean Seas, for example, allowed an increase in tuna fishing quotas for those area in 2014.

Ocean Garbage Patches

There are few more unsightly natural scenes than a beach cluttered with litter and debris. Unfortunately, much of that debris is not from local sources, but it may have come from hundreds or thousands of miles away. Humans have long assumed that, because the oceans are so big, their disposal of liquid and solid waste into them would, ultimately, have no impact. Ships, from small fishermen to large cargo and cruise ships, have historically spewed litter, oil and sewage directly into the oceans, and many continue to do so today.

Urban settlements and coastal industrial activities have similarly used the oceans as open sewage pits. Plastics, in particular, can be fatal to marine life that encounter them and require many decades to decay. Areas with the worst concentrations of plastics tend to be major coastal tourist destinations (Europe, the U.S., the Caribbean and Indonesia). In addition, remote areas of both the north Pacific and north Atlantic oceans have developed massive garbage patches, where small and large fragments of plastics accumulate and circulate a thousand miles from the nearest coastline.

Marine Protected Areas

Since the mid-2000s several very large marine protected areas have been established in the Pacific Ocean in which fishing is now either banned or severely limited. The largest of these is the Pacific Remote Island Marine National Monument, which encompasses 490,000 sq mi (1,271,500 sq km; about the size of Spain and a little less than twice the size of Texas) between the Hawaiian Islands and Samoa. International Maritime law allows countries to control up to 200 miles from their territories and the U.S. used its seven Pacific island and atoll territories, inhabited only by temporary military personnel, to create the protected area. The national monument includes over 130 biologically rich underwater "sea mounts."

Other major marine protected areas around the world include:

- Northeast Greenland National Park (Greenland, Denmark)—357,917 sq mi (927,000 sq km)
- Chagos Marine Protected Area (British Indian Ocean Territory)—247,105 sq mi (640,000 sq km)
- Phoenix Islands Protected Area (Kiribati)—157,626 sq mi (408,250 sq km)
- Papahānaumokuākea Marine National Monument (Hawaii, USA)—138,997 sq mi (360,000 sq km)

- Great Barrier Reef Marine Park (Australia)—133,205 sq mi (345,000 sq km)
- Galapagos Marine Reserve (Ecuador)—51,352 sq mi (133,000 sq km)

The goal of these protected areas is to allow for the natural migration of large pelagic fish to enable their recovery and protection from overfishing. Marine ecotourism, including whale watching and scuba diving, are secondary opportunities offered by these protected area.

Consumers and Ocean Protection

Because the oceans are so large, they are beyond the human sensory capacity to comprehend. How individual consumers can reduce their impact on the oceans is equally challenging. Some of the things that you can do to ensure a rich ocean resource for the world include (source: National Geographic):

1. Reducing your carbon footprint to reduce the amount of carbon dioxide in the atmosphere that is being absorbed by the oceans. Atmospheric carbon is making the oceans warmer and more acidic. Reducing your oil and electricity consumption directly reduced your carbon impact on the environment.
2. Reduce your chemical footprint, much of which ends up in our rivers and oceans. Reduce your use of plastic that contributes to oceanic garbage patches, especially plastic shopping bags that are hazardous to marine life. Dispose of chemicals, paints and medicines safely—never pour them into a toilet or open drain because they will very easily end up in our rivers and oceans. Limit your use of pesticides and fertilizer (and foods that are grown with them), as well as toxic household cleansers.
3. More challenging is being aware of which fish are being overfished and not purchasing them in restaurants and stores. Make sustainable seafoods your preference. You can find out which fish are sustainable and which are not through the internet.
4. Vacation in environmentally friendly beach and ocean resorts. Ask how their sewage and swimming pool water (with high chemical content) is disposed and if the fish in their restaurants are from sustainable sources. Stay away from fragile dunes, do not disturb plants and wildlife, do not take or buy products (including jewelry) made from coral, shells and other marine animal parts. Do not litter; participate in beach and waterway litter cleanup efforts.
5. Support organizations that help to protect our oceans.

Taking at least some of these actions will not only help the oceans, but will also benefit life throughout the many varied and fascinating geographies over our endlessly evolving planet Earth.

CHAPTER 4 STUDY QUESTIONS & ACTIVITIES

Name ____

Human Mobilities

1. Identify and describe two examples of human mobility that shaped the historical development of one or more of the subregions of Asia and Oceania.

____

____

____

____

2. Identify and describe the role of human mobility in the contemporary societies of one or more of the subregions of Asia and Oceania.

____

____

____

____

Tourism Destinations

3. Describe the contemporary international image of one or more of the subregions of Asia and Oceania. How do these image impact the potential of the regions as tourist destinations?

____

____

____

____

4. Select one or more of the subregions of Asia and Oceania and describe how the countries in the subregion might use their cultural and physical resources to improve their global image and attractiveness.

____

____

____

____

Sustainable Environments

5. Identify and describe tine major environmental challenges of one or more of the subregions of Asia and Oceania. What challenges do the countries in the region face in addressing these issues?

_____

_____

_____

_____

6. Identify and describe the major social and economic development challenges of one or more of the subregions of Asia and Oceania. What challenges do the countries in the region face in addressing these issues?

_____

_____

_____

_____

Instructor Questions (answer any additional questions from your course instructor in the space below)

_____

_____

_____

_____

CHAPTER | 5

The Americas, North & South

5.1 ANGLO NORTH AMERICA

5.2 LATIN NORTH AMERICA

5.3 SOUTH AMERICA

Chapter 5 Online Resources (http://wrgeography.com/5.html)

CHAPTER 5 OVERVIEW QUESTIONS

HUMAN MOBILITIES

- What are the different ways that the Americas (North and South) have been shaped by migrations from the other parts of the world?

TOURISM DESTINATIONS

- How do the touristic images of the regions and countries of the Americas vary based on their geographies, histories and modern challenges?

SUSTAINABLE ENVIRONMENTS

- What are the contemporary relationships between the natural and human geographies that existed in pre-Columbian times and those that have be introduced post-1492 in the Americas?

INTRODUCTION

This chapter covers the continents of North America and South America. North and South America, together with Australia and New Zealand, are known as "The New World", as opposed to Eurasia and Africa, which are known as "The Old World", as defined from a European colonial powers perspective. In some parts of Eastern Europe, the idea of the Americas as "colonized lands" is so strong that the two are considered as a single continent in geography textbooks. In physical terms, however, the North American land mass and the South American land mass comprise two distinct continental bodies that are just barely touching one another in the country of Panama. Each of these continents has considerable internal diversity, both in physical and human geography. However, for this chapter, we only divide North America into two sections and leave South America, which is among the smallest continents, as a single entity.

There are actually two widely used definitions of the term *North America*. The first is based on physical geography and defines the North American continent as all of the lands north of the South American continent. The second is based in cultural geography and refers to the mostly English-speaking cultural region of the United States, Canada and the island of Greenland (a territory of Denmark). This cultural region is also referred to as *Anglo North America*, and it is covered in the first section of this chapter. The term Anglo North America is used primarily to distinguish it from the mostly Latin speaking areas to the south. It is not entirely accurate given the large Francophone region of Quebec in Canada, Danish Greenland, and the small islands of Saint Pierre and Miquelon, which are a French territory off the coast of Newfoundland, Canada.

The mostly Spanish-speaking region of the North American continent encompasses lands and islands south the United States. *Latin North America* includes three major subregions: Mexico, Central America and the islands of the Caribbean Sea. In addition to the language differences, the long border between the United States and Mexico marks one of the few places in the world where a highly industrialized country is directly adjacent to an economically developing country. The Caribbean is only partially Spanish speaking, with major French-, English- and Dutch-speaking islands, as well. Although geographers have traditionally referred to this region as Middle America, the second section of this chapter is called Latin North America, to better distinguish it from Anglo North America. Because physical geography does not always follow political or *linguistic boundaries*, some of the physical geography of Latin North America is discussed in the Anglo North America section.

The geographic term *Latin America* usually refers to all of Latin North America (covered in section 5.2 of this chapter), plus all of South America. Although South America is almost twice the size of Australia in land area, it is still considerably smaller than Asia, Africa and North America. The continent of South America has several significant subregions, including the Andes Mountains, the Amazon Basin and the Pampas in the south, and its major cultural division is between Portuguese speaking Brazil and the other countries where Spanish and Native American languages are spoken. These are all discussed in the third section of this chapter.

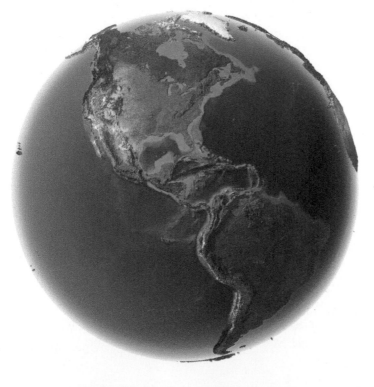

◀ *North America and South America from space, showing their major physiographic landforms and waters. (Note: This image exaggerates elevations to make them more prominent than they would appear on the actual planet.)*

5.1 | ANGLO NORTH AMERICA

Anglo North America Geography Issues and Resources (http://wrgeography.com/anglo-namerica.html)

Physical Geography of Anglo North America

Physiography refers to the shape of the surface of the earth. This is known more popularly as *topography*, although the formal definition of topography also includes vegetation and built structures on the land (i.e., everything that you might see on a topographic map). The physical landscape of Anglo North America is comprised of the following physiographic regions (see Map, page 338):

1. *The Interior Lowlands and Coastal Plains*, which from the far north to south consist of the Arctic Coastal Plain, the Canadian Shield, the Central Lowlands and the Great Plains, and the Gulf (of Mexico) and Atlantic (Ocean) Coastal Plains.
2. *The Appalachian Mountain System*, which includes the Appalachia and Ozarks regions in the American South, and extending northward to the mountains of New England and the Canadian Maritime Provinces.
3. *The Western Mountains and Basins*, which extend along the entire Pacific Ocean length of the continent, starting in Central America and Mexico, and including the Rocky Mountains, the Pacific coastal mountain ranges and the basins and plateaus that lie between these mountain systems all the way up through Alaska.

The Interior Lowlands and Coastal Plains

Over half of Anglo North America is flatland. The largest expanse of this flatland is located between the Appalachian Mountains in the east and Rocky Mountains in the west. This vast *Interior Lowland* region stretches from the Gulf of Mexico in the south to the northern tip of Canada. Mexico's Yucatan Peninsula is the southernmost outlier of this lowland region. The southern and eastern portions of the Interior Lowlands in the U.S. are made up of the Gulf Coastal Plain (along the Gulf of Mexico) and Atlantic Coastal Plain (which wrap around the Appalachian Mountains and gradually slope into the sea). Florida is part of both of these coastal plains.

The Canadian Shield

Underlying most of the eastern half of Canada and most of Greenland is the Canadian Shield, which is centered on Hudson Bay, the largest bay in Anglo North America. The Canadian Shield contains some of the oldest exposed rock on the surface of the earth, created 4 billion years ago in the Precambrian Period. While the Canadian Shield is almost entirely in Canada and Greenland, geologically important extensions of it lie in northern Michigan, Minnesota and Wisconsin. In New York, the Adirondack Mountains are part of the Shield. The metamorphic processes that created the Canadian Shield resulted in zones that are rich in metallic mineral deposits that were once deep inside some of the tallest mountains that the earth has known.

Those mountains were eroded long ago. More recently, during the great ice ages (ending about 10,000–12,000 years ago), vast sheets of snow extended out of the Hudson Bay area, covering much of lowland Canada and the Northeastern United States. Successive periods of glacial cover removed most of the sedimentary material that once covered the Shield, exposing its valuable ores. Canadian Shield iron from northern Minnesota and Canada was one of the main resources allowing for the development of the American and Canadian heavy manufacturing industries (including the car industry) around the Great Lakes in the early 20th century.

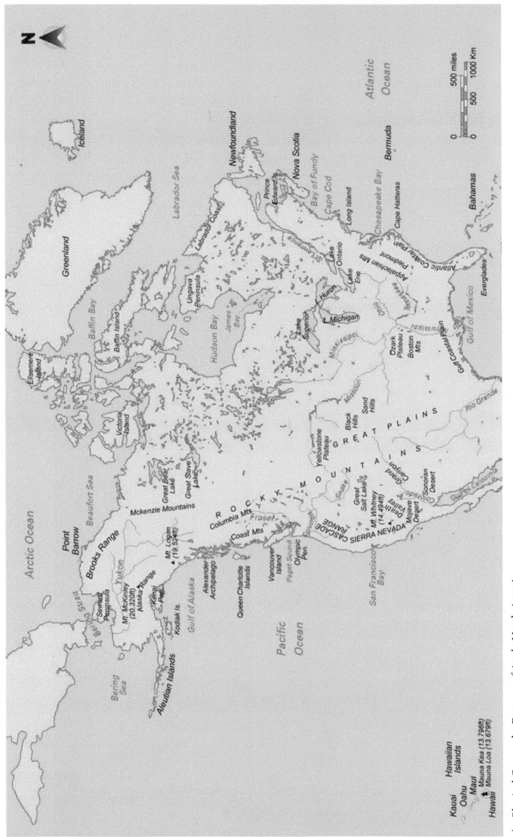

▲ *Physical Geography Features of Anglo North America*

The Great Lakes

A string of lakes and other water bodies surrounds the edge of the Shield to the south and west. The Great Lakes, Lake Winnipeg, Great Slave Lake and Great Bear Lake are among these and are the largest lakes on the North American continent. They, along with the St. Lawrence River and Lake Champlain Lowland (between New York and Vermont), and the Mackenzie River in the far north, demarcate the boundary between the Canadian Shield and surrounding higher elevation areas. Skinny spruce and aspen forests vegetate most of the Canadian Shield, with the exception of the far north, which has a treeless *tundra* vegetation. The tundra line also marks the southern limits of continuous *permafrost* soils. (South of the tundra line, permafrost exists, but it is discontinuous and not in a solid mass.) The permafrost zone have been in recent year along with an overall warming of global climate patterns.

The Great Lakes are situated on the border between the Canadian Shield to the north and the Central Lowlands to the south. The location of the lakes reflects eroded layers of sedimentary rock that encircle a large bulge in the earth's crust, centered on the state of Michigan. Continental glaciers that were over 5,000 ft (1500 m) thick carved out the weaker layers of rock that were then filled in with water when the glaciers last receded. Glacial action has also affected the river drainage pattern of the Great Lakes and upper Mississippi, Ohio and Missouri Rivers. The path of the Missouri and Ohio Rivers roughly marks the southernmost reach of continental glaciers during recent ice ages.

The Great Lakes have a significant impact on the regional climate of the upper Midwest. Winds that cross the lakes pick up moisture, which increases precipitation (including snowfall) on the eastern side of each lake. The lakes also have more moderate air temperatures than the surrounding land (similar to the oceans), which makes for longer growing seasons on lands that border the eastern sides. As a result, the Niagara Peninsula, north of Lake Erie and surrounded by three of the Great Lakes, is the major fruit-growing region in Canada.

The Great Plains

The Great Plains rise in elevation from near sea level at the outlet of the Mississippi River to just over a mile high (1 mile = 5,280 ft; 1609 m) in Denver at the foot of the Rocky Mountains. The Great Plains are composed of layer upon layer of sedimentary rock and soil that was washed out from the Rocky Mountains to the west. Rivers such as the Platte, Missouri and Red rivers parallel one another as they flow from west to east down this gentle grade.

While most of the Great Plains is flat and undistinguished, it does contain some distinct physical anomalies. In Nebraska, a belt of *Sand Hills* has been formed by wind and the outwash of the last continental glaciers. In South Dakota, however, water erosion has created a vast *badlands* landscape of deeply cut and barren soils. The one major mountain system that interrupts the evenness of the Great Plains is the Black Hills of South Dakota, which rise to over 7,000 ft (2134 m). Here, old crystalline igneous rock breaks throughout the surface in a dome-like swelling that is believed to be an outlier of the Rocky Mountain system. The oldest and largest goldmine in the United States, as well as the famous Mount Rushmore presidential carvings, are both located in the Black Hills.

The Great Plains becomes narrower as it extends into Canada and gradually disappears in northern Alberta, north of Edmonton (the capital of Alberta), where the prairie grasslands give way to a belt of mixed coniferous and deciduous forests. Eventually, these become spruce, fir and aspen forests, also known as *taiga*, which dominate the cold northern latitudes all the way to Alaska, both on the Canadian Shield to the east and the Rockies to the west, and beyond to northern most Europe and Asia.

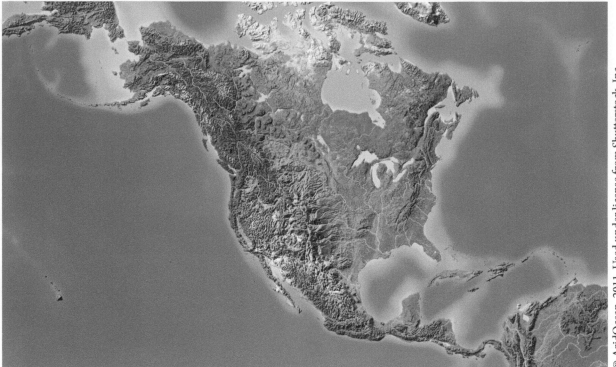

▲ *Physiographic map of the North American continent showing the pattern of major surface features and water bodies.*

The Coastal Plains

Further to the south, the Gulf Coastal Plain slides into the Gulf of Mexico, while the Atlantic Coastal Plain borders the Atlantic Ocean. This is the lowest and flattest part of the Interior Lowlands. These plains are also among the youngest exposed land in Anglo North America. When the Rocky Mountains were first beginning to rise some 70 million years ago, almost all of the Gulf and Atlantic Coastal Plains were under water. The coastal plains were gradually built up by unconsolidated (and easily eroded) sediments washed out from the Appalachian Mountains, the Central Lowlands and the Great Plains. They are now being threatened by rising sea levels and more frequent hurricanes, both of which are resulting is a loss of land.

Along much of the coastline are swamps (wet forests), marshes (wet grass lands) and lagoons, along with sand bars and reefs. These characterize a transition zone between land and sea. Florida lies entirely within the Gulf and Atlantic Coastal Plains. Its highest elevation is only 345 ft (105 m), and mostly of its land is sandy with large swamp and marsh areas. The southern tip of Florida is dominated by the 5,000 sq mile (12,950 sq km) Everglades marsh. A shallow continental shelf extends up to 250 miles from the coast line to the underwater edge of the North American continent. Coastlines that have a broad coastal zone and wide continental shelf are known as *trailing coastlines*, because they are often found on the coast that is trailing behind a migrating continent. The North American continent has been moving westward, toward the Pacific Ocean (the leading coastline) and away from the Atlantic Ocean (the trailing coastline).

Interior Lowland Climate

The region north of Hudson Bay remains at freezing and below throughout most of the year, with very short summers when high temperatures reach about 45°F (7°C). (Only interior Greenland averages below freezing all year round, although even that is starting to change as its ice sheets melt from global warming.) The major air mass influencing this region is the dry and cold continental polar system. This air mass is weaker in the summer, but it becomes strong in the winter months. Annual precipitation is under 10 in. (25 cm) for the entire Canadian Shield, except for the areas to the east and south of the Great Lakes. The southern two-thirds of the Canadian Shield experience mild summers (50°F to 70°F; 10°C to 21°C) and cold (below freezing) winters.

The climate throughout the southern eastern lowlands, below the Great Lakes area, is humid subtropical with summer rains. Temperatures are moderate in the winter and hot in the summer. Annual rainfall is over 40 in. (102 cm) for the entire coastal plains area, with over 60 in. (152 cm) in the Everglades. This drops to 20 in. (51 cm) further to the west on the Great Plains. Rainfalls below 20 in. typically indicate desert conditions, and the Great Plains was originally known as the *Great American Desert* by early European immigrants.

The major air mass that influences the climate of the Southern Interior Lowlands comes from the Gulf of Mexico. It brings warm, moist tropical air in the summer months, which is when most of the precipitation falls in this region. This air mass is weaker in the winter when occasional strong Arctic air masses reach deep into the Southeastern United States, destroying the region's fruits and vegetables. Vast areas of needle leaf evergreens make the Southeastern U.S. one of the two major timber-producing regions in Anglo North America (the other being in the U.S.'s Pacific Northwest and Canada's British Columbia). Further to the north, temperatures are cooler and growing seasons are shorter, though summers can still be quite warm due to the continental climatic conditions of the interior of Anglo North America.

The Mississippi River and its tributaries drain the entire Central Lowlands (known as the Midwest in the United States) and most of the Great Plains, including the Missouri and Ohio Rivers. Good soils, flat land and navigable rivers made the Midwest an easy region to cross and settle in. Large grain shipments, coal (from the Appalachian Mountains) and iron ore (from the Canadian Shield) could also be transported easily on these waterways. These attributes contributed to making the Midwest and Southern Ontario the agricultural heartland and the heavy industry core of the United States and Canada.

The Appalachian Mountain System

The Appalachian Mountain system comprises a band of mountains that runs from Canada's Newfoundland in the northeast to just north of Atlanta, Georgia, in the American southeast. The system can be divided into four sections:

1. A Northeastern section that covers northern Maine and the Maritime Provinces of Canada, and which continental glaciers have eroded heavily;
2. A New England section, which has been somewhat less eroded by glaciers and has some steep sloped mountains and a jagged, glacier-carved coastline;
3. The Appalachians proper, culturally known as Appalachia, and which runs from Pennsylvania to Alabama and includes the Allegheny and Cumberland Plateaus; and
4. The Ozark Plateau and the Ouachita Mountains (pronounced waw'-chi-taw), which were the southernmost part of the Appalachian system, but were cut off from the rest of the range by the Mississippi River.

The Appalachian Mountains were once the principal mountain system in North America, rising as high as the Rocky Mountains today. They were created by the intense uplifting and folding of sedimentary rock as North America shifted and slid against Europe and Africa some 570 million years ago during the formation of the Pangaea supercontinent. This was caused by the tectonic force of swirling molten magma inside the earth that causes continental and ocean plates to move about on the earth's surface. Starting about 200 million years ago, North America began to break away from Europe and Africa, forming the Atlantic

This view of the Appalachian Mountains at Harper's Ferry, West Virginia, illustrates the rolling hills effect of millions of years of erosion on what was once a very high mountain range.

Photo: Dallen J. Timothy

Ocean. As a result of this process, a portion of the Appalachian system was split, with one part being in North America and another part ending up across the Atlantic in Northern Ireland, Scotland and Norway.

Although the Appalachians mostly stopped growing after the continents separated from one another, they are still the dominant mountain range in the eastern U.S. Continuous erosion has brought most of the peaks down to under 5,000 ft (1524 m) in elevation. Narrow valleys, steep hillsides and thin soils have resulted in low population densities throughout this mountainous region, and they have kept its cities relatively small. This has also been true of the Ozarks and the Ouachitas. Metallic minerals (especially copper and iron ore) have been largely mined out of the Appalachians, though coal still remains in abundant supply.

The Maritime Provinces and New England

A portion of New England and most of the Maritime Provinces of Canada were subjected to mountain-building processes that involved huge intrusions of igneous rock (known as batholiths). The hard rock underlying this land, much of which is granite, has experienced considerable scraping and erosion from continental glaciers and today closely resembles the landscapes of the Canadian Shield. The St. Lawrence River and the Gulf of St. Lawrence form a water boundary between the northern Appalachians and the Canadian Shield of Quebec and Labrador. Two north-south trending mountains dominate the Appalachians in New England. The Green Mountains in the west (mostly in Vermont) are lower in elevation. Their peaks have been rounded due to continental glaciers that once covered them. The higher White Mountains (mostly in New Hampshire) include Mount Washington, the tallest peak in the Northeastern United States at 6,288 ft (1917 m).

Even though the New England part of the Appalachian Mountains is close to the coast, the climate is continental because of the west-to-east flow of air. Cold polar air masses from the Canadian Shield regularly blow across the northern Appalachians region, bringing winter snow all the way to Boston and New York. The climate is generally one of mild summers and long, cold winters in the Canadian portions, with hot summers and shorter winters in the southern parts of New England.

Because large amounts of the earth's surface water were stored in ice during the glacial advances, sea levels were about 450 ft (137 m) below their current levels. Much of the North American continental shelf was exposed and a land bridge existed to Asia. Because rivers seek to erode until they reach sea level, the lower sea levels allowed rivers to cut deeper valleys, especially in coastal areas. When sea levels rose, these valleys were filled in, creating Chesapeake Bay (Susquehanna River), Delaware Bay (Delaware River) and New York Harbor (Hudson River). Bays created by the filling in of river valleys are called *estuaries*. The result is that the coastline of the northeastern U.S. has many bays, peninsulas and islands, which contributes to giving North America the second longest coastline among the world's continents, after Asia. These bays also made excellent harbors for seafaring boats.

Appalachia

Starting in Pennsylvania, the *Appalachians* proper are found. Portions of this part of the Appalachians formed the edge of a carboniferous era swamp that covered the interior lowlands of North America some 300 million years ago. These old swamp lands formed the basis for the rich coal deposits found throughout this part of the Appalachian system. The transition between the Appalachians and the Central Lowlands of the U.S. Midwest is sometimes called the *interior low plateaus*. These consist of the Allegheny Plateau in Pennsylvania and the Cumberland Plateau further south. They are, for the most part, sandstone underlain by limestone, which creates a karst topography with limestone caves. The largest cave in North America is Mammoth Cave in Kentucky's Cumberland Plateau. When the sandstone cover is removed by erosion, limestone basins are created, such as the Blue Grass Basin of Kentucky and the Nashville Basin of Tennessee. Both have rich soils and good farmland.

The Blue Ridge Mountains are in the southern part of the Appalachians. These mountains reach an elevation of over 6,000 ft (1,829 m) in the Great Smoky Mountains National Park, along the North Carolina and Tennessee border. *The Smokies* are the highest mountains east of the Rocky Mountains, with Mt. Mitchell reaching 6,684 ft (2,037 m). To the west, the mountains are heavily folded ridgelines that are parallel to one another and are known as the *ridge and valley region*. The north-flowing Shenandoah River and the south-flowing Tennessee River run through this area. Most of the Appalachians proper experience a hot and humid summer and cool to cold winter.

The Ozarks

The Ozark Plateau and Ouachita Mountains consist of a group of low mountains sometimes called the *interior highlands*. On the southern edge of the Ozark Plateau are the Boston Mountains, the peaks of which are exposed granite rock. To the south, across the Arkansas River, are the Ouachita Mountains. They are a folded belt that resembles the ridge and valley area of the Appalachians. At its highest points this region is under 2,700 ft (823 m). The climate of the Ozarks and Ouachitas is similar to that of the surrounding warm and humid lowlands. This is in part because these interior highlands are an east-west trending system, which benefit less from altitudinal zonation and rain shadow impacts of air systems that also trend from west to east, at least in comparison to the north-south trending ridgelines of the southern Appalachian mountains.

Earthquakes in the Eastern United States

Earthquakes are more commonly associated with the Pacific Coast of Mexico, the U.S. and Canada. However, some of the strongest earthquakes ever to occur in the lower 48 states of the U.S. (excluding Alaska) were in the areas around New Madrid, Missouri, in 1811 and 1812 (7.7 magnitude). Other devastating quakes have occurred in Charleston, South Carolina (7.3 in 1886), and New York City's Jamaica Bay. Geologists estimate that there is a very high probability of a destructive earthquake occurring in the eastern United States within most of our lifetimes. Because earthquakes are so infrequent in this region though, none of the buildings have been constructed to resist tremors, as they have in California, and predictions are very difficult to make.

The Western Mountains and Basins Region

The Western Mountains and Basins comprise about a third of Anglo North America and most of Mexico and Central America. The major subsections of this diverse region are:

1. The Rocky Mountains in Canada and the United States, which extend to the Sierra Madre ranges in Mexico;
2. The Intermontane Basins and Plateaus region, which is located between the Rocky Mountains and the Pacific Coast ranges, and includes interior Alaska and central Mexico; and
3. The Pacific Coast Mountain and Valley region, including the Alaska Range and the Aleutian Islands in the far north and Baja California in Mexico to the south.

The Western Mountains were created by the rapid speed with which the North American continent moved westward, colliding into and over the Pacific Ocean Plate. About 80 million years ago, this mountain-building uplifted the western half of the continent where a long series of sedimentary beds lay evenly spread over the ancient continental floor. This disturbance was accompanied by volcanic activity, known as the Laramide Revolution, and resulted in a large uplift of the sedimentary beds, folding and faulting, and metamorphic changes. In some places, the uplift was followed by the erosion of the sedimentary layers and the exposure of the underlying igneous rock. In other places, the older layers were preserved. The result is a great mix of sedimentary, igneous and metamorphic rock across western North America, along with a diversity of landscapes.

The *Intermontane (or Intermountain) Basins and Plateaus* region was actually the first portion to be uplifted, creating a large mountain range through present-day Nevada some 80 million years ago. The Rocky Mountains were formed 60 million years ago. Today, the Front Range of the Rockies rises a mile above the Great Plains at Denver, Colorado. Mountain building in this region continues today, causing earthquakes along the Pacific Coast, down to the mountains of Central America.

The Rocky Mountains and the Sierra Madres

The main part of the Rocky Mountain system lies in Canada (the Canadian Rockies) and in the United States from Montana and Idaho (the Northern Rockies) to Colorado (the Central Rockies) and New Mexico (the Southern Rockies). Disconnected outliers of the Rockies include the Brooks Range in northern Alaska and the Sierra Madre Occidental (west) and Oriental (east) in Mexico. Although of different origin, the highlands of Central America, south of Mexico, share a similar geographic role as the more northern mountain systems.

Climate in the Rocky Mountains is generally cold in the winter and mild in the summer. Highland climates, however, are difficult to classify because of the great variation in micro-conditions between valley floors and mountain peaks. In addition, south-facing slopes often receive much more sunshine and are very different from colder, north-facing slopes in their climate and vegetation. Latitude plays a major role, with the mountains located farther to the north being much colder and having more permanent snow and active alpine glaciers than those to the south.

Intermontane Basins and Plateaus

West of the Rocky Mountains (as well as south of the Brooks Range in Alaska) lies the Intermontane Basins and Plateaus region. This diverse area was created by crustal faulting, volcanic activity and intense recent erosion and downcutting by rivers. The basic form of this region is that of a series of high-elevation plateaus. These are older mountains that have eroded into low-lying hills, internal drainage basins with sand dunes and flat playas, as well as the deepest river valleys on the continent.

The Cold Triangle

In Alaska, this area forms the vast drainage basin of the Yukon River, with its rolling hills and wide, braided rivers. The Yukon River originates on the Yukon Plateau (a higher-elevation area) in Canada's Yukon Territory. It flows through the *Cold Triangle*—an area enclosed by Fairbanks in Alaska, Snag in Canada's Yukon Territory and Good Hope in the Northwest Territories of Canada. This area regularly has the coldest temperatures in North America, with -60°F to -80°F (-51°C to -62°C) cold periods lasting for weeks at a time. Dawson, on the Yukon River, and the Klondike Region, bordering Alaska and Canada, were major gold mining districts in the late 1890s that lie at the center of the Cold Triangle. Summer high temperatures in this area are typically in the 70s F (low 20s C), although temperatures in the 90s F (mid-30s C) are also common. In the summer, the melted marshlands of the far north become infested with mosquitoes.

The great extremes of summer and winter temperatures at higher latitudes (closer to the poles) are due to the earth's tilted axis. Because the earth moves around the sun on a tilted axis, the northern hemisphere faces toward the sun in June and looks away from the sun in December, thereby having shorter days and receiving more indirect rays from the sun. These extremes become most pronounced closer to the poles. North of the Arctic Circle, Alaska and Northern Canada's summer solstice has 24 hours of sun, while the winter solstice has 24 hours of cold darkness. At the other extreme, the length of days in the Caribbean Sea remains about the same all year round, and average winter temperatures are only a couple of degrees less than average summer temperatures.

In Canada, a deep fault line separates the Rocky Mountains from the lower mountains of the Interior Basins and Plateaus and the Coast Mountain range further to west. The broad trench that these faults create serves as the headwaters for the Columbia, Fraser, Peace and Yukon Rivers. The Fraser Plateau comprises much of central British Columbia, situated between a narrow band of the Rockies and the Coast Mountains. It is made up of large areas of extrusive igneous lava. The entire plateau is drained by the Fraser River, which enters the sea just south of Vancouver.

The Columbia Plateau

The Columbia Plateau is another vast area of lava rock that underlies most of the state of Washington and portions of Oregon and Idaho. It is drained by the Columbia River and the Snake River. Like the Fraser Plateau, the Columbia Plateau was created when giant fissures opened up and poured forth 200 ft (61 m) thick lava flows. Over time they created a 50,000 sq mile (129,500 sq km) area of old lava several thousand feet thick. The Columbia River Basin, also known as the *Inland Empire*, lies in eastern Washington at the center of the Columbia Plateau. Along with the Great Sandy Desert in southeastern Oregon, it is the driest area in the Pacific Northwest region of the United States (less than 10 in. (25.4 cm) of rain a year). This is mostly because it lies in the *rain shadow* of the volcanic Cascade Range. The large rivers that flow through this desert region provide water that makes it one of the richest wheat-growing areas in North America. Hells Canyon, on the Snake River between Oregon and Idaho, is also in this area and is the deepest canyon in the United States. It forms a V-shaped canyon as the river cuts straight down through the old lava rock.

The Great Basin

The Great Basin is centered on the state of Nevada, though it also includes areas of neighboring states. It is the largest of the internal drainage basins in the Intermountain West. During the last ice age it was filled with large lakes that have since evaporated, often leaving behind much smaller salty remnants, such as the *Great Salt Lake*. The landscape of the Great Basin consists of ancient mountains that have been eroded away into widely dispersed smaller ridge lines. Their eroded sediments now form flat expanses and out-wash plains. This landscape is known as the basin and range and extends well into northern Mexico.

Rainfall is generally under 5 in (13 cm) throughout the basin and range deserts of the American West and northern Mexico. This is due to a high-pressure air mass that tends to dominate this area, as well as the rain shadow effect of the Pacific Coast mountain system in California. Irrigated fields rely on water from the plateaus and higher peaks of the western mountains. On the higher mountains there is a well-pronounced increase in precipitation, and one can often travel from desert to alpine conditions within 100 miles (160 km). Groundwater is another important source of irrigation water, although it is becoming more difficult to reach as water tables drop due to over use. Most of the rainfall in this region comes in the summer months, from summer monsoon moisture originating in the Gulf of California and the Gulf of Mexico.

The Colorado Plateau

One of the most prominent features in the Intermontane Basins and Plateaus region is the Colorado Plateau, which is centered on the four corners where Colorado, New Mexico, Arizona and Utah meet. It is made up of sedimentary rock that was uplifted at about the same time as the Rocky Mountains. Today, the Colorado Plateau has an average base elevation of around 5,000 ft (1524 m). As the Colorado Plateau lifted upward from a former shallow sea bed, the Colorado River carved down through it to form the Grand Canyon. Unlike Hells Canyon, the different layers of sedimentary rock that form the Colorado Plateau created a multi-colored, stepped canyon of mesas and rock monuments.

The Pacific Coast Mountains and Valleys

The Pacific Coast of Anglo North America is very diverse. Along it are found the highest peaks and the most active volcanoes on the continent. Like the rest of the mountainous west, the Pacific Coast was shaped by plate tectonics. In the south, the *Baja California* peninsula (including the city of Los Angeles, California) is actually a portion of the Pacific Ocean Plate, which is currently sliding northward against the North American Plate. The granite peaks of the Sierra Nevada Mountains in California are an exposed intrusive igneous (mostly granite) formation, while the Cascade Range further to the north is made up of extrusive igneous (lava) rock. Farther to the north, Canada's Coast Mountains and the Alaska Range in Alaska are intrusive igneous formations, similar to the Sierra Nevada. The Coast Mountains of British Columbia are the world's highest coastal mountain system, while the Alaska Range contains the highest mountains on the continent, including Mt. McKinley (also known as Denali) at 20,322 ft (6194 m).

Subduction

Both the intrusive and the igneous rock that comprise the higher Pacific Coast mountains come from the melt-ing of the Pacific Ocean sea floor plates that have been pushed under the North American Plate in a process known as *subduction*. The melting of the ocean plates results in molten rock squeezing upward through cracks in the North American plate. This created the Alaska Range, Canada's Coast Mountains, the Cascade Range and the Sierra Nevada Range. (The Rocky Mountains and the Sierra Madre Range were also formed this way.) As these mountains were uplifted, the area immediately to their west was sunken to form a depression. The depression originates as the Gulf of California in Mexico and continues as the Central Valley in California, the Willamette Valley in Oregon, the Puget Sound lowland in Washington, the coastal straits and Inland Passage of British Columbia and the Alexander Archipelago in the Alaska Panhandle (southeast Alaska).

◀ *This view of Napa Valley, California, shows the fertile farms and surrounding peaks of the Pacific Coast's mountains and valleys. The Napa Valley is especially well known for its vineyards and wineries.*

Photo: Dallen J. Timothy

Today, a small ocean plate (the Juan de Fuca Plate) continues to subduct under Oregon and Washington, keeping the Cascade Range volcanically active. Subduction and active volcanism are also still occurring in Alaska from about Anchorage out through the Aleutian Islands. In Mexico, California and British Columbia, however, subduction has mostly stopped, and instead the North American Plate is sliding sideways against the large Pacific Ocean Plate. This causes lateral (or strike-slip) faulting and frequent earthquake activity in most of the Coast Range of California, as the Pacific Plate moves northward at about 2 in. (5 cm) a year.

The low-lying Coast Range of California, Oregon and Washington (which is different from the Coast Mountains in Canada) formed by crumpling the edge of the North American continent. The Coast Range continues as Mexico's Baja California peninsula in the south (though Baja's geologic origin is different from that in California, Oregon and Washington) and as Vancouver Island and Queen Charlotte Island in British Columbia to the north. The Alaska Peninsula, which contains more active volcanoes than anywhere else on the continent, and the Aleutian Islands, can also be considered an extension of this coastal edge. Unlike the *trailing coastline* of the U.S. eastern seaboard, the west coast is a *leading coastline*, which is characteristically steeper and shows signs of emergence, or rising up from the sea, with older coastlines now hundreds of feet up slope in some areas.

West Coast Rain Shadows

The two north-south trending mountain systems on the Pacific Coast (the Coast Range and the Sierra Nevada/Cascade Range) create two rain shadow patterns. The first and smaller rain shadow occurs in the inland valleys of California, Oregon and Washington and the coastal channels of British Columbia. Thus, Sacramento, Portland, Seattle and Vancouver (British Columbia) all lay in a warmer and less rainy rain shadow area compared to the mountains to their west and east. A more significant rain shadow exists east of the Sierra Nevada, the Cascade Range and Canada's Coast Mountains. This rain shadow allows the naturally occurring deserts of the United States-Mexico border area to be extended up to eastern Washington state, while in Canada the dry polar desert that exists over the North Pole is extended southward deep into Canada by the same rain shadow process behind the Coast Mountains and Rocky Mountains. A dry rain shadow is also found in central Alaska, where the east-west trending Alaska Range forms a moisture barrier to the northern interior.

The Climate of Anglo North America

The rings of high- and low-pressure air masses encircling the globe shape the weather and climate of the North American continent in different ways. The *eastern half of Anglo North America* is primarily influenced by interactions between cold and dry continental air coming down from the high-pressure North Pole, and from warm, humid air rising from the equator and the Caribbean Sea. The warm tropical air brings most of eastern North America's rainfall in the summer months. This is true for both Anglo and Latin North America. The humid air spins around the *Bermuda High* pressure air system, which circulates in a clockwise direction in the Atlantic Ocean, approximately around the island of Bermuda. The cold polar air keeps the winter months frigid and comparatively drier than in summer in the eastern areas of Canada and the United States. Winter precipitation occurs when humid, low-pressure systems move in from the Pacific Coast and intermix with the polar air mass (see Map, page 351).

For most of Anglo North America, the predominant wind pattern is from west (the Pacific Ocean) to east. For Latin North America, the predominant wind pattern is from east (the Atlantic Ocean) to west. Smaller local exceptions to these patterns sometimes occur, and the whole system of air masses moves northward in the summer months and southward in the winter months. As the warmer air masses move northward, they warm up the Atlantic Ocean east of the Caribbean Sea.

Hurricanes

Starting in June each year, warm and humid air rises from the Atlantic Ocean off the coast of Africa and then moves toward the Caribbean Sea. Occasionally these air masses become tropical storms and hurricanes. The hurricane season lasts through the end of November, when the warm air masses shift further southward. Unfortunately, the hurricane season occurs during the summer holiday season for North Americans, which can cause vacation trip cancellations in the Caribbean and on the coastal plains when inclement weather is forecast.

The Rocky Mountains and the prevailing west-to-east wind pattern (from the Pacific) keep the tropical and polar air masses that affect the eastern United States from entering the western half of the continent most of the time. Occasionally the polar air mass makes its way into the Pacific Northwest in winter, and the tropical Caribbean air mass regularly affects Mexico and the United States' Southwest in the summer months.

Western North America is affected by a dry high-pressure air mass in the southern portions (the southwestern United States and northwestern Mexico) that creates desert conditions that extend to the Canadian border due to the mountain rain shadows, as mentioned above. Further to the north, an upper latitude humid low-pressure system brings year-round humid weather to the Pacific Northwest all the way up to southern Alaska. Central and northern Alaska and Canada are under the influence of the dry polar air mass that affects eastern parts of the continent in the winter months.

California's Mediterranean Climate

California is famous for its Mediterranean climate. This climate exhibits a distinct summer dry and winter wet pattern, which is different from that found anywhere else in North America. This climate transitions into a desert climate on the northwest coast of Mexico, and then to a summer-wet tropical climate in southern Mexico and Central America. North of California, the Mediterranean climate gradually changes

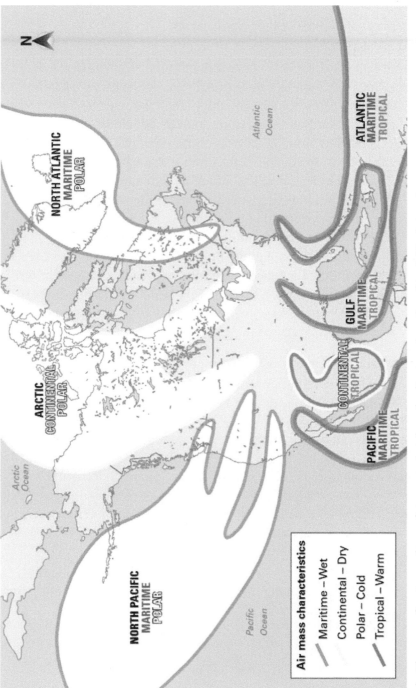

▲ *The source, direction, and type of air masses that shape the climate and weather of Anglo North America.*

to the *Marine West Coast* climate on the coast from Oregon to southern Alaska. The Marine West Coast climate is wet all year round, although, like the Mediterranean Climate, it also has a winter peak. Because the moisture hitting the west coast comes off the ocean, it has a moderating maritime influence on the coast and coastal snow is very uncommon, except in northern British Columbia and Alaska.

Human Geography of Anglo North America
Pre-Columbian Settlement

Migrants from Asia are generally believed to have arrived in the New World at least 12,000 years ago, at the end of the last ice age (though some archaeological finds suggest possibly pushing the earliest crossing date back to 45,000 years ago.) At the time Europeans arrived, the area that today constitutes the United States contained perhaps 5 million Native Americans, with an additional 2 million in Canada, including Inuits. The Caribbean Islands had from 500,000 to 1 million people, and Mexico and Central America had 25 to 30 million inhabitants. In Anglo North America, the native peoples were sparsely scattered, with major concentrations in the Central Valley of California, the Pacific Northwest, the U.S. Southwest and portions of the U.S. South. Most of their societies were technologically primitive in comparison to Western Europe, though some developed highly-organized social systems. However, their population densities were never large enough to allow for sustained high civilizations.

Native American is a general term used in the United States that applies to all pre-European contact people in North America, including Hawaiians and Inuits. In Canada the preferred term is *Aboriginal*, or *First Nations*. *American Indian* is the term used in the United States to refer to most of the pre-European contact people in the North America, excluding Hawaiians, Inuits and several northern ethnic groups that are more closely related to ethnic groups in Arctic Asia and Europe.

Native American Tribes

The tribe was the basic unit of social organization for American Indians in North America. A tribe is any group of people who share the same distinct customs and language and who believe that they have

◀ *These pre-Columbian ruins of a Native American pueblo at Tuzigoot National Monument in central Arizona were home to between 200 and 300 people approximately 1000 years ago. Archeological sites, along with contemporary American Indian settlements, attest to their adaptations to the arid lands of the southwestern USA.*

Photo: Dallen J. Timothy

descended from a common ancestor, whether real or mythological. At the time the Europeans arrived, the tribes of North America spoke several hundred languages and led a variety of lifestyles, from settled agriculturalists (such as the Hopi) to nomadic hunters and herders. Some of these groups got along very well (such as those belonging to the Iroquois Federation); others, however, did not.

Common European diseases, especially smallpox and measles, killed large numbers of Native Americans because they had no resistance to them. In the early years of colonization, Europeans hotly debated whether or not Native Americans were humans or animals. Even when they were finally declared to be human, they were often treated inhumanely. Large numbers were forced to relocate as Europeans came to dominate the continent, and many were killed trying to protect their homelands in the Indian Wars in the United States (1865–1885). Many Indian battlefields and routes of forced migration are now commemorated as 'dark' heritage sites and promoted as tourist attractions to commemorate this difficult part of U.S. history. Indian place names , however, are common throughout the United States and Canada. The word Canada itself may have come from the Iroquois word *Kanata*, meaning town or village.

Land Ownership Differences

One of the major differences between the European colonialists and the Native Americans was how each group viewed ownership of land. In Native American societies, land was owned *communally* by the tribe. An individual had the right to a particular parcel of land only as long as he or she worked that parcel. As soon as the work stopped, it went back to communal ownership. The concept of land as a private commodity that could be bought or sold did not exist in any of the civilizations of the New World, with the possible exception of the feudal land ownership among the Aztecs in Mexico. European settlers brought with them the concepts of private land ownership (*freehold tenure*), the right to do whatever one wishes with one's land (*fee simple ownership*) and the ability to buy and sell land (*property deeds*). These concepts were fundamental to the founding values of the United States and Canada.

For the most part, the northern European settlers in the New World preferred to ignore Native Americans as much as possible. The cultural gap between the two groups was considered too large to bridge, and the preferred means of dealing with the situation was to remove Native Americans to isolated and undesirable lands that became known a *reservations* (reserved against settlement by Euro-Americans). Today, many Indian reservations have been found to contain valuable mineral deposits and have demonstrated considerable development potential.

Native Americans constitute less than 1 percent of the population of the United States and a little more than 3 percent of Canada's population. Their absolute numbers are only slightly larger than when Europeans first arrived. In the United States they are mostly settled in the western states, with especially large concentrations in Arizona, Oklahoma, South Dakota and California. Large non-Indian native populations reside in Alaska and Canada's Nunavut territory (mostly Inuit people), and in Hawaii (Hawaiians).

European Colonial Settlement

Columbus landed in the Bahamas in 1492. In the 1500s, colonial settlement in the New World was primarily from Spain (to Mexico) and France (to Canada). In the 17th century, increasing numbers of settlers were starting to arrive from the British Isles and Germanic Europe, settling in the British colonies in the United States and Canada. Soon after the American Revolution (1776-1778), 60 percent of the Americans in the 13 colonies were from the British Isles, 20 percent were from German-speaking Europe, including the Netherlands (Germany did not yet exist), and 20 percent were from Africa who had been brought to the Americas as slaves. Today, approximately 40 percent of the population of the United States consists of descendants of German immigrants, which is more than from any other national origin.

The French were the first Europeans to settle in Canada, settling in *Acadia* (modern-day Nova Scotia and New Brunswick) in 1605, and founding the city of *Quebec* in 1608. The British arrived about a decade later and the Quebec area changed hands between the two several times until 1760 when the British took control of Montreal. In 1763, the Treaty of Paris gave all of French Canada to the British. At that time there were an estimated 60,000 French settlers in Canada.

British Canada

After the British took full control of Canada, some of the French-speaking Acadians migrated to the New Orleans, Louisiana, area, where they are today known as *Cajuns*. A few years later, British loyalists, fleeing the colonies after the American Revolution, settled in the new colony of New Brunswick. After the United States gained its independence from Britain, large numbers of British immigrants started to settle in Canada, eventually coming to predominate over the French-speaking inhabitants who are today centered in the province of Quebec with substantial secondary populations in New Brunswick and Ontario as well. In addition, there is still today a significant French-speaking Acadian population in the U.S. state of Maine.

The Confederation of Canada was established on July 1, 1867, and it included today's provinces of Ontario, Quebec, Nova Scotia and New Brunswick. Later, Manitoba (1870), British Columbia (1871), Alberta (1905), Saskatchewan (1905) and Newfoundland (1949) joined the Confederation. Britain transferred its jurisdiction over the Yukon Territory and the Northwest Territories to Canada in the 1880s. In 1999, the territory of Nunavut was created from a portion of the Northwest Territories as a homeland for the Inuit people. This large area, however, only has a population of a little more than 30,000.

The Industrial Revolution

The industrial revolution had a major impact on European migration to Anglo North America. The industrial revolution started in England in the late 1700s and entered Northern Europe in the early 1800s. In each area where it took hold, a similar pattern of development occurred. Death rates declined, populations exploded, massive migration from rural areas to industrial cities took place and poverty increased dramatically. The Irish and British were the first to come to North America to escape the squalor of England's industrial cities as well as rural poverty and deprivation. They were followed by the Germans and the Dutch when industrialization occurred in Northern Europe.

Southern and Eastern Europe experienced the industrial revolution at the end of the 19th century, leading to the migration of large numbers of people from these areas to Anglo North America (and to southern South America). The United States population was 14 times larger in 1900 than in 1800 (growing from 5.3 million to 76.2 million). The impact on Canada was more limited because its potential for large-scale settlement was less. Massive European migration to North America ended with the outbreak of World War I in 1914 and has been very low since then.

Post-Industrial Era

Since about the 1950s and 1960s, North America has entered a postindustrial era in which tertiary (service-sector) employment has replaced secondary (manufacturing) employment as the leading source of jobs. This is not the case for many other countries in the world, where manufacturing continues to employ more people than any other sector. Europe and Japan have only recently begun to transform into postindustrial economies. In the postindustrial economy, both industries and workers are highly mobile and less place-bound. Because of this, many companies and employees have moved from the old industrial core near the Great Lakes to higher amenity *Sun Belt* states in the U.S. South and on the east and west coasts. At the same time, the postindustrial economy is characterized by lower paying service jobs (compared to manufacturing), higher rates of mass-consumerism, very low savings rates by employees and higher job turnovers. All this makes for less security in employment and in life for the postindustrial North American population.

Population Distribution

Today, the overall North American settlement distribution shows three distinct geographic patterns (see Map, page 357):

1. The Northeast Manufacturing and Financial Center, which includes the northeast coast, the Great Lakes region, southern Ontario and Quebec, is the most densely settled region of Anglo North America. It has the largest concentration of major metropolitan areas, even after a couple of decades of migration that have favored the Sun Belt. This is still a highly productive manufacturing region and the center of financial services for both the United States and Canada. Ethnically, the rural parts of this region are of English and French heritage in the east and in Canada, and of Germanic and Scandinavian heritage west and south of the Great Lakes. Urban populations have larger concentrations of Southern and Eastern Europeans, as well as of blacks that migrated from the South after the U.S. Civil War (1861-1865) and more recent Spanish-speaking arrivals.

2. The Agricultural Interior Lowlands, which include the U.S. Midwest, the Great Plains, and much of the U.S. South are characterized by a very even distribution of population and cities over a largely flat terrain. This distribution models a classic spatial pattern in human geography known as *Central Place Theory*. This theory says that the service area of different size settlements distributes them in an even grid-like pattern across the landscape. The density of this settlement distribution pattern becomes increasingly spread out as one moves into Canada's northern areas. Ethnically, the European settlers of the Southeastern United States, including the southern Great Plains, were predominantly of *Scots-Irish* descent (these were Scottish migrants to British Northern Ireland, who then migrated to the American Colonies). On the northern Great Plains and in the American Midwest, they were mostly of Germanic and Scandinavian descent, with some Russian settlement areas and significant French areas in Canada.

3. The Dispersed Western Settlements exhibit a pattern that is highly dependent on the location of natural resources, especially water and minerals. Water-enabled agriculture was the basis for most of the largest cities in the interior of western North America. Other cities, mostly smaller in size, were based on mining in the past and recreation today. Port locations were important for Pacific Coastal cities. Vast areas with low populations exist between these larger western settlements. This pattern extends up the western mountains system into Alaska. Ethnically, western North America is a great mix of diverse populations. American Indians are a major population group here, along with immigrants and their descendants from Europe (Utah is mostly of English descent). Middle America (especially Mexico) and Asia (China, the Philippines and Japan are the main source countries). Los Angeles alone is the largest Mexican city outside of Mexico, the largest Guatemalan city outside of Guatemala and the largest Iranian city outside of Iran. San Francisco is the largest Chinese city outside of Asia.

Regions and Tourism of Anglo North America
The Northeastern United States

The eastern United States is perhaps best known for large cities. New York City, Boston, Washington, Philadelphia and Baltimore are important international gateways and are favored destinations for domestic tourists. New York City is America's definitive city. A major transportation hub, gateway and destination in its own right, New York is known for glamorous shopping, ethnic foods and neighborhoods,

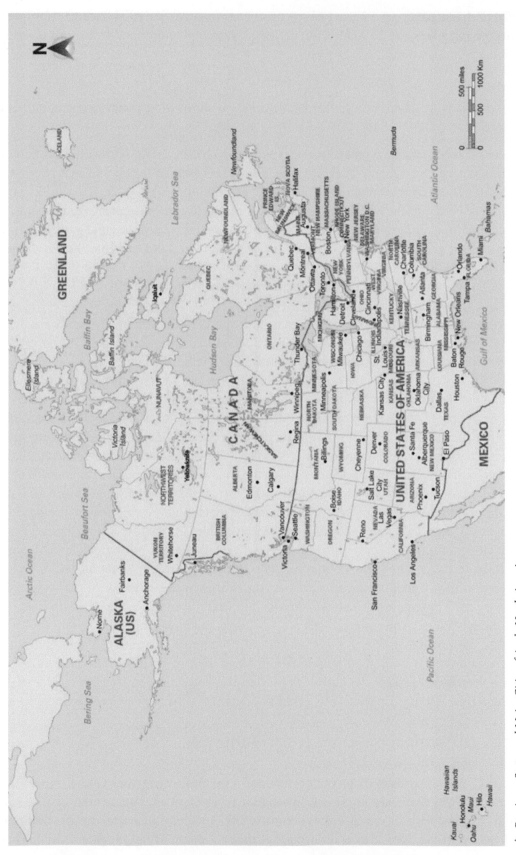

▲ *Provinces, States and Major Cities of Anglo North America*

ISSUES AND INSIGHTS   Urban Imaging

A major feature of Canadian and American cities is the use of urban imaging as a major development tool. Contemporary imaging strategies are typically policy responses to the social and economic problems associated with the changing nature of cities due to globalization and economic restructuring. Issues include de-industrialization, urban renewal, increased gaps between haves and have-nots, multi-culturalism, and social integration and control. The principal aims of imaging strategies are to:

- attract tourism expenditures;
- generate employment in tourism and related industries;
- foster positive images for potential investors in the region, often by reimaging previous negative perceptions;
- provide an environment that will attract and retain the interest of professionals and white-collar workers, particularly in clean service industries such as tourism and communications;
- increase public spirit and solidarity by making communities feel good about themselves.

Imaging processes in urban areas are characterized by some or all of the following:

- the development of a critical mass of visitor attractions and facilities, including new buildings/ prestige/flagship centers (e.g., shopping centers, stadia, sports complexes and indoor arenas, convention centers and casino development);
- the hosting of hallmark events (e.g., Olympic Games and the Grand Prix) and major league;
- sports teams;
- the development of urban tourism strategies and policies often associated with new or renewed organization and the development of city marketing and promotional campaigns;
- the development of leisure and cultural services and projects to support the marketing and tourism effort (e.g., the creation and renewal of museums and art galleries and the hosting of art festivals, often as part of a comprehensive cultural tourism strategy for a region or a city).

Such measures are widely applied in North America. Toronto bid, and failed, to host the Summer Olympic Games, while Vancouver hosted the 2010 Winter Olympics. Cities compete to host major league franchises in hockey, football, baseball, basketball and soccer. Downtowns, especially near waterfronts, such as in Baltimore and Vancouver, become focal points for urban redevelopment. Tourism is integral to such processes as it serves as both a justification for development strategies as well as a source of capital. The core focus of imaging is to attract tourists in the short term and migrants, capital and businesses in the longer term.

Sources

Hall, C. M. 2005. *Tourism: Rethinking the social science of mobility*. Harlow, UK: Pearson.

Page, S., and C. M. Hall. 2003. *Managing urban tourism*. Harlow, UK: Pearson.

---

its city skyline and the theater district. The most devastating terrorist attack on U.S. soil took place in downtown Manhattan on September 11, 2001. While this event essentially annihilated tourism in New York, and virtually the entire United States, for some time, the industry has recovered, and Ground Zero, the site of the World Trade Center's twin towers, and the new memorial constructed there, have become a major tourist attraction.

Boston is an old city and home to many places of historical significance in the American Revolution and the years leading up to it. It has one of the nation's most visible urban heritage trails and an important maritime history. The United States' capital, Washington DC, is a unique political destination popular among Americans who desire to know their nation's political history. Between the 1960s and 1980s, Baltimore's old waterfront was redeveloped into a thriving tourism, service and recreational area. Today, it has one of the best-known waterfront attractions in the country and is used as a model for other cities around the world that are revitalizing their derelict waterfronts. Philadelphia is home to many patriotic icons, including the Liberty Bell and Independence Hall, which contribute to its heritage appeal for American tourists.

Most of the nation's colonial heritage is located in the eastern U.S., and many of the earliest settlements have become important attractions. These include reconstructed living heritage villages, such as Old Sturbridge Village, Plimoth Plantation, Mystic Seaport and Colonial Williamsburg. In the New England states, maritime heritage overshadows other types of patrimony. Lobster and cod fishing are synonymous with Maine. Shipyards, fishing villages and marine aquariums are important in the tourist milieu, as are quaint villages with churches and autumn colors. Skiing dominates some mountainous

Photo: Dallen J. Timothy

◀ *The National September 11 Memorial and Museum in New York city commemorates the terrorist attack in 2001 and is situated on the site of the destroyed World Trade Center Twin Towers. That attack had significant impacts on international relations, as well as on travel and tourism around the world through major increases in security measures and concerns.*

communities in New York, Vermont and New Hampshire. The famed witch trials of Salem, Massachusetts, and cranberry agriculture on Cape Cod are important regional heritage icons as well.

Outside cities, the coastline and mountains are the backdrop for most recreational and tourist activities. The rugged coast of Maine in the north, and the beaches further south, in Massachusetts, Connecticut, New York, New Jersey, Virginia, Delaware, Pennsylvania and Maryland, have long been important destinations. The nascent popularity of eastern seaboard resorts, such as Coney Island, New York; Atlantic City, New Jersey; and Ocean City, Maryland, were made even more popular and accessible in the mid-1800s by the expanding rail lines, which attest to the coastline's importance. The Appalachian Mountain chain is a focal point for much nature-based and recreational tourism. Several national parks and recreation areas are located in the Appalachians, but the most obvious is the famed Appalachian Trail, which extends 3,500 km from northern Georgia to northern Maine and traverses 14 states. Thousands of people hike the trail or portions of it every year, spending anywhere from a single afternoon to several months.

The South

The South is a popular destination region, known for cuisine, slave heritage, Civil War battle fields, beaches, business travel and amusement parks. Florida is the region's most popular destination for tourists and the second most visited state in the country. Florida is home to sandy beaches, a semi-tropical climate, interesting wildlife, swamplands and Disneyworld, besides dozens of other mega theme parks. Miami is a major international gateway for flights from the Caribbean and Latin America and home to a large transnational Spanish-speaking population. Most of the U.S.-based Caribbean cruises depart from Florida.

New Orleans

New Orleans, Louisiana, is another important urban destination that has heritage appeal. The city is known for its music, festival (Mardi Gras), and spiritual and slave-related traditions. The New Orleans area is still a stronghold of old French Acadian culture (Cajun) with its culinary, musical and mystico-religious connotations. The city is also an important cruise destination and point of departure. In August 2005, Hurricane Katrina hit land in New Orleans, destroying thousands of homes and public buildings, killing nearly 2,000 people and resulting in tens of billions of dollars in damage. Thousands of residents were relocated as thousands more outsiders were mobilized to assist in recovery efforts. It was the costliest and second deadliest hurricane in United States history, and, as would be expected, devastated the area's tourism industry, along with most other industries.

Since 2005, the old historic quarter has recovered and tourists are once again arriving. One manifestation of dark tourism surrounds the events of Hurricane Katrina. Today, Katrina tours are offered by several tour companies. These include visits to devastated neighborhoods and destroyed levies. These tours have received a fair amount of criticism because some people perceive them as exploiting the suffering of the people of New Orleans for profit.

Civil War Heritage

Plantation and slave heritage, together with the *Civil War legacy* in the southern states, has received a great deal of attention in recent years among scholars and tourists, particularly African-Americans, as plantations, slave cabins, museums, and the homes of prominent African-Americans have been restored and promoted as tourist attractions. In the past, these sites have kept hidden the true conditions under which slaves lived and worked, but public cries for truth in heritage interpretation have begun to effect more realistic portrayals of slaves' life and death.

Although the United States is ashamed of its slave past, it is beginning to come to terms with it through history, storytelling, heritage management and site interpretation. Thus, the *whitewashed South* is seen as becoming more balanced in its treatment of history. Hundreds of thousands of Civil War enthusiasts, not all of them Americans, travel to the South to involve themselves in battle re-enactments or simply to visit the battlefields and other sites associated with this 1860s conflict.

The southeastern coasts are home to many beach and golf resort developments and vacation homes. Offshore gambling, also known as casino cruises, have become popular in southeastern states where high-stakes gaming typically is not allowed. These casino cruises embark from ports in Georgia and Florida, travel just outside of United States territorial waters (twelve nautical miles) and float as passengers enjoy an afternoon or evening of gaming.

Photo: Dallen J. Timothy

◄ *The pirate Jean Lafitte started this blacksmith shop as a front to his pirate activities in the late 1700s. Today it is one of the few remaining buildings of true French colonial architecture in the French Quarter (most of which is more Spanish influenced). The French culture of Louisiana included both settlers from France and Acadians (also known as Cajuns), which are migrants from Canada. The French-Acadian culture of southern Louisiana gives it a sense of place that is clearly distinct from the rest of the country and provides a cultural context to the annual Mardi Gras celebrations.*

The Midwest

The middle portion of the United States possesses the largest and most voluminous rivers and lakes in the country. The tourism and recreational role of these water bodies is clear. Sailing, fishing, water skiing and jet skiing are popular pastimes on the Great Lakes of the upper Midwest. Several of America's largest and most important industrial cities developed on the shores of the Great Lakes, including Chicago, Detroit and Cleveland, owing to the navigability of the lakes in relation to the St. Lawrence River and the transportation of industrial products. These cities are all important international air gateways, airline hubs and highway and railroad exchanges. The region's great rivers (Mississippi, Missouri and Ohio) have played a vital role in the history of industrial growth and transportation in the United States.

Previously an important method of transportation, riverboats on the Mississippi and Ohio Rivers have become attractions, not just for their nostalgic value, but for their role in the gambling sector. Several states do not permit gaming on land, but they do on water. Once the riverboats leave dock, the casinos open their tables and slots, navigate up and down the rivers and close up shop before returning to the riverbank. *Indian gaming* is also important on reservations in the Midwest, and several of the Indian reservations in places like Minnesota were among the first to establish casinos following the passing of the IGRA.

Because the upper Midwest states have a long history of heavy industry, the region has become known for its industrial heritage. This sub-theme within heritage tourism focuses on factories, waterfronts and dockyards, mines, canals and other infrastructure and enterprises associated with the history of manufacturing or resource extraction. This region is best known for steel mills and automobile assembly plants, as well as associated museums, such as the Henry Ford Museum and Greenfield Village in Michigan.

Agriculture

Among the most prominent elements of human geography in the Midwest is agriculture. The agricultural heritage includes dairy products, corn, wheat, oats, soy beans, pork, beef, cotton, fruits and vegetables. Like wineries in Northern California, the cheese factories of the upper Midwest attract many people to visit the farms, observe the production process and sample and purchase the final product. The ethnic heritage of the Midwest includes Swedes, Finns, Norwegians, Dutch, Germans, Czechs, Poles and Irish who arrived throughout the 19th and early 20th centuries.

These two heritages, agriculture and ethnic migration, have resulted in a vast array of food-based and ethnically-oriented festivals and events. The Turkey Festival, the National Cherry Festival, Strawberry Days, the Watermelon Festival, Pork Days, Platteville Dairy Days, the Sparta Butterfest and the Great Wisconsin Cheese Festival are only a few examples of the hundreds of annual community events that commemorate the region's agricultural importance. Similarly, from a cultural perspective, Tulip Time, the Wilber Czech Festival, Finnfest, German Folk Fest, the St. Patrick's Day Parade and Pub Crawl, Hostfest, Midsommar Festival Swedish Days and Polish Fest are an important part of the immigrant landscape that draws visitors from around the U.S. and Canada, as well as from the homelands of these diaspora peoples. Nearly every city, town and other settlement, no matter how small, boasts some kind of agricultural fair or ethnic festival to increase its tourism potential and assert its heritage identity.

The Mountain West and Southwest

The western and southwestern regions of the U.S. are endowed with high mountains, grasslands and barren deserts. This varying physiography has laid the foundation for the highest concentration of national parks and monuments in the United States. Over 67 percent of the territory of the state of Utah, for example, is owned by the federal government, much of it in national parks and other protected uses. Bryce Canyon, Carlsbad Caverns, Glacier, Grand Canyon, Grand Teton, Joshua Tree, Rocky Mountain, Saguaro,

Yellowstone, Yosemite and Zion are among the most popular national parks in the U.S. and located in this western region. Likewise, tucked in the Rocky Mountains system are many of the country's finest ski resorts and exclusive mountain getaways. Colorado, Utah and Idaho contain the most famous and prominent ski destinations in the country.

Traditional Culture

The themes of most cultural attractions in this region are Native American lifestyles and archeology, cattle ranches, cowboys, mines and ghost towns, and the Spanish-American borderlands. Indian ruins are featured prominently in many National Park Service properties, and several reservations today invite tourists to their dances, food-ways and music. The region's ranching traditions have given rise to numerous dude ranches and guest-oriented cattle drives where visitors are permitted to assist in feeding and caring for animals and working on large farms. Ranching, cowboys, mining and ghost towns are all interrelated heritage features of the west. Mining locations, such as Tombstone, Arizona, and Virginia City, Nevada, epitomize old stereotypes where cowboys and outlaws ruled the Wild West. These communities have based their tourism industries on their Wild West cowboy image, although their degree of authenticity has often been questioned and has resulted in significant public debates in recent years.

In addition to defunct mines which attract heritage tourists, active mines in Arizona, Colorado, Montana, Nevada, Utah and Wyoming have become salient industrial tourism attractions. They offer guided tours and souvenir shops that are merely incidental to the mines' primary function and are seen as a public relations tool to demonstrate the important role of mining in everyday life to the public and to show that they are not negatively affecting the environment.

Spiritual Tourism

A distinctive form of heritage tourism is spiritual tourism. Hundreds of Native American sites and natural places in the western states have been identified as important New Age spiritual attractions, where

◀ *Telluride ski resort in the Rocky Mountains of Colorado. The old mining town of Telluride is located in the u-shaped glacier-carved valley below the ski area. Dozens of world-class ski resorts like Telluride have developed in the Rocky Mountain states and attract millions of skiers every year.*

Photo: Alan A. Lew

nature worshippers and other animists travel to experience the powers of Mother Earth. Most notable of these is Sedona, Arizona, known to some worshippers as the New Age capital of the world. Several holy sites associated with the Catholic Church have become significant religious destinations in Arizona and New Mexico. Salt Lake City, Utah, is the headquarters of The Church of Jesus Christ of Latter-day Saints and a chief destination for the church's 15 million strong global population.

The southwest, particularly southern Utah, Arizona, southern New Mexico and western Texas, has become a highly desirable winter destination for *snowbirds* from the north—retirees who spend the warm summer months in the north and the cold winter months in the southwest. The population of the greater Phoenix region of Arizona, for example, swells by approximately 300,000 between December and March each year. Spending by these temporary residents is estimated to inject over a billion dollars into the local Arizona economy each winter. Snowbirds in Arizona and Texas often travel to nearby Mexican border towns for medications, dentures, dental services and physician care that are a fraction of their cost in the U.S.. Other travelers appreciate the warm winters of the southwest, which have become a haven for golfers and sun, sea and sand tourists, minus the sea, but with mega-resort development.

The Pacific States

This region includes California, Oregon, Washington, Alaska and Hawaii. Contrary to most popular images of Alaska and Hawaii, people can snow ski in Hawaii and lay on the warm beaches of southern Alaska. Alaska's main draw is its wild and remote setting, separated from the rest of the United States by Canada. It is possible to drive from the mainland United States to Alaska through British Columbia and the Yukon, but the majority of people arrive by air and cruise ship. The state's unique selling propositions are its arctic climate, glaciers, abundant wildlife, scenic mountain parks and unique native cultures. Dog sled races, skiing, ice fishing and snowmobile racing are popular winter activities, and the state's cities—Juneau, Fairbanks and Anchorage—are becoming more popular business travel and special events destinations.

Hawaii (or Hawai'i)

Hawaii, while not in the top ten destination states, is a special place in the minds of most Americans and a desirable destination. The newest state in the union, it is one of the country's favored wedding and honeymoon destinations, owing to its tropical climate, beaches, lush rainforest and many resorts. In addition to domestic visitors, Hawaii is a favorite destination for Japanese and Korean tourists, whose primary goal on the island is to shop. From an academic perspective, Hawaii is often cited as an example of how tourism (beginning with early explorers, traders, missionaries and whale hunters) can change a destination in both positive and negative ways. Waterfalls, volcanoes, vegetation and dramatic coastlines are natural foundations of the tourism sector, while its World War II heritage is appealing to most Japanese and American visitors, especially veterans of that war. Agritourism is important in Hawaii, as visitors tour pineapple plantations and banana farms. Hawaii Volcanoes National Park is one of the few places in the United States where visitors can get close to molten lava and see volcanic processes in action.

As already noted with Alaska and Hawaii, nature plays a critical role in tourism in the far western United States. Volcanic activity, including the destructive 1980 Mount St. Helens eruption, is a fact of life along America's Pacific Rim. Many mountains and craters through Washington and Oregon are volcanic in origin and play a major role in tourism as hiking venues, ski resorts and parklands. The countless islands in Washington's Puget Sound are focal points for recreational second homes. The entire coast from British Columbia to central California is a major linear attraction for people traveling along its length. This is an important corridor for camping trips and scenic drives.

California

California is the country's most visited state, and tourist expenditures are highest there. Amusement parks are a trademark of southern California, with Disneyland leading the way since its opening in 1955. The creation and success of Disneyland has spurred the development of *theme parks*, especially Disney theme parks, around the world. While European colonization was taking place on the east coast, Spanish explorers and missionaries were penetrating deep into the territory that is today northern California. Historic missions and churches are what remain from these explorer days. Other heritage resources include mining ghost towns and Hollywood studios where many movies and television shows have been filmed. *Film-induced tourism* is important for the Los Angeles region; it includes studio tours, famous film locations and tours to stars' homes. Near San Francisco, Sonoma Valley and Napa Valley are world-renowned wine-growing regions, and their products have been distributed throughout the world for many years. This heritage of viticulture forms the backdrop for a thriving *culinary and wine tourism* sector in northern California.

California's largest cities are important international ports of entry for air and sea travelers. Many cruises along Mexico's west coast originate in Los Angeles or San Diego, and San Francisco is a major port of departure for cruises to western Canada and Alaska. Los Angeles, the second largest city in the U.S., is a major conference center, shopping venue and urban destination. The San Ysidro area of San Diego abuts the Mexican border town of Tijuana. Tijuana and San Diego are closely linked by history, culture, economics and modern transportation. San Diego's promotional literature includes cross-border visits to Tijuana, expressing the role of the Mexican city in greater San Diego tourism. Sea World and the San Diego Zoo are San Diego's most visible icons. As the spiritual capital of the Left Coast, San Francisco has distinguished itself as a significant *gay and lesbian tourist destination*. The city hosts many of the world's best attended gay events, including the annual San Francisco Pride Event. Chinatown is also a salient feature of San Francisco. It is one of the earliest and largest Chinatowns in North America and is popular for its food selection, shops and interesting cultural landscape.

Canada

As already noted in this section, North America is a vast territory that transects many climatic, topographic and vegetative zones. Canada, geographically the second largest country in the world, exemplifies this pattern. The diverse physical and human geographies of Canada give the country its tourist appeal. In the west, tourism is based largely on mountains, aboriginal cultures and European settlement. In the mid-portion of Canada, tourism focuses on indigenous people and agriculture-based heritage. Ontario and Quebec are seen as the heartland and original points of European discovery and settlement, and they are well-established cultural heritage destinations, especially pertaining to Canadian national heritage and Franco-Canadian identity. East of Quebec, in the Atlantic Provinces, tourism is largely oriented toward the Atlantic Ocean and its resources, such as fishing, and the cultural landscapes created by maritime trade, fishing villages, rugged coastal and forest scenery, and European settlement. The far north is home to vast expanses of arctic wilderness, giving it a significant nature-based tourism appeal.

Like its southern neighbor, Canada faces many aboriginal (First Nations) issues in relation to tourism. Gaming on First Nations lands has been a focal point of tourism development in recent years, and *aboriginal land rights* have been a primary focus in legislation within and outside of tourism debates.

Canada and the United States have long been major trading partners and traditionally have had interdependent economies. One aspect of this has been the phenomenon of *cross-border trade and shopping*. For decades, Canadians have crossed into the United States to shop, but in the early 1990s, when the Canadian dollar gained significant strength against the U.S. dollar, the phenomenon grew to incredible heights, with some 60 million Canadians visiting American border communities in 1991 for the express purpose of shopping. Reasons for this growth among Canadians were favorable exchange rates, lower taxes in the United States, a wider variety of products south of the border and higher-quality goods and services. American border towns thrived on this deluge of people and money coming from the north, and new *retail landscapes* were created to cater to their northern neighbors' desires.

However, by the mid-1990s, the Canadian dollar lost much of its competitive value, making it more expensive to shop abroad. With this turn, a reverse pattern occurred as Americans began shopping north of the border for items that became a better value with the strength of the U.S. dollar. A primary form of United States-Canada *border tourism* is pharmaceutical purchasing, where Americans cross to buy prescription medications for a fraction of their cost in the United States. Meanwhile, the weakening of the U.S. dollar since the year 2000 has once again reversed the shopping migration in favor of the Canadian shoppers traveling to the U.S..

British Columbia

Canada's westernmost province is British Columbia (BC), which possesses some of the most notable mountain scenery in North America. There are many national and provincial parks in British Columbia that attract millions of visitors each year. Skiing is especially popular, and many world-class resorts have been built since the 1960s. The Okanagan Valley is also growing in significance as a wine and food tourism destination, as are Vancouver Island and the Gulf Islands that lie between Vancouver Island and the mainland. Canadian native culture is an important tourist draw on Vancouver Island, along with the rainforests that the natives depended on for centuries. *Urban tourism* is concentrated in Vancouver, the province's largest city, and Victoria, its capital. Both cities are noted for the quality of their lifestyles. Regular ferries ply between Vancouver Island and the mainland, and the populated southwest corner of British Columbia is in many ways more connected to the United States than to the rest of Canada. Tourism in BC's sparsely populated far north is based on natural areas with their accompanying fishing and hunting. British Columbia is also an important cruise destination. Many west coast cruises from San Francisco and Seattle stop in Victoria or Vancouver on their way to Alaska.

The Prairie Provinces

Alberta, Manitoba and Saskatchewan comprise the Prairie Provinces of Canada, although the westernmost portion of Alberta lies in the Rocky Mountains. These provinces are the *breadbasket* of Canada, producing most of the grains and other agricultural products for the entire country and for export abroad. There is also a unique migration history, particularly from Ukraine and other Eastern European countries, which has left its footprint on the social and physical environment of the region. Like the U.S. Midwest, these ethnic and agricultural heritages have led to the growth of many ethnic and food-based festivals.

Likewise Parks Canada and the individual provinces maintain a large number of cultural sites that highlight the human past of the region. Native Canadians still live in reservations in the region and have begun utilizing their heritage as a tourism product and capitalizing on Canada's gaming laws that allow casinos on First Nations reservations. Cattle ranching history also plays an important role in the region's tourism industry. The Calgary Stampede, for instance, was first held in 1912 and has since become the largest rodeo in the world.

The Prairie Provinces reach from the United States border northward into the subarctic and taiga northlands, including the vast Canadian Shield. The northern half of the provinces is sparsely populated and strewn with lakes, rivers and vast forests, providing many outdoor opportunities for recreational fishers and hunters. Some of the world's largest national parks lie in northern Canada. The far north has little by way of road and transportation infrastructure, although it is possible to drive the entire length of

◀ *A walk across the Capilano Suspension Bridge is a popular tourist experience in Vancouver, British Columbia. The natural scenery and mountain landscapes within easy travel distance of Vancouver, a major gateway to Canada and one of the country's largest cities, make the area very attractive to foreign and domestic tourists.*

Alberta and into the Northwest Territories. Many of the water bodies are favored by recreationists because of their relative isolation; floatplanes are the most commonly used mode of long-distance transportation.

Several international attractions lie astride the United States-Canada border in this region. The International Peace Garden on the border of Manitoba and North Dakota (USA) is a botanical park divided by the international border but united as a monument to peaceful relations between the two countries. It is one of the most popular attractions in Manitoba and the most significant attraction in North Dakota. Against the Saskatchewan-North Dakota border are a few provincial parks and other protected areas. The Waterton-Glacier International Peace Park is located on the border of Alberta and Montana (USA), and it comprises Glacier National Park, in the U.S., and Waterton Lakes National Park, Canada.

Ontario

Ontario is the home of Canada's capital, Ottawa, and its largest city, Toronto. The cities are the hallmark of Ontario's tourism industry, with thriving business and stock market exchanges, theaters and arts, and ethnic communities. Toronto is home to strong Chinese, Italian, Greek and Portuguese ethnic communities that add appeal to the city with food, festivals and culture. The outbreak of *SARS* (severe acute respiratory syndrome) had a drastic impact on Toronto's tourism sector in 2003. Dozens of cases were reported in the city, leading to thousands of canceled bookings and an estimated loss of half a billion dollars in tourism revenue. Even though the situation was largely under control and the outbreaks were limited to hospitals, the situation was exacerbated by the World Health Organization's release of a worldwide warning against non-essential travel to Toronto.

Ontario's vast rural areas have become trendy destinations as well, especially former agricultural villages in the southwest that have been transformed into popular *tourist-shopping villages*. St. Jacobs is one of the most notable. It is the center of Mennonite Country, where old-order Mennonites still dress, farm, drive carriages and perform their traditional religion, language and agricultural practices. Another notable rural activity in Ontario is *wine tourism*, which is mostly concentrated in the Niagara Falls region. Famous for its unique ice wines, the Niagara wine country has become popular for winery and vineyard tours, and its products are exported all over the world.

Second home ownership is a salient phenomenon in various parts of the province, particularly on lakeshores around Georgian Bay and Muskoka. Canadians from all over the country and even Americans, when their dollar is strong, come to the region to purchase vacation properties in the woodlands and on the lakeshores of Ontario. The rural, agricultural landscape of the area is an important part of heritage as well, which comes in many forms. Battle sites of the War of 1812 along the Niagara and St. Lawrence Rivers appeal to history buffs, while colonial and indigenous heritages are exemplified in outdoor living museums. Canadian national heritage is preserved in an array of museums and monuments in Ottawa and Toronto.

Niagara Falls

Niagara Falls is probably the province's best-known landmark. The United States-Canada border bisects the falls, with nearly all of Horseshoe Falls located in Canada. The town of Niagara Falls, Ontario, is one

◄ *These grain silos in Manitoba are a vital part of the prairie heritage and demonstrate the importance of agriculture (primarily grain) in the Prairie Provinces. Like many areas of North America and the world, agricultural heritage and rural heritage in general are becoming a more appreciated part of a region's overall tourism product.*

Photo: Dallen J. Timothy

of Canada's premier destinations, and it has been the subject of much geographical inquiry into how tourist destinations develop into measurable zones of intensity. By all accounts, Niagara Falls, Ontario, is more appealing to tourists than its cross-border counterpart, Niagara Falls, New York. The U.S. side of the falls has been described by several observers as a backdoor to the United States, where power plants, factories and heavy industry dominate, and the community is old and in a state of disrepair. The Canadian side, however, is seen as Canada's front door, which is reflected in the falls and riverfront being designated a provincial park. It is well groomed and more inviting than its U.S. neighbor. In spite of this, the Ontario town is often criticized as being too touristy and over developed with souvenir shops, cheap restaurants, tacky museums and other peripheral attractions.

Quebec

Since the mid-twentieth century, there has been a movement in the Canadian province of Quebec to separate itself from Canada and become an independent country. Through the 1960s, 1970s and 1980s, this movement gained momentum as more and more French-speaking Canadians became interested in sovereignty for their homeland. They felt their language and culture were not receiving the recognition and prioritization they deserved in the federal model of Canada. Several referenda have been held over the years wherein the public has been able to voice its vote on whether or not to secede from Canada. While there has yet to be a majority yes vote in favor of secession, support appears to be growing with each successive referendum, which might mean that in future efforts, a majority vote could lead the province to declare independence from Canada.

If the separation ever comes to fruition, it will have important implications for tourism. Namely, questions about whether or not Quebec will continue to use the Canadian dollar and the Canadian national banking system will need to be addressed. Citizenship, passports and the right to cross borders into Canada and the U.S. will also be salient points of negotiation. Likewise, the establishment of *customs and immigration* posts on Quebec's new international boundaries and its potential membership in NAFTA would need to be addressed. These are only a few of the many issues that would need to be faced that reflect directly on tourism, if Quebec were to go its own way from Canada.

Montreal and Quebec City

Montreal and Quebec City are the two centers of tourism in the province. Quebec City, the provincial capital, is one of the oldest European settlements in North America, and its old town (Vieux-Quebec) was inscribed on the *UNESCO World Heritage List* in 1985. Heritage is a major resource in Quebec City; many of its buildings date from the 17th century, and the city has a distinct European ambiance. Its position on the St. Lawrence River has facilitated economic growth for hundreds of years and today allows cruise ships to visit. Montreal is Canada's second largest city and an important heritage destination. The old town, harbourfront and Olympic Park are important components of Montreal tourism. Scenic French-Canadian rural landscapes, where little English is spoken, contribute to the appeal of the province as a destination. Skiing is a popular winter pastime near Quebec City in the Laurentian Mountains, primarily for domestic tourists.

The Atlantic Provinces

While the four Atlantic provinces of Canada (Nova Scotia, New Brunswick, Prince Edward Island and Newfoundland and Labrador) are among the least affluent areas of Canada, they are among the most scenic, culturally rich and desirable destinations in the country. Rugged coastlines, thick forests, lakes and rivers, lighthouses and small fishing villages are trademarks of the region. Newfoundland and Labrador is the easternmost province and home to several unique national parks. Iceberg sighting and whale watching are popular tourist activities, and the UNESCO-designated L'Anse aux Meadows, the site of the first European (Viking) landing and settlement in the Americas, typically draws some 30,000 people a year.

Prince Edward Island (PEI) is a good example of a *literary heritage* destination in North America. The province, located off the coast of New Brunswick and Nova Scotia, is known as the location of the *Anne of Green Gables* books and television series. It is especially popular among Japanese, Canadian, American and European tourists. Until the mid-1990s, the only way to arrive in PEI was by boat or air. However, in 1996, the 13-km long Confederation Bridge was finished to link PEI with New Brunswick and the rest of Canada. As another alternative means of travel, the new bridge has resulted in a growth in visitation levels since its completion. PEI markets itself as North America's Island Experience, and it offers biking and hiking trails, national park walks and picnics, windsurfing and lighthouses.

Nova Scotia and New Brunswick are perhaps Canada's best examples of maritime and fishing heritage. Small fishing villages, many of which have been the settings for TV series and motion pictures, attract visitors to their dockyards and shipyards, shopping zones, seafood restaurants and lighthouses. Many of the historic communities still survive largely on fishing and tourism. Nova Scotia is also a hub of genealogy and family history research, for the area was one of the primary immigration points in Canada and all of North America. The province estimates that approximately 25 million North Americans can trace their roots back to Nova Scotia since the 16th century. The province is home to Scottish, Acadian (French Canadian) and indigenous Mi'kmaq cultures that are still manifest in everyday life and form the foundation of heritage tourism. Ethnic festivals, music, food, dance and folklore are important components of tourism in these provinces, and New Brunswick is endowed with thick forests and rivers that have become a focal area for scenic drives and historic routes.

The Northern Territories

In addition to its ten provinces, Canada is home to three territories: Nunavut, the Northwest Territories and the Yukon Territory. Most of the area's population is located in a few medium-size towns and cities

◀ *Plenty of snow and ice remain in the summer months in the highest areas along the border region between the state of Alaska and Canada's Yukon Territory. This region was a major mining destination in the late 1800s. Its dramatic natural scenery is home to several national parks on both sides of the border today.*

Photo: Alan A. Lew

(e.g., Iqaluit, Yellowknife and Whitehorse) that serve as territorial capitals and regional trade centers. The rural population is scattered throughout the north in small, isolated villages where travel is done exclusively by airplane, and where residents supplement their grocery stores with seal, walrus, polar bear and caribou hunting. Owing to physical limitations and lack of demand, few roads exist, most of them extending only a few kilometers outside of regional towns. Travel in the far north is extremely expensive, with even short flights often costing thousands of dollars. As a result, most northern travel is done by residents themselves, who sometimes receive government travel assistance, and affluent tourists who desire to experience the vast Arctic national parks, indigenous Inuit culture, or fishing and hunting.

Inuit (Eskimo) culture is appealing to many people, and winter guests can enjoy dog-sledding and snowmobiling tours, as well as igloo stays. In the Yukon Territory around Dawson, gold rush heritage is heavily promoted, and people travel there to try their own hand at gold panning. One type of tourism that is starting to develop in the coastal northern territories is seasonal cruise tourism. As a result of climate change smaller 'expedition' cruises and even some larger cruise ships are able to travel further north. However, such tourism often places substantial pressures on small settlements. Nevertheless, tourism is seen as one of the few means of economic development apart from mineral exploitation in the far north.

ISSUES AND INSIGHTS   Tourism Issues in Anglo North America

Following World War II, tourism in North America grew quickly. The continent today is one of the largest tourism receiving and generating regions of the world, together with Western Europe. The United States and Canada (as well as Mexico) are regularly ranked by the World Tourism Organization (UNWTO) in the world's top 20 destinations. Furthermore, the United States has been the largest tourism earner for many years. Five primary characteristics account for North America's global tourism position. The first is its sheer size. The continent extends over dozens of physiographic, climatic and botanical zones, which provide a wide range of climates and weather patterns, land- forms, flora and fauna, natural landscapes, cultures and patterns of human habitation. This varied environment is the foundation for the growth of much of North America's tourism sector.

The second factor is rapid advances in travel innovations. Post-World War II air travel improvements and larger transportation networks (including an expanded high way system) have facilitated higher levels and more efficient forms of travel and opened up new destinations. Similarly, computer-based technology has put travel planning in the hands of individuals who had in the past relied more heavily on travel agents and tour operators. Approximately half of all online dollars spent in the U.S. are for travel and tourism products and services. Fourth, since the 1950s, the economies of Canada and the United States have grown rapidly, with a relatively low level of unemployment and inflation, enabling many more people to travel than before.

NAFTA
North America has been a fertile laboratory for cross-border alliances. In 1989, the Canada-United States Free Trade Agreement (FTA) was launched to start reducing trade barriers for goods and services between the two countries. This coalition was instrumental for tourism because it addressed tourism in detail. It effected various regulations pertaining to air travel, collaborative marketing efforts and increased cross-border travel, and it eliminated restrictions on the value of services that may be purchased in the neighboring country. In 1989 and 1990, additional bilateral treaties were signed between Canada and Mexico and the United States and Mexico that pertained to tourism and that facilitated cross-border travel, cultural exchanges, binational human resource education, common promotional efforts and mutual collaboration in achieving World Tourism Organization (UNWTO) standards and practices. On January 1, 1994, the FTA was replaced by the North American Free Trade Agreement (NAFTA), a tripartite accord between Canada, Mexico and the United States. Its primary objectives were to eliminate duties and other trade tariffs in stages over the next two decades. Unlike the FTA, tourism was not an explicit concern of NAFTA, but it did deal with issues affecting tourism, such as temporary entry, trade in services, monetary investments and telecommunications.

Shopping
Shopping has emerged as a significant tourist activity in North America, focusing largely on outlet shopping centers, mega-malls and shops in border regions. Cross-border shopping is important in the Canada-United States and Mexico-United States borderlands. Since the 1990s, shopping centers have been built in Mexico near the U.S. border in an effort to dissuade Mexicans from outshopping to the United States. However, this has not stopped Mexican consumers from crossing into the United States to purchase household items, such as clothing, food and supplies. Americans shop in Mexican

border towns for pharmaceuticals, alcohol and tobacco products. Shopping tourists are attracted to the U.S. for lower prices and wider varieties of goods from much further afield than Mexico, and include shoppers from Brazil and China who save money on both high and low end products even after taking into account the cost of air fare.

Modern shopping malls have popped up all over North America since the end of World War II. The first mega-mall, and still the largest mall in North America, the West Edmonton Mall in Edmonton, Alberta, Canada, was built in the 1980s and expanded in the 1990s. Following suit in the 1990s, the Mall of America was constructed in Minneapolis, Minnesota, and is North America's second largest mall. These centers are tourist destinations in their own right, offering playgrounds and theme parks, zoos, restaurants, bowling alleys and skating arenas, travel agencies, car rentals, hotel accommodations, currency exchange booths, cinemas and swimming pools, as well as other recreational and financial services. West Edmonton Mall and Mall of America package tours are offered in Japan and Europe. Shopping enthusiasts fly into Edmonton or Minneapolis, spend two or three nights at the mall and return to London or Tokyo. Malls and other shopping centers in North America no longer serve only a utilitarian function; they have become entertainment complexes and vacation enclaves.

National Parks

North America's national park systems are among the most inclusive and expansive in the world. Yellowstone National Park was the world's first national park (in 1872) and a precursor to the U.S. National Park Service (NPS). Parks Canada and the U.S. NPS, and their precursors, have managed hundreds of natural and cultural sites in their respective countries. Most of these sites in the eastern portions of the U.S. and Canada focus on maritime history and the colonial past, while western properties tend to spotlight natural environments and indigenous cultures. National Park Service properties in the United States fall under several categories, including battlefields, battlefield parks, battlefield sites, historic parks, historic sites, lakeshores, memorials, military parks, monuments, parks, parkways, preserves, recreation areas, reserves, rivers, scenic trails, seashores, and wild and scenic rivers. Each has its own definitional criteria. Mexico has a complex system of approximately 123 national parks and protected natural areas under the administration of several national-level conservation agencies.

American Indian Gaming

Native American cultures have play an important role in the U.S. tourism product, especially in the West. However since the 1990s, American Indian cultural as a tourism resource has been superseded by Indian reservation gaming. According to treaty, federally controlled American Indian reservations are sovereign nations and not legally part of the states in which they are located. Within this legal framework, the Indian Gaming Regulatory Act (IGRA) was passed by the United States Congress in 1988 to allow various forms of gambling, including high-stakes casinos, to be developed on Native American lands that are under the control of the federal Bureau of Indian Affairs. Casinos could be developed on federal reservations (not state reservations) within the boundaries of states where any form of gambling was already allowed. Thus, several states that permitted lotteries or sports betting suddenly became hosts to Native American casinos. Since 1990, hundreds of casinos have been built on Indian reservations in 28 states, with most gaming revenue coming from California and Oklahoma. Location is crucial, however, with Foxwoods and the Mohegan Sun, two of the world's largest casinos, being located on two small Indian reservations in the state of Connecticut that have ready access to the large metropolitan market extending from Boston to New York.

Post-9/11 Tourism

Since the 2001 terrorist attacks on New York and Washington DC, there has been a tightening of controls on the US-Canada and US-Mexico borders for fear of terrorism. The United States sees Canada and Mexico as having more relaxed immigration policies, and it has set out to protect itself from potential threats via the north and the south. On September 11, 2001, immediately following the attacks, all travel into and out of the United States was halted, both United States borders were sealed and all airports and harbors were closed. This immediate sealing of the border had a critical impact on Canada and Mexico, and the intensification of border controls has created additional economic pressures for both neighbors, who are heavily dependent on U.S. trade and tourism for their economic well-being. Biometric controls (e.g. eye retinal scans and fingerprinting) have been initiated at entry points for non-US citizens, and stricter passport and visa policies have been introduced.

The most recent policy to affect travel between Canada, the U.S. and Mexico was the change in the U.S. government's passport regulations, known as the Western Hemisphere Travel Initiative (WHTI). Previously, Canadians and Americans did not require a passport to visit each other's countries, and Americans have not required a passport to visit or return from Mexico or to take Caribbean cruises. As of January 2007, however, all travelers who fly into or out of the U.S. must possess a valid passport. Travel to U.S. overseas territories (American Samoa, Guam, Northern Marianas, Puerto Rico and the United States Virgin Islands) is not considered international, and passports are not required. In 2009, the same requirement to present a valid passport or other enhanced proof of citizenship at ports and border crossings was extended for people traveling to the United States by car, boat and foot. Canada and Mexico vehemently argued against this new policy, suggesting that it would harm cross-border trade and tourism, and increase border delays.

Other Countries in Anglo North America

Most people are unaware that there are other countries located within the geographical reaches of North America. As noted earlier in the chapter, Saint Pierre et Miquelon (population 7,000), a territory (overseas collectivity) of France, is a small group of islands off the coast of Newfoundland and the last remnant of France's North American possessions. Cod fishing has traditionally been the economic mainstay of the islands, but with the forced reduction in fishing activity in the 1990s and Cod Wars with the U.S., the islands have become more dependent on tourism. As part of France in North America, Saint Pierre et Miquelon is becoming a popular cruise stopover and shopping destination. Canadians are the islands' biggest market; they come to purchase French products (e.g., perfumes), French food and French lifestyle without having to travel to Europe.

Greenland

Greenland, the world's largest island, is located off the northeast coast of Canada. The island is a self-governing territory of Denmark and home to a large Inuit population that still depends on fishing and hunting for much of its subsistence. Most travel to Greenland is undertaken by Danes arriving from Denmark, although there are air connections to Iqaluit, Canada, and Iceland, and many visitors arrive in east Greenland on short trips from Iceland. Cruise ships stop in Greenland's western communities in conjunction with visits to Canada's Arctic islands.

Much of Greenland is an impenetrable ice cap, though it too is warming in recent years, with the population occupying the green coastal zones. Passenger ships ply coastal waters, and most remote communities are serviced by regularly scheduled flights. In the winter, visitors can sleep in a hotel constructed completely of snow and ice, play winter ice golf and tour by dog sled. During the non-winter months, tourists tour the fjords and small coastal islands, and gaze at migrating whales and icebergs. Viking archeological sites and Inuit culture are the basis of Greenland's cultural heritage.

Bermuda

Although Bermuda is not officially part of North America (it is better classified as a mid-Atlantic island), it is often considered part of North America, being 620 miles (1,000 km) off the U.S. coast. It is better connected to the U.S. than to Europe. Bermuda (a territory of the United Kingdom) is an expensive destination and geared toward an affluent clientele that stays in apartments, bed and breakfasts, or beachfront resorts. To reduce tourism's ecological and social impacts, there are no rental cars on the island, so travelers must use scooters, taxis, or horse-drawn carriages for local transportation. Bermuda resembles its British colonial cousins in the Caribbean in terms of architecture, population, culture and festivals, and it has become a popular port of call for cruises from the United States.

5.2 | LATIN NORTH AMERICA

Current Latin North America Issues and Resources (http://wrgeography.com/latin-namerica.html)

Physical Geography of Latin North America

Latin North America consists of three major geographic subregions: (1) the country of Mexico, (2) Central America (also known as Meso-America) and (3) the Caribbean Sea (also known as the West Indies). Central America includes the countries of Guatemala, Belize, Honduras, El Salvador, Nicaragua, Costa Rica and Panama. As noted in the introduction to this section, physical geography transcends the boundary between Anglo and Latin North America, so there is some overlap of the discussion about Latin North America's physical geography with that in the previous section. In particular, the subregions of Latin North America share many of the physical geographical characteristics of the southern portions of Anglo North America. The two major physiographic regions of Latin North America are (see Map, page 380):

1. The Mountains and Basins, which extend south from Anglo North America's Rocky Mountains and Basins and Plateaus, and
2. The Caribbean Sea and Gulf of Mexico, including the large islands of the Greater Antilles and the smaller islands of the Lesser Antilles, which together are known as the West Indies.

The mountains of Mexico and Central America are closely related to the Rocky Mountains. Both were formed by the collision of the North American continent with several plates that make up the Pacific Ocean floor in this region. Mexico's Sierra Madre Occidental (west) and Sierra Madre Oriental (east) are extensions of the Rocky Mountains into Mexico, while the Sierra Madre Del Sur (south) transitions to the high lands of Central America, and then down to the Andes Mountains in South America. The Sierra Madre Occidental and Oriental are heavily folded (like the Appalachians), while the Sierra Madre del Sur is mostly igneous (volcanic). All of these mountain ranges are part of the Alpine Orogeny, which is the current geologic period of mountain building that started about 80 million years ago. The islands of the Caribbean, from southeast Cuba and Hispaniola southward, are also part of this Alpine Orogeny and are geologically connected to the highlands of Central America and the northern Andes in South America.

Plate Tectonics

The northern Mexico portion of the North American continent is sliding laterally against the Pacific Ocean Plate, similar to the situation in California. The major difference here, however, is that Baja California, which is on the Pacific Ocean Plate, along with the city of Los Angeles in the United States, is slowly breaking away from North America as it moves northward. Some models predict that it will completely break away from the North America mainland in the geologic future. South of Baja California there is a separate ocean plate, the Cocos Plate, that is subducting directly under southern Mexico and the entire Central American region. This is similar to the Juan de Fuca Plate in the United States Pacific Northwest, except that the number of active volcanoes and frequency of earthquake activity is much greater on the southwest coast of Mexico and Central America than in Oregon and Washington.

The Basin and Range landscape of Nevada and Arizona extends into northern Mexico with its high-elevation plateaus, older mountains that have eroded into low-lying hills and internal drainage basins with sand dunes and flat playas. This area between the Sierra Madre Occidental and the Sierra Madre Oriental is known as the Mexican Plateau as it increases in elevation toward central Mexico, around Mexico City. The Mexican Plateau is the heart of Mexico. It is about 700 miles (1,130 km) long and ranges from 4,000 to 8,000 ft (1,220 to 2,440 m) in elevation. Most of the Mexican Plateau is arid, with climatic conditions similar to the U.S. Southwest. Rainfall, however, increases further south toward Central America, where the summer rains from the equatorial intertropical convergence zone (ITCZ) are more pronounced.

Like the Great Basin to the north, most of the drainage basins on the Mexican Plateau drain internally and have no outlet to the sea. Mexico City, one of the world's largest cities (over 22 million people), is located in one of these basins. A few large rivers drain from the plateau into the Gulf of Mexico in the east and the Gulf of California in the west. A chain of extinct volcanoes is located in the vicinity of Mexico City, including Citlaltepetl (also known as Orizaba) which, at 18,700 ft (5,700 m), is the highest peak in Mexico. South of Mexico City, the Sierra Madre Occidental and Sierra Madre Oriental converge in a jumbled mass of mountains called the Sierra Madre del Sur.

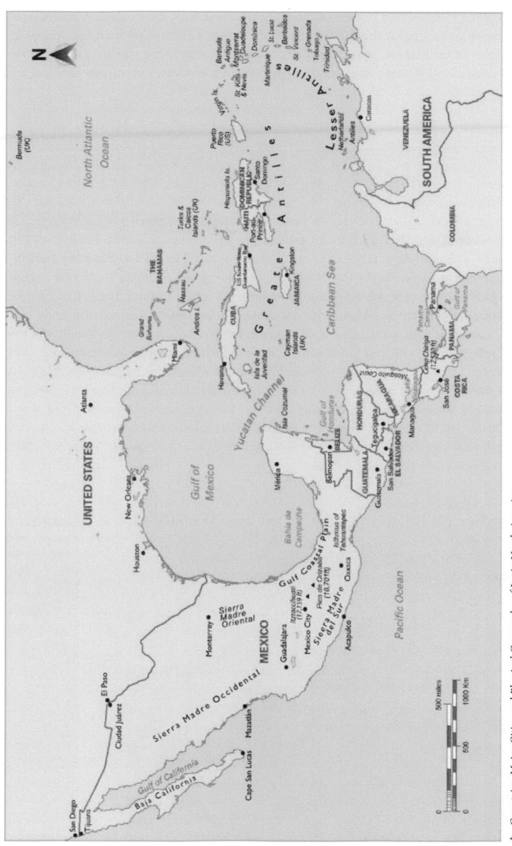

▲ *Countries, Major Cities and Physical Geography of Latin North America*

South of the Sierra Madre del Sur is a break in the mountains known as the Istmo de Tehuantepec. This isthmus is 135 miles across, and some use it to define the boundary between Mexico and Central America, because the climate, geology and culture of the region to the south are distinct from those of both Mexico and Anglo North America to the north. Here, the Mexican state of Chiapas, with its large Mayan Indian population, borders the country of Guatemala. The highlands of Central America share the west-to-east trending mountain system that created the Greater Antilles islands in the Caribbean Sea (Cuba, Hispaniola and Puerto Rico). These mountains of folded sedimentary rock, uplifted by the collision of South America and North America, extend along the southern coast of the Yucatan Peninsula, under the Caribbean Sea and over to the Greater Antilles and then down the Lesser Antilles islands.

The Caribbean

The Caribbean islands, also known as the West Indies, stretch in an arch from Cuba and the Bahamas to the north coast of South America. They are divided into the larger islands to the north and west, known as the Greater Antilles, and the smaller, more volcanic islands to the southeast, known as the Lesser Antilles. This chain of islands separates the Caribbean Sea from the Atlantic Ocean. Northwest of the Caribbean Sea is the Gulf of Mexico, which is enclosed by Mexico's Yucatan Peninsula, the island of Cuba and the Florida Peninsula. The Gulf of Mexico is an extension of the North American Interior Lowlands and the Gulf Coast Plain, which includes much of northeastern Mexico, Mexico's Yucatan Peninsula, most of Cuba, the Bahama Islands and Florida. All of these areas have very low elevations and, along with the lower-elevation islands, are seriously threatened by rising sea levels due to global warming.

The Caribbean islands were formed by the complex collision of several small ocean floor tectonic plates that were pushed together by the northward movement of South America into North America starting about 70 million years ago. Some of the islands are former sea beds, while others are volcanic in origin. Earthquakes and volcanic activity remain a frequent occurrence in the region, which occasionally result in major disasters.

◄ *The highlands of Central America were formed primarily through plate tectonics. These mountains in Costa Rica are volcanic in origin and are covered by cloud rainforests. They abound in wildlife and vegetation and create an attractive venue for agriculture and rainforest ecotours.*

Photo: Dallen J. Timothy

The Greater Antilles

The islands of the Greater Antilles were once a single mountain range that was forced upward by plate tectonics and was connected to the highlands of Guatemala and Honduras in Central America. Today, it includes the islands of Cuba (the largest island in the Caribbean), Jamaica, Hispaniola (containing the countries of Haiti and the Dominican Republic), Puerto Rico (a commonwealth territory of the United States) and the Virgin Islands (which are divided between a British part and an American part). The highest elevation in the Caribbean is Duarte Peak in the Dominican Republic, at 10,400 ft (3,170 m). All of the Greater Antilles islands have mountainous centers surrounded by relatively broad coastal plains. Coral reefs are also common off the coasts of these large islands. The valleys of the ancient mountain range are now submerged below sea level, with the deepest place in the Caribbean being the Cayman Trench (25,200 ft,; 7680 m) deep) between Cuba and Jamaica.

The Lesser Antilles

The Lesser Antilles lie to the southeast of the Greater Antilles and generally mark the outer edge of the North American continental shelf. The Lesser Antilles have two types of islands. An inner, western ring of larger islands (from Saint Kitts and Montserrat in the north to St. Vincent and Grenada in the south) are more mountainous and volcanic (known as high islands). An outer, eastern ring (including Anguilla, St. Martin, Barbuda and Antigua in the north, and Barbados in the south) consists of low islands that are mostly limestone sitting atop the stubs of eroded volcanoes and other rocks.

The Lesser Antilles are also divided into the Leeward (meaning downwind) Islands in the north and the Windward Islands in the south that face closer to the ITCZ rains. The Lesser Antilles are more volcanic because they are located in a subduction zone facing the Atlantic Ocean. Other islands are also considered part of the Caribbean, but they are not part of the Antilles. These include the Bahamas (which are low-lying limestone and coral reef islands on the continental shelf to the southeast of Florida), Trinidad and Tobago and the Netherlands Antilles (that are geologically part of South America) and Bermuda (a British territory that is actually a mid-Atlantic island). The Anglo North America section discussed Bermuda.

Climate of Latin North America

Throughout Latin North America the predominant wind movement is from east (Atlantic Ocean) to west, which is the opposite of the predominant pattern in Anglo North America. The major influence on the region is the ITCZ, which is a belt of low pressure and rainy weather that roughly follows the equator around the globe. The ITCZ moves northward, following the sun, in North America's summer months. It brings more rain throughout Latin North America, as well as hurricanes. In the winter months, the ITCZ moves southward toward the Amazon Basin in South America. This is the dry season for all of Mexico. Central America and the Caribbean receive less rain at this time of the year, though some precipitation still occurs throughout the season, and temperatures remain very warm, producing rainforest vegetation.

In addition, many of the taller and larger islands of the Caribbean are affected by altitudinal zonation and rain shadow effects, with cooler peaks, more rain on their windward sides and arid conditions on their leeward side. These conditions make the winter season in the Caribbean and Central America ideal for holiday makers in the northern latitudes of Anglo North America and Europe who are looking for warmer weather.

In the far north of Latin North America, northern Mexico has dry and desert-like conditions throughout the year, with cooler winter temperatures similar to the border areas of the United States. The high elevation and low latitude of the Mexican Plateau give Mexico City mild weather conditions throughout the year. In the far south of Latin North America, the tropical rainforests are denser, though much of them have been cut for forestry products. Precipitation is also more common throughout the year on the southernmost Caribbean islands, close to South America.

Hurricanes

Starting in June each year, the collision of warm air masses off the coast of West Africa in the Atlantic Ocean gives rise to low pressure air systems that move toward the Caribbean Sea. About 100 of these swirls become tropical disturbances, and about a dozen of those become tropical storms. When the winds in these storms reach 74 miles per hour (119 km per hour), they are classified as Category 1 hurricanes. (The same phenomena in several other parts of the world are called typhoons and cyclones.) When a hurricane hits land, it quickly loses its source of energy-the warm waters of the tropical and subtropical ocean and seas-and dies out, though winds can remain quite strong far inland. The hurricane season lasts through the end of November, when the warm air masses shift further southward.

Hurricane Alley is the name given to the Caribbean region just south of the Greater Antilles and up through the Gulf of Mexico. That is because it receives more tropical summer storms and hurricanes than anywhere else in the Americas. At least six hurricanes a year take this path, typically affecting the Yucatan Peninsula and making landfall in Texas or Louisiana, though they could also land in Mexico or Florida. Alternatively, some hurricanes approach North America to the north of the Greater Antilles and then turn up the east coast of Florida, sometimes making landfall as far north as New York City or New England.

A large hurricane can have devastating impacts on human settlements and the natural flora and fauna of the places in its path. Because of the nearness of South America to Africa, only the northernmost portions of that continent are ever affected by hurricanes (which are called cyclones in South America). Unfortunately, the hurricane season occurs during the summer holiday season for North Americans, which can cause trip cancellations when inclement weather is forecast. Hurricanes are also found in the Pacific Ocean side of North America, and on rare occasions they hit Hawaii and bring strong winds and rains to Baja California in Mexico and southern California.

Photo: Dallen J. Timothy

◄ *Five years after Hurricane Ivan passed directly over Grenada in 2004, the damage was still visible throughout the island, including to this historic house in St. George. Several important government and heritage buildings were destroyed, along with hundreds of individual homes. Ivan was the tenth most powerful hurricane ever to have been recorded in the Caribbean, and it nearly destroyed the communities on the island country of Grenada.*

Human Geography, Countries and Tourism of Latin North America
Mexico

The country of Mexico is officially known as Estados Unidos Mexicanos (United Mexican States). It has a population of approximately 120 million (2014 est.), more than half of whom live in the center of the country on the Mexican Plateau. The more arid north and the tropical south are sparsely settled, except for several large cities on the United States-Mexico border. In recent decades, Mexico has experienced large migrations from the countryside to the cities. Two out of three Mexicans now live in cities, and while some agriculture is still practiced, the petroleum industry and tourism have become the country's two most important economic sectors. The name Mexico comes from Mexica, which was an alternative name for the Aztecs. Prior to European contact, Mexico City was the center of the Aztec civilization, one of the most advanced in the New World. The Aztec culture dominated central Mexico from the 14th to the 16th centuries. The Aztecs had a complex culture in which feudal-like lords paid tribute to a central authority, often referred to as the Aztec emperor.

Aztec Civilization

Like many early civilizations, the central rulers were spiritually responsible for the predictability and success of the annual agricultural cycle. To do this, the Aztecs developed a very precise calendar, which had 365 days in a year, combined with a 260-day ritual cycle. Together, these two cycles created a 52-year cycle that was similar to a century in the modern calendar. The Aztec cities demonstrated highly advanced architectural design, irrigation technology and social organization. Their capital city of Tenochtitlan was built on islands in Lake Texcoco. Today, the lake is gone and the valley is the site of Mexico City.

Hernan Cortes landed in Mexico in 1517 and named it New Spain. Montezuma II, the Aztec emperor at the time Cortes landed, thought that Cortes was the god-king, Quetzalcoatl. Cortes used this confusion to overthrow Montezuma and conquer the Aztecs. The Spanish brought smallpox and other communicable diseases to Mexico, which killed up to 90 percent of the estimated 25 to 30 million Indians who inhabited Mexico and Central America at the time of European contact. For the next 300 years New Spain was ruled by Spain. In 1810, Mexico declared independence from Spain, and, after a long war, it finally gained its full independence in 1821. At that time, all of the southwestern United States from Texas to California was part of Mexico. Growing United States migration into these areas prompted the U.S. government to try to purchase them from Mexico, which Mexico refused. The United States then instigated the Mexican-American War (1846-1848), which resulted in Mexico's loss of about a third of its territory to the United States.

Mayan Civilization

In addition to the Aztecs, there were many other Indian groups living throughout Mexico at the time of Spanish arrival. Remnants of the Maya civilization (that peaked in the seventh century CE in Guatemala) are prominent to this day on Mexico's Yucatan Peninsula and in the southern state of Chiapas. Zapotec Indian communities abound in the Oaxaca area and in the remote Sierra Madre del Sur mountains south of Mexico City. Unlike Anglo North America, racial mixing was much more commonplace between the Spanish and Indians in Mexico. The resulting mestizo racial group forms the bulk of Mexico's population today. Mestizos in Mexico tend to have much more Indian blood than European blood because of the comparatively large numbers of Indians who lived there at the time of European contact.

Today, Mexico is considered one of the more successful countries in the developing world, and compared to the countries further south in Central America. However, in comparison to its neighbors to the north, Mexico remains poor, and large numbers of Mexicans leave the country each year both

Photo: Alan A. Lew

◀ *This all-inclusive resort is at Cabo San Lucas at the southern tip of the Baja Peninsula in Mexico. Coastal resorts have long been a part of the tourism product of Mexico, especially on the Pacific Coast in the west. East-coast Caribbean resorts have surpassed the west-coast destinations in popularity since the 1980s.*

legally and illegally to seek their fortunes in the United States and Canada. The number of illegal crossers apprehended by United States border officials ranges from 1 to 1.7 million a year. It is difficult to know how many illegal crossers are not apprehended.

Within Mexico, the most traditional lifestyles tend to continue in isolated rural areas and in Indian peasant societies. Smaller Spanish colonial towns are rich in their historic architecture, and some have become major tourist attractions, such as Oaxaca and Talaquepaque. The large urban centers of Mexico City, Guadalajara, Monterrey and Ciudad Juarez are modern metropolitan regions, each with over two million people. The result is a wide range of incomes, life styles, wealth and poverty, on top of a tremendous cultural and environmental diversity.

Tourism in Mexico

Most of Mexico's international tourists come from the United States and Canada, but many also visit from Europe, Asia and other parts of Latin America. Mexico is best known for beach resort-based tourism on the west and east coasts, particularly in the resort destination of Cancun (a purpose-built resort community) and Cozumel in the east, and in Los Cabos, Mazatlan, Acapulco and Puerto Vallarta on the Pacific Coast. Cruise tourism is also a vital component of the tourism industry in Mexico. Cruises on the west coast typically leave from Los Angeles or San Diego on trips three to seven days in length along Baja California and reaching as far south as Acapulco and Puerto Vallarta. On the Caribbean side, cruises frequently call in Cancun and Cozumel.

Another form of tourism that has become more popular in Mexico is nature- based tourism. This is particularly so in the rainforests of the southern states, where a large variety of endemic species of plants and animals can be found and where large tracts of rainforests still dominate much of the natural landscape. Most images of ecotourism, or of nature-based tourism, are geared toward tropical regions, with rainforest trekking as the major activity. However, in Mexico, a significant level of nature-oriented tourist activities also take place in the mountains, canyons and desert areas of the central and northern regions, focusing primarily on the country's widespread system of national parks and nature preserves.

Cultural heritage tourism is also a popular niche in the Mexican tourism industry. This specialized form of tourism is based primarily on the built heritage of Mexico's rich indigenous and colonial past. Many of the largest cities of Mexico are home to important built heritage areas of historic zones, where centuries-old Spanish colonial architecture stands among modern skyscrapers and busy streets in the form of churches, government buildings and stately homes.

Indigenous Heritage

The country's indigenous heritage is known worldwide for ancient cities and archeological sites. As mentioned above, the Aztecs built large cities in Mexico's central region, and Mayan ruins and cities dominate in the south. Many of these sites have been designated UNESCO World Heritage Sites and are among the country's most salient tourist attractions. Chichen Itza, the best-known ancient Mayan city, recently was elected to the list of the New Seven Wonders of the World, which will likely result in increased visitation. Another important form of cultural heritage that receives considerable tourist attention is Mexican cuisine. The food is the highlight of many trips, and Mexican culinary heritage has spread to other countries and influenced some of the largest fast food chains in the world.

Border Towns

In spite of the importance of tourism in Mexico's coastal regions, large cities, natural areas and heritage places, the country's most significant tourist region in terms of total visitor arrivals and money spent has long been its northern border with the United States. Mexico's border towns have been in statistical terms its most important tourism asset. In fact, the Tijuana-San Ysidro port of entry is the busiest border crossing in the world. These communities, most of which were originally cattle trading centers, became popular tourist destinations in the early 1900s, beginning with the prohibition of alcohol and its production and consumption in the United States. Unable to purchase and consume alcohol at home, many Americans traveled to Mexican border towns to drink where it was legal. As this phenomenon grew, so did other exotic vices, including prostitution and gambling. Some of these activities have become less important among visitors to the border towns (e.g., prostitution), but others continue to thrive.

Today, the most important border tourism activities on Mexico's northern frontier include drinking, gambling, purchasing pharmaceutical items, plastic surgery at purpose-built hospitals, and the outpatient services of dentists and physicians. These last three activities have become more popular in recent years with the high cost of medicine and health care in the United States. All of these activities have grown as a result of legal and pricing differences on opposite sides of the border.

Maquiladoras

Throughout the 20th century, the reputation grew of Mexico's border towns as lewd and dirty places. In an effort to change the public view of the border towns, the Mexican federal government started a border development program in the early 1960s, in an effort to stimulate tourism and expand the borderland economies. As part of this measure, the government cleaned up border towns, razed old buildings, poured sidewalks, remodeled shops and parks and built welcome archways. Additionally, the Mexican government created a special economic zone near the border to entice foreign-owned industries to set up shop just inside Mexico. Assembly plants and factories (maquiladoras) were built by U.S. automobile manufacturers, small appliance companies, pharmaceutical companies, and clothing manufacturers. By locating just inside Mexico (typically between 50 ft and a few miles), foreign-owned companies could pay lower wages to Mexican workers, and the national government provided many tax and location incentives.

The maquiladoras have become among the largest employers in northern Mexico, and people from around the country often give up rural life for the booming border cities to work in manufacturing, where they earn as much as two or three times their wages in the interior of Mexico. This phenomenon has resulted in waves of relatively affluent Mexicans traveling across the United States border to shop on same-day or overnight trips.

Mexican Emigration

In 2013, there were some 36 million Americans of Mexican descent and direct migrants from Mexico living in the United States. One of their preferred tourist activities is to travel back to Mexico to visit relatives. Mexicans are socially connected by strong family ties, which often determine where they will take their vacations. Visiting friends and relatives (VFR) tourism is therefore one of the most significant forms of tourism in Mexico.

Tourism Challenges

Because of its widespread appeal and diverse resources, Mexico is one of the most important tourist destination countries in the world, regularly ranking in the UNWTO's to 20 international arrival countries. But history has not always been kind to Mexico. In 1994, NAFTA took effect and started the process of eliminating trade barriers between the United States, Canada and Mexico. NAFTA was an unpopular treaty among many of Mexico's southern indigenous people, who still eked out a living from traditional agriculture. It meant that they would have to compete with more efficient and inexpensive farm products from the United States and Canada. As a result, on January 1, 1994, an indigenous uprising broke out, supported by the Zapatista National Liberation Army, against the national government and its military forces. Thousands of people were displaced and many were killed. A ceasefire was later signed, and today the Zapatistas have moved toward more peaceful means of negotiation. This event dramatically affected tourism to all parts of Mexico, particularly the south, and it placed the country on the travel warning lists of its major markets. While relations between the Mayas and the government are normalized, the conflict still lingers in the tourism image of Mexico, and there is still a military presence that has an effect on tourism.

Also affecting tourism in recent years have been drug-related murders and kidnappings in Mexico near the northern border. Along with crimes against tourists, these have prompted the U.S. State Department to issue warnings against travel to the region. By 2009, the U.S. media and government branded Ciudad Juarez, adjacent to El Paso, Texas, and long one of Mexico's most significant border tourism destinations, the most dangerous city in the world. In 2010, there were more than 3,000 murders in the city, and the federal government has increased police and military presence in the city to nearly 8,000 troops to help deal with the growing drug trade and related violent crime. While few tourists were targeted or injured in the drug war, the violence scared the once faithful American market away, resulting in a significant decline in tourists in Ciudad Juarez and other formerly popular cities such as Tijuana, which has also seen a good deal of drug-related violence since 2005. Since 2012, however, conditions have improved dramatically with Mexican government efforts. With far fewer murders and kidnappings, tourism in many of Mexico's border cities has begun to recover, notably in Ciudad Juarez and Tijuana.

◀ *In this photo. Negates, Mexico is on the left and Nogales, Arizona is in the USA on the right. These twin towns have had a long history of complementary border-related tourism development. Shopping, pharmaceuticals, medical services, auto repair shops, and prostitution are all located on the Mexican side within 300 meters of the border crossing.*

Photo: Dallen J. Timothy

Central America

Central America comprises the southern tip of North America, between Mexico and South America. As mentioned above, there is some logic to placing the northern boundary of Central America at the Isthmus of Tehuantepec in Mexico, because of the tropical climate and Mayan cultural influences that roughly start there. Excluding Mexico, the countries of Central America include Guatemala, Belize, El Salvador, Honduras, Nicaragua, Costa Rica and Panama. Together, the land area of these seven countries is about one-fourth the size of Mexico, and about half the size of Colombia to the south. Sometimes Panama is included in South America instead of Central America because it used to be part of Colombia.

Central America separates the Pacific Ocean from the Caribbean Sea and the Atlantic Ocean. Costa Rica and Panama form an *isthmus*, which at its narrowest point in Panama is only about 30 miles (50 km) wide. All of the countries of Central America are close to one or both of these large water bodies. The Caribbean coastal lowlands of Central America are closely tied to the Caribbean Sea and share many cultural similarities with the Greater and Lesser Antilles. A major characteristic that they share is the almost complete extinction of native populations and the dominance of immigrant peoples from Europe and Africa. One major exception to this is on the Mosquito Coast, which still has native Indian groups who live very traditional lifestyles. However, much of the Caribbean east coast of Central America has very low population densities and is thickly covered with tropical rainforests.

Tropical rainforests and high mountains are found throughout Central America, though the forests have been heavily cut or burned for agriculture. These conditions have created many isolated and re-mote areas within this relatively small geographic territory. The interior mountains and the western coast of Central America still maintain strong Indian and Spanish colonial influences, much like southern

Mexico. It is here where the majority of the Central American population lives. Because the mountains are cooler, the higher slopes are the preferred places to live, with most of the major cities lying between 3,000 ft (900 m) and 8,000 ft (2400 m) in elevation. Overall, the east coast of Central America is very lightly populated, the West Coast is moderately populated, and the interior highlands are heavily populated, and even overpopulated in some parts.

Mestizos

Overall in Central America, about 20 percent are pure Indian, while 60 per cent are *mestizo* (European-Indian mix, also called *ladinos* in Guatemala). *Mulattos* (black and European) make up 5 percent of the population, and *zambos* (Indians and blacks) make up 1 percent. Pure European ancestry is claimed by 12 percent of the people in Central America. Costa Rica is the only country in Central America with a majority of its population claiming white European ancestry. It had a very small Indian population in colonial times, and it was not of particular interest to the Spanish. Most of its population was attracted to the country in the 19th and 20th centuries, as Costa Rica appeared to have a more stable government and economy than the rest of Central America. Even today, Costa Rica is often recommended as an inexpensive retirement haven for Anglo North Americans. Blacks are proportionally greater in Belize, which was a British Colony until 1981, and in Panama, where many West Indians were brought to construct the Panama Canal (late 1800s and early 1900s).

Independence

In 1821, the colonies of Central America declared their independence from Spain, and with Mexico's support they gained full independence in 1823. In that year, they formed the United Provinces of Central America, which was a federal, democratic republic, much like the United States. This union, however, collapsed by 1840, when countries sought different policies on trade and governance. (However, the flags for four of the countries of Central America still share the colors and stripes from the original federal republic.) Many other attempts at union have been made since 1840, but none has been successful. A Central American Parliament was established in 1991 and continues to meet, though it only has advisory status. The Dominican Republic is part of the Parliament, but Costa Rica has so far refused to join.

Central America has suffered significant economic problems since the countries gained their independence from Spain. Their primary export is fruit, and they have been highly dependent on trade with the United States, as well as on U.S. development assistance and private investment. Poverty remains widespread, with wealth isolated in the hands of a few. Insurgencies and civil wars have been a problem in Guatemala, Honduras, El Salvador and Nicaragua. Civil unrest has calmed down in recent years in most of the countries, allowing them to develop their significant eco-tourism potential. The quieter country of Costa Rica has long been one of the leading ecotourism destinations in the world, and Belize is developing a similar reputation.

Tourism in Central America

Although the countries of Central America have considerable potential for tourism development, only about half have experienced notable tourism growth. In addition to many issues related to underdevelopment, one of the primary reasons for this has been political conflict and security problems. The most successful tourism destination countries in this region are Belize, Costa Rica and Panama, with Guatemala and Honduras having seen some success as well. El Salvador, Honduras, Guatemala and Nicaragua have been plagued by political and security problems (i.e., civil wars) for the past quarter century, although these have calmed down considerably in recent years. Nonetheless, these countries have had a difficult time recovering from their tainted images of fighting guerillas, military conflicts and rampant crime.

The countries of Central America have similar tourism products: rainforests, colorful cultures and beachfront developments. Mayan ruins and the lifestyles of indigenous people today form the basis for much of the region's tourism. Mayan civilizations in the northernmost countries of the region appeal to throngs of outsiders each year. In countries such as Panama and Costa Rica, the success of indigenous involvement in tourism has been cited throughout the world as being among the most sustainable, where the natives have a strong voice in, and benefit from, the development of tourism. Nature provides another important resource for the industry, primarily rainforests, beaches and volcanoes. Because of the combination of these cultural and natural elements, more and more Caribbean cruise itineraries have begun to include stops in Central America, such as Belize City, Roatan (Honduras), Puerto Limon (Costa Rica) and the Panama Canal.

Supranational Alliances

There are several supranational alliances that also focus on Central America, such as the Central American Economic Integration (SIECA, members include Costa Rica, El Salvador, Guatemala, Honduras and Nicaragua); the System of Central American Integration (SICA) (formerly the Organization of Central American States); and SICA's related Central American Commission of Directors of Migration (OCAM). Like other free trade agreements, SIECA aims to improve trade between member countries and facilitate the transport of goods across national boundaries, as well as eventually to establish a customs union between its five member states. SICA's goals include regional economic development, the promotion of democracy, environmental protection, establishing a common customs union (trading block) to enhance trade with other non-party countries, and the protection of human rights.

Together these two organizations, comprised of all seven Central American countries, work together to ease border crossings for people from within the region. For non-Central American tourists, however, border crossings have not been streamlined as they have for locals, although the two organizations jointly issue a common immigration card for foreigners arriving in any of the member states. While tourism is not openly a significant priority in these agreements, many of their other priorities, such as environmental protection, border-crossing formality alleviation, free trade and transportation, have a significant bearing on tourism.

Pan-American Highway

Another issue of significance in this region is transportation. In 1923, an agreement was reached at the Conference of American States to build a single transportation corridor (Pan-American Highway) that would link all of the Americas from Alaska in the north to Argentina and Chile in the south, some 16,000 miles (25,800 km). This goal has been met, except for a nearly 60 mile (100 km) break in eastern Panama and northern Colombia known as the Darien Gap—a swampy, rainforest-covered region with high biodiversity value that is under the control of dangerous drug smugglers and rebel guerrillas. The area's physical features have long prevented this last stretch of Interamericana from being developed, and now that adequate technology exists to overcome these barriers, precarious security conditions continue to inhibit the project. The countries of Central America have good and comprehensive bus transportation systems. Nearly all cities and towns, and many rural areas, are served by coach routes and minivan services. Unique to this region is the international company, TICA Bus, with routes from southern Mexico through all countries, except Belize, as far south as Panama City.

Belize

One of the most developed countries in Central America, in terms of tourism, is Belize. It has a well-established tourism image, particularly among the American, Canadian and Western European markets.

One of the appeals of Belize is its population's ability to speak English, owing to its political history as a former British colony (known as British Honduras). It gained independence in 1981, and it has flourished since then as an important destination, despite strained relations with neighboring Guatemala over their common border and claims by Guatemala that Belize is Guatemalan territory. Belize is home to the *Belize Barrier Reef* (also known as the Great Maya Reef or the Mesoamerican Reef), the second largest barrier reef system in the world after Australia's Great Barrier Reef. In 1996, the Belize Barrier Reef was inscribed as a UNESCO World Heritage Site. The reef also extends into Mexico in the north and Honduras in the south. This natural resource is home to thousands of aquatic plant and animal species and provides much of the appeal of Belize as a tourist destination. Small islands on and near the reef (e.g., Caye Caulker and Ambergris Caye) are the center of the country's resort facilities and diving operations. Many of the smaller islands are rented out for honeymoons and other significant events.

Mainland-based tourism focuses on the remnants of ancient Mayan civilizations, such as Altun Han and Lamanai, and on ecotreks through the rainforests of the country's southern and western regions. Most of the Mayan ruins in Belize have yet to be fully excavated. Belize City, the country's primary international gateway, is now a regular stopover on Caribbean cruises. Ecotourism and archeology-based heritage tourism, as well as limited diving and snorkeling, are among the most popular on-shore cruise excursions. Belize has been quite successful in withstanding pressures to welcome the establishment of large, foreign-owned and brand-name lodging and fast food services. This is despite the importance of tourism in the national economy, and although many of the lodging establishments and souvenir shops are owned by expatriates. In the remote south, Belize has also been noted as being a socially and ecologically friendly tourist destination with the development of village-based and community-operated lodging and tour services.

Guatemala

Civil war plagued Guatemala from 1960 to 1996, and it still has salient repercussions for the country's economy and society today. In spite of the country's 1996 peace accord, between 2000 and 2006 there were several reports of foreign tourists being raped, murdered, carjacked or robbed, and bouts of violence continue to plague certain parts of the country. These images have been accentuated by media reports

◀ *On Caye Caulker, a sand bar island located on Belize's barrier reef system, it is easy to arrange diving and snorkeling tours. Water sports are one of Belize's most important tourist activities, and Caye Caulker is considered a budget destination for people who desire a reef experience, Maya ruins and jungles are found in the interior of the mainland portion of Belize.*

Photo: Dallen J. Timothy

and have been a significant impediment for tourism development in Guatemala. Nonetheless, the country has potential for tourism growth. Its primary appeal lies in indigenous (Maya) heritage, both living and ancient. Languages, handicrafts, folklife, mixed Mayan and Catholic religious celebrations, costumes and food are important parts of the cultural product today. Several towns and regions in the highlands of Guatemala have become significant cultural destinations, including the famed Mayan market towns of Chichicastenango, Panajachel and Santiago Atitlan.

Built heritage is an important part of tourism in the country. Tikal in the northeast is an important national park and one of the best examples in Central America of pre-Columbian Mayan civilization. Antigua and Quetzaltenango are important towns for many reasons in Guatemala, but their tourism value lies in their exemplary Spanish colonial influence. In addition to cultural heritage, Guatemala has a rich and diverse natural heritage, which it is beginning to exploit for tourism purposes by offering rainforest-based nature tours and adventure activities. It lags behind its neighbors Mexico, Belize, Costa Rica and Panama in the area of ecotourism though. This is, in part, because some of the most pristine areas of the country have also been bastions of conflict between guerilla fighters and the military, and many of those areas are still risky to visit.

Honduras, El Salvador and Nicaragua

Traditionally, Honduras suffered from an under-exposed image in the world of tourism rather than the political conflicts plaguing its neighbors. The primary constraint to tourism in Honduras today, however, is violent crime. Since the 2009 coup d'état, drug trafficking, crime and human rights abuses have increased dramatically, earning the country the distinction of having the highest murder rate in the world. In spite of the current situation, in common with Belize, Mexico and Guatemala, Honduras has a rich Mayan past that has long formed the basis of much of the country's heritage tourism product. Honduras also has several national parks and protected areas where nature-based tourism was beginning to develop before the security problems emerged in 2009-2010. Perhaps the most popular region of Honduras for tourism today, however, is the Bay Islands off the Caribbean coast of the country. Roatan, particularly,

◀ *Ancient Maya pyramids extend above the rainforest canopy in Tikal National Park, Guatemala. The Mayas were one of the most advanced pre-Columbian civilizations in the Americas and are known for their scientific abilities and construction prowess. The ruins of many Mayan cities dot the landscapes of Guatemala, Belize, Honduras and southern Mexico.*

Photo: Dallen J. Timothy

attracts water-loving tourists to its growing beach-based resorts and accommodations. The Bay Islands are the safest part of the country for tourists.

Like Guatemala and Honduras, El Salvador and Nicaragua have seen a good share of political conflict and instability in recent decades, which have hampered their efforts to develop tourism. Efforts are underway by both countries to develop tourism, but the world, particularly their main potential market (i.e., the United States), is still a bit reluctant. Between 2010 and 2015, in conjunction with the increasing violence in neighboring Honduras, robberies, muggings, and carjackings of tourists in El Salvador increased dramatically, as did the national murder rate, resulting in many governments publishing warnings cautioning their citizens to avoid traveling there.

In 2006, Nicaragua held national elections wherein Daniel Ortega, a former adversary of the United States, was re-elected president. In the 1970s, he was a leader of rebel guerillas who overthrew the government, and after coming into power he adopted a Marxist- socialist form of government, which the United States opposed. In fact, the United States funded efforts by Contra rebels to oppose the Ortega (Sandanista) government. In 1990, he was ousted from office but returned again as president in November 2006. Since his recent re-election, Ortega has met with some foes of the United States, including Iran, and declared his hostility toward the United States. After years of economic recovery and growth, the future of tourism in Nicaragua will no doubt be affected by inter-American politics.

Costa Rica

Like Belize, Costa Rica has a well-developed tourism sector and a positive image abroad. Costa Rica has experienced the least degree of political turmoil in Central America and is one of the most affluent countries of the region. This has assisted in establishing a successful tourism industry that is well-known throughout the world and is commonly cited as an example of the congruent relationship between sustainable development and tourism, particularly among the native inhabitants of the mountains and rainforests. Forest-based ecotourism is the country's primary tourism product, focusing on numerous cloud forests and volcanoes in the Central and Guanacaste regions. Between San Jose and the Panamanian border are a series of less-accessible national parks that are popular among wilderness adventurers. Puerto Limon on Costa Rica's Caribbean coast is becoming an ever more popular cruise port city, and beach resorts are well established on both the Caribbean and Pacific coasts. Like the rest of Central America, Costa Rica is seen as a good value for North Americans in that airfares and accommodations are relatively inexpensive, and the country is well connected with transportation options. For these and other reasons (e.g. good health care), Costa Rica has become a fashionable home base for retirees and second home owners from the United States and Canada.

Panama

Panama as a destination is growing in popularity. Since the 1990s there has been considerable growth in the number of beach resorts on the Caribbean and Pacific coasts, to supplement the already popular San Blas Islands. Many tourism service providers are still small scale and locally owned and operated, but the country has seen the growth of expatriate ownership of small hotels, guesthouses, restaurants and souvenir shops, as has happened in Costa Rica and Belize. There is also a movement toward large-scale resorts, many of which are and will be owned and operated by large multinational corporations.

Of all the countries of Central America, Panama has the highest level of rainforest cover (ca 40 percent), and it is dotted with amazing volcanoes. These forests and mountain regions are home to a vast array of flora and fauna. Based on these resources, Panama has become a destination for nature buffs with a notable growth in nature-based/ecotourism since the late 1980s. Several national parks exist in the northern and southern portions of the country, including the UNESCO-listed World Heritage Area of La Amistad, which is an international park comprised of two adjacent parks in Panama and Costa Rica.

Panama is home to thousands of indigenous Amerindians, who also have a significant stake in tourism enterprises. Their material and non-material culture is an important base for Panama's tourism product, although there have been significant conflicts in recent years between the indigenous people and the government in the utilization or conservation of natural resources. The best-known tourism feature of Panama, however, is undoubtedly its famous canal, built by the French and Americans in the early twentieth century. The Panama Canal and the adjacent Canal Zone functioned as though they were part of the United States until they were gifted back to Panama by President Carter and returned to full Panamanian control in 1999. It is estimated that more than 14,000 ships pass through the canal each year, including cargo ships and cruise ships. Many cruises in the eastern Caribbean include the Panama Canal on their itineraries.

The Caribbean Sea and Gulf of Mexico

The islands of the Caribbean Sea (or West Indies) include 25 separate political entities, including sovereign countries, overseas departments (French provinces) or colonies, and territorial dependencies. The largest island and country is Cuba, with more than ten million people. Many other islands are very small, and some are uninhabited. Some islands are independent, while others are still governed by colonial rulers. They all share their insularity in common (they are islands), a colonial heritage, a former *plantation economy* (mostly sugar) and a history of African slavery.

Colonialization

Spain, France, Britain, the Netherlands, Denmark and the United States have all fought for control of various islands in the Caribbean Sea during the 17th, 18th and 19th centuries. Islands were captured during wars and traded during peace negotiations. Because they belonged to different colonial rulers, there has never developed a strong sense of cohesion among the islands, as there was in Mexico and to a degree in Central America. The official languages of the islands tended to be that of their colonial rulers, and economic ties were often stronger with the colonial homelands than with neighboring islands. Today, the Spanish-speaking territories (such as the Dominican Republic) have closer ties to other Spanish-speaking countries of Latin America, but the other islands and territories do not.

At the time Columbus arrived in the Caribbean, there were perhaps 500,000 to 1 million Indians living there, with the main tribes being the Ciboney, the Taino and the Carib. Of these, only a couple of hundred Caribs survive in the Caribbean region today, though there are some Ciboney in Florida. The main cause of their death was the diseases brought by the Spanish. Common English language words that are of Carib origin include hammock, iguana, hurricane (after the Carib god of evil) and maize. To replace the native population, large numbers of African slaves were brought to the Caribbean to work on the sugar plantations established by the Europeans. Today, approximately half the population of the Caribbean is of African descent, and about half is of European descent. The number of mulattos (mixed black and white) is very small, although they comprise 75 percent of the population of the Dominican Republic. On Cuba and on Puerto Rico, 60-70 percent of the population is of white European descent, though there is also considerable mixing. Elsewhere, blacks form the largest racial group.

Impacts of Slavery

Although slavery ended more than 150 years ago, society in the Caribbean continues to be highly stratified based on race and culture (African v. European). This is especially true on the Spanish-speaking islands. On the English, French and Dutch islands, mulatto and black leaders have been more likely to hold political power. Jamaica (formerly British) has had a strong black national identity that has formed and fostered a multi-racial society. Neighboring Haiti (formerly French) has had a much more difficult time fostering racial equality under its black leadership. (The United States has invaded Haiti several times in the 20th century to enforce peace in the country.) Trinidad and Tobago also has a large Asian Indian population, brought there by the British, who have made demands for greater political participation.

Today, the sugar plantations are almost completely gone, except in Cuba and a few in Barbados. The mainstays of the Caribbean economy are tourism and some mining, oil and natural gas drilling. The Caribbean is home to the largest cruise ship tourism industry in the world, and luxury beach resorts there cater to Anglo North Americans and Europeans. These often stand in stark contrast to impoverished local residential areas. Most Caribbean islands are too small to develop diversified economies and are dependent on only one or a few export products. They are also highly dependent on subsidies and aid from their current or former colonial rulers. Because the islands are so dependent on imports for almost everything they consume, the cost of living is very high and subsistence lifestyles are difficult. Puerto Rico, a United States commonwealth territory, is the wealthiest entity in the Caribbean, though its per capita income is well below that of the mainland United States.

Tourism in the Caribbean

The Caribbean basin is one of the most popular tourism regions in the world. It is especially popular among Americans, Canadians and Europeans for its warm climate, beautiful beaches and seaside resorts. It is the most popular cruise destination in the world, with many ships plying its waters every day. Because of the region's varied colonial history, led by the British, French, Spanish and Dutch, the islands are rich in cultures that are a patchwork of African and European celebrations, foods, music, religions and languages. Many of the islands have become independent countries, while others are still under colonial rule from the United Kingdom, France and the Netherlands. Nearly all of the islands have well-developed tourism industries, with most of them being heavily dependent on tourism for their economic livelihoods. Some, in fact, are dangerously over-dependent on tourism, which creates economic hardships when hurricanes blow through, or when economic conditions in their market areas decline. Hurricanes are a major threat to the entire Caribbean region, and every year at least some of the islands are negatively affected by these tropical storms and their aftermath, sometimes wiping out an island's economy in one storm. In these cases, it is important for government leaders and other decision makers to try, inasmuch as possible, to diversify the local economies, such as through agriculture, manufacturing or other service industries aside from tourism.

Caribbean Supranationalism

Several supranational alliances have also formed in the Caribbean, but unlike the situation in North, Central and South America, the Caribbean treaties are more concerned with tourism. The Caribbean Community (CARICOM), the Organization of Eastern Caribbean States (OECS) and the Association of Caribbean States (ACS) are the most significant of these and include most of the region's independent countries and several members from Latin America as well. In tourism terms, these trade alliances particularly deal with air transportation, facilitating inter-island flows of people, trade in goods and services, economic growth and human resource development. The ACS has the highest number of member countries and is deeply involved in tourism-oriented treaties that deal with sustainability, air transportation and regional marketing. The OECS also has adopted a common currency (East Caribbean Dollar) among

◄ *Cruise ship docked at St. George, Grenada, The cruise sector is the most pervasive form of tourism in the Caribbean. Hundreds of cruise ships ply the waters of the Caribbean every year, and it is the most popular cruise ship destination in the world in terms of ships, ports of call and passengers. There is considerable debate about the negative environmental and social impacts of cruise tourism, as many small islands depend almost entirely on the cruise sector for their economic survival.*

Photo: Dallen J. Timothy

its member nations to simplify inter-island transactions. Member nations are Anguilla, Antigua and Barbuda, and the British Virgin Islands (BVI), which uses the United States dollar, Dominica, Grenada, Montserrat, St. Kitts and Nevis, St. Lucia, and St. Vincent and the Grenadines.

The Caribbean Tourism Organization was formed in 1989 as a non-political international development agency with offices in the Caribbean and North America. It is not a supranational alliance in the strictest sense, because it is not imbued with political power to legislate. It works closely with the ACS and is responsible for marketing, human resource development, tourism research and information management, statistics collection, technical assistance, product development and consulting services through the entire region. Although the tourism product is at one level fairly uniform throughout the entire region, the following sections attempt to elucidate the uniqueness of each island and country and highlight their commonalities with other islands.

Bahamas, Turks and Caicos, and the Cayman Islands

While the Bahamas and Turks and Caicos border the Caribbean Sea, they are officially located in the Atlantic Ocean. Nonetheless, the two archipelagos are culturally and environmentally very similar to their Caribbean brothers and sisters, and in general terms and for most tourism purposes they are classified as Caribbean islands. The Bahamas, Turks and Caicos, and the Cayman Islands, have a very similar product—sun, sea and sand—and all three island groups are heavily dependent on tourism. In the Bahamas, tourism employs over half the workforce directly and much more indirectly (e.g., through construction for tourism); it makes up some 60 percent of the country's gross domestic product (GDP).

The Caymans depend heavily on tourism for nearly three quarters of their GDP and foreign exchange earnings. Nearly 50 percent of the workforce of the Turks and Caicos is involved in tourism and other services, and tourism is a significant portion of the local economy. All three entities are havens for offshore banking and insurance, and all three produce small amounts of fruit and fish for local consumption and export. In 2003, more than 68,000 companies were registered in the Cayman Islands alone, many of which were banks, mutual funds and insurance agencies. In these islands, the banking and insurance industries stimulate a great deal of business travel as well.

Historical evidence suggests that Christopher Columbus first set foot in the New World on San Salvador Island in the Bahamas. A monument has been erected to commemorate the event, and the Bahamas Ministry of Tourism has worked hard to promote the site as one of the country's pre-eminent heritage locations. Cruise tourism is especially important in the Bahamas and the Cayman Islands. The Bahamas are one of the heaviest cruised archipelagos in the world, while the focus of tourism in the Turks and Caicos is beach resorts. The Bahamas lie only 50 miles (80 km) off the coast of Florida and have become a convenient, yet foreign, destination for Americans who desire a quick trip overseas that is not far from home. Many beach resorts have been built on Grand Bahama, New Providence, Andros and Great Abaco to cater to this specific niche.

Cuba

Cuba is perhaps one of the most interesting island situations in the Caribbean. Until the late 1950s, Cuba was a major destination for Americans and Canadians. However, as a result of the island's socialist revolution in the mid-twentieth century that resulted in the eventual placement of a communist government in power, relations with the United States soured, while relations with the USSR improved. To have a communist country in such close proximity to the United States was a threatening proposition for the Americans, and the situation escalated nearly to the brink of war on a few occasions in the 1960s. Nonetheless, a non-armored approach was eventually taken, and in an effort to try to overthrow Fidel Castro and his state-socialist system, the United States enacted strict economic sanctions against Cuba, which are still in effect today. These sanctions translate into embargos against the purchase and use of Cuban products by Americans and travel by United States citizens to Cuba. The exceptions to the travel ban are journalists, sport teams and religious service missionaries. Occasionally, Cuban- Americans are allowed to visit relatives on the island, although even this last function was severely curtailed by the Bush administration in 2004. Nevertheless, there were some changes in travel policy under the Obama administration. U.S. federal regulations restrict travel to Cuba to licensed travelers engaged in certain specified activities. In 2009, the United States announced the lifting of restrictions on family travel and remittances to Cuba while in 2011, the United States announced regulatory changes to increase purposeful travel including religious, cultural, educational, and people-to-people travel and allow all U.S. international airports to apply to provide charter services to Cuba (as of August 2014 19 airports were authorized).

Despite the travel ban, tens of thousands of Americans travel to Cuba each year to enjoy the island's rich Latin culture and natural landscapes. Historic Old Havana is a significant tourist destination and was listed by UNESCO as a World Heritage Site in 1982. In addition, the island boasts many luxury coastal resorts with tours to the rainforest interior. Cuba is a popular destination for Canadian, Mexican and Spanish tourists, and while their own government prohibits visits to the island, Americans are welcomed by Cubans with open arms, as they tend to be big spenders. However, as there are no direct commercial flights or ferry services between the United States and Cuba, people must travel via a third country, such as Canada, the Bahamas, Jamaica or Mexico. If caught visiting Cuba, United States citizens can be fined thousands of dollars and jailed, although this is rare.

Jamaica

The population of Jamaica, like most Caribbean islands, is primarily of African origin, reflecting the former thriving slave trade on the island and in the region. Jamaica's slavery took place mostly on sugarcane plantations, but with the abolition of slavery there in 1834, the slaves were freed and many of them became small-scale farmers. The island's independence from the United Kingdom in 1962 furthered the self-determination of the people and paved the way for Jamaica to become one of the most successful tourism destinations in the Caribbean. Jamaica, unlike several of its regional counterparts, is not overly dependent on tourism. While tourism is a very crucial part of the economy, the island has a more diverse economy than many of its neighbors. Activities such as gypsum, limestone and bauxite (aluminum) mining, rum production, agriculture (tropical fruit, vegetables, sugarcane, coffee), and cement and chemical production lend considerable diversity to the Jamaican economy.

An extremely popular cruise and all-inclusive resort destination, Jamaica depends heavily on visitation from the United States, Canada and Western Europe. Kingston, the capital, is not the central focus of tourism, but it is one of the country's two primary air hubs, the other being Montego Bay. Jamaica is unlike most of the other islands as well in its diversity of attractions. While sun, sea and sand are the principal appeal, slavery-related historic sites, Christopher Columbus monuments, colonial architecture, reggae music and food form a solid foundation for cultural heritage tourism as well. In addition, Jamaica has capitalized on recent growing demand for *agritourism*, wherein people visit agricultural landscapes, plantations, farms and processing plants. Several large papaya and banana plantations are involved in tourism, together with rum (made from sugarcane) estates where visitors can taste and see the production process. Likewise, coffee plantations are a growing element of agritourism in Jamaica, and the island's Blue Mountain coffee has become a popular drink among visitors and an essential export item. River cruises complete with crocodile sightings, jungle walks and waterfall climbs are the most typical nature-based products of Jamaica.

Hispaniola: Dominican Republic and Haiti

The island of Hispaniola is home to two countries: the Dominican Republic (DR) and Haiti. Haiti, one of the poorest countries in the world, has significant potential for tourism development; however, it has experienced some notable political problems in recent years that have prevented the successful growth of the industry. Coups d'état have been a common occurrence, with successive governments put into power (sometimes with United States' assistance) and systematically removed by the military. Likewise, in 2005, political unrest in a few provincial towns led to even more decreases in tourist arrivals and the placement of Haiti on several countries' travel warning lists. Most tourism in Haiti today is comprised of cruise stopovers in Port au Prince and Labadee (a fenced-off area developed specifically for cruises and inaccessible to local residents), as well as cross-border day visits from resort areas in the Dominican Republic.

The massive earthquake of January 12, 2010, wrought much change to Haiti. Cruises ceased, as did land-based tourism. Most of the infrastructure of the capital, Port au Prince, was destroyed, leaving many hotels and homes in ruin. It crushed the historic presidential palace, damaged the airport and upfolded many of the streets and main roads. Rescue workers from around the world arrived quickly on the scene, and once the airport became functional, "*disaster tourists*" began arriving to view the destruction for themselves. Within a few weeks of the disaster, cruises once again began to call at the port of Labadee.

Haiti's neighbor, the Dominican Republic, is among the best holiday values in the Caribbean. It appeals foremost, like most islands in the region, to American and Canadian tourists, although it also receives considerable numbers of arrivals from Europe and other parts of Latin America. It has not experienced the same turmoil as its neighbor, which has resulted in a more stable tourism economy that has grown considerably in the past quarter century. Most tourism in the Dominican Republic centers on beach resorts in the areas of Puerto Plata, Sosua and Santo Domingo. Cruise ship stopovers on the north coast supplement this tourism. The country is also attempting to capitalize on its Spanish colonial heritage to enhance its traditional sun, sea and sand product.

United States Caribbean Territories

The United States has three territories in the Caribbean: Puerto Rico, the United States Virgin Islands (USVI) and Navassa Island (uninhabited). While Guantanamo Bay is occupied and controlled by the United States military, it is still under Cuban sovereignty and therefore part of that nation's territory. Many non-military international visitors to Guantanamo Bay have not travelled there voluntarily. Tourism flourishes in Puerto Rico and the USVI, but obviously not on Navassa Island or in Guantanamo Bay. Both Puerto Rico and the USVI are unincorporated territories, which means that they belong to the U.S., their residents are United States citizens, but not all the rights set out in the constitution apply to them (e.g., voting in certain elections). They are able to elect representatives in the United States Congress, however, and enjoy most of the benefits of all other United States citizens.

USVI

The USVI were purchased from Denmark in 1917 as part of an effort to prevent German submarines from taking over the islands in World War I. Since that time, tourism has thrived on the islands, and many luxury resorts have been built to service the needs of a primarily American market. The United States government now requires passports for everyone traveling to and from the Caribbean. However, the USVI and Puerto Rico are exempt from these requirements since they are part of the United States. Both territories emphasize this fact in their promotional literature and websites. Some observers believe this new regulation will grow tourism in the two territories since non-passport bearing Americans have been seen to divert their Caribbean vacations from other islands in the region to the USVI and Puerto Rico.

St. Croix, St. Thomas and St. John have become popular USVI destinations, especially for cruise and beach resort tourism. Water-based activities and honeymoons are especially popular forms of tourism in the USVI and gaining popularity. Virgin Islands National Park on St. John is a popular attraction for nature enthusiasts, and the USVI hold the unique distinction of being the only place in the United States where cars drive on the left side of the road.

Puerto Rico

The United States gained possession of Puerto Rico in 1898 from Spain during the Spanish-American War. There is a political movement in Puerto Rico toward independence from the United States, led by the nationalist Puerto Rican Independence Party, through a peaceful electoral process. Other groups have lobbied to achieve full statehood as America's 51st state. The third movement is to keep the status quo—the Commonwealth of Puerto Rico, an unincorporated organized territory of the United States. Puerto Rico is an integral part of Caribbean tourism. Many cruises embark from the Port of San Juan, and the island is a primary air gateway into the United States from islands across the Caribbean Basin. Spanish colonial heritage (forts, government buildings, plantation mansions) forms a significant part of the island's heritage base, together with a rich tradition of Latin food, music and dance. Because of its large size (by Caribbean standards), the island is endowed with a rich variety of landscapes that have long formed the basis for tourism—white sandy beaches, rainforests, mountains and historic cities.

Anglo Eastern Caribbean

Tourism generates nearly half of the total economy of the British Virgin Islands (BVI), although services in general comprise over 90 percent of the islands' GDP. The BVI is perhaps best known for sailing and diving. Most people fly into Tortola, the main island, rent boats and sail casually through the BVI's over 36 green islands. Like BVI, Anguilla is still a colony of the United Kingdom and best known for white sandy beaches. The island caters to an upscale market, but it is not accessible directly from North

◀ *Isla Verde in San Juan, Puerto Rico, is the largest of the island's many beach resort areas, Puerto Rico and the U.S. Virgin Islands are popular tourist destinations for Americans, who do not require a passport for travel to the two territories. As such, they provide a domestic travel alternative to many of the other islands in the Caribbean.*

Photo: Alan A. Lew

America or Europe, which adds a sense of seclusion that simultaneously attracts and wards off potential travelers. Connections by air are only from nearby islands (e.g., Sint Maarten, Puerto Rico, Antigua, St. Kitts and the USVI) and boat services from St. Martin. Antigua and Barbuda, and St. Kitts and Nevis are both independent countries in the eastern Caribbean and both provide similar tourist experiences. They are fundamentally resort-based tropical destinations where tourism focuses on the beaches, diving, golf and honeymoons, with a little exploring and colonial heritage mixed in. Although both St. Kitts and Nevis are part of the same country, each has its own separate tourism authority, which is uncommon for the Caribbean.

The British colony of Montserrat was a rather successful, if low key, tourist destination that boasted of beautiful coral reefs and a fascinating volcano. In 1995 and 1997, however, the Soufriere Hills Volcano erupted twice, killing at least 20 people and forcing the evacuation of nearly 8,000 residents to other islands. The eruption essentially buried the capital, Plymouth, and resulted in the designation of the southern half of the island a volcanic exclusion zone, which is generally inaccessible to everyone but scientists. In 2010, further pyroclastic flows covered the remains of the original airport.

These disasters essentially ruined tourism in Montserrat, but since that time the industry has started to recover. Many residents have returned, and the Tourist Board is once again actively promoting the island. Capital city functions and the tourism industry have been relocated to the northern end of the island out of immediate danger. While snorkeling and scuba diving are still the foundation of Montserrat's tourism industry, the volcanic eruptions have also resulted in a new product—disaster tourism—where people visit the island to see the volcano and its destruction from safe distance lookout points.

Dominica

Dominica is unique in the eastern Caribbean and differs from its neighbors in three fundamental ways. First, it is the only island with a remaining indigenous population. There are approximately 3,000 Carib Indians still living in Dominica. Second, tourism and other services amount to less than 50 percent of the GDP; Dominica's economy depends more on agriculture and manufacturing than it does on tourism.

Finally, the focus of efforts by the Ministry of Tourism, Industry and Private Sector Relations is heritage (indigenous and colonial) and ecotourism. Unlike other Caribbean islands, most of Dominica is still covered by rainforest, which provides an important backdrop for trekking trails, waterfalls and river walking.

Like several of their Leeward and Windward Island neighbors, St. Lucia and St. Vincent and the Grenadines are volcanic islands with green mountains and hills that lend themselves well to nature-based tourism. Both countries specialize in diving and water sports, yachting, beaches and eco-adventures. The Creole culture, a mix of African and European cultural traditions, such as music, language, food and religion, are also an important part of tourism. Although both countries rely heavily on tourism, agriculture is an important economic mainstay. St. Lucia is home to one of the very few UNESCO World Heritage Sites in the Caribbean: the Pitons National Park.

Grenada

In addition to bananas, cocoa, spices, citrus, avocados, sugarcane and vegetables, Grenada's economy is nearly three quarters dependent upon diving, boating, cruises and wedding/honeymoon tourism. This dependence has placed the country in a vulnerable position, especially given its location in the Windward chain of islands. Grenada has seen a larger number of tourism ups and downs than most other islands in the region. In 1979, for example, a series of coups by the leftist Maurice Bishop and the New Jewel Movement overthrew the young country's government and established a socialist system that thrived on improving standards of living, health care, education and self-reliance. The new government did, however, have a Marxist slant; it supported many Soviet policies and established close ties with Cuba and other socialist states. Under Bishop's regime, opposition media were censored and free elections were canceled, which swiftly caught the attention of the United States.

On October 19, 1983, hardline Marxists in the New Jewel Movement led another coup that ousted and executed Bishop and his advisors. With the hardline communist government in control for only six days, the United States, with assistance from Jamaica, Barbados and a handful of other Caribbean countries, invaded Grenada on October 25, 1983, overthrowing the government and establishing free elections. This significant event in the history of the Caribbean had serious consequences for tourism in the short term, although the industry has since recovered, and Grenada has turned this dark period of history into a national holiday and tourist attraction. Another major blow to Grenada's tourism occurred in 2004 with the passing of Hurricane Ivan, which was especially destructive for Grenada and its neighbors. Dozens of people were killed, some 90 percent of all homes were destroyed or heavily damaged, and much of the tourism infrastructure, including the cruise ship port, was destroyed. While not as powerful as Ivan, Hurricane Emily hit Grenada the following year, doing additional damage to an island already struggling to recover. Tropical storms are a significant threat throughout the entire region, but Grenada has paid a heavy cost in recent years.

Barbados

Barbados is a low-lying Windward Island and one of the few in the Caribbean with petroleum and natural gas resources. In addition to sugar, rum, cotton and vegetable production, tourism forms a significant part of the economy based on festivals, culture, sport, cruises and golf resorts. Barbados remains a popular destination for leisure travel. Nevertheless, the Barbados Tourism Authority actively markets itself as a meetings and conferences (MICE) destination in an effort to boost sluggish growth in tourism. It offers good access, beautiful weather and scenery, first-class facilities, quality hospitality and, more recently, an interesting cultural heritage.

With the emancipation of slaves in 1834, the British colonialists brought South Asian laborers to Trinidad and Tobago to work the sugarcane industry, which traditionally has formed the backbone of

◀ The Pitons are two volcanic peaks covered in rainforest and reaching far above the coast on island of Saint Lucia. One of only a handful of UNESCO World Heritage Sites in the Caribbean, they are one of the most visited natural areas in the Caribbean basin.

Photo: Dallen J. Timothy

the nation's economy. As a result of this late 19th and early 20th century labor migration, the population of Trinidad and Tobago remains predominantly South Asian Indian. Petroleum and chemical products account for approximately half of the GDP, followed by tourism and agriculture. Trinidad and Tobago is the largest producer of oil in the Caribbean and as a result it also has one of the highest standards of living and per capita incomes in the region.

Trinidad and Tobago

Trinidad and Tobago is faced with a unique situation, shared with St. Kitts and Nevis, and Antigua and Barbuda in the Caribbean Basin—that is the notion of twin-island states. Scholars and other observers have commented recently about conditions of twin-island states, although Tobago has received the majority of attention given its separatist desires that have made world headlines. Tobago, Barbuda and Nevis argue that they are treated by their larger national partners as second-rate regions, lacking in political power and public representation at the national level, and ignored in economic development efforts, particularly in tourism. This has resulted in strained relations between the larger islands and the smaller islands and a growth of independence movements in all three of the smaller island partners. While separatist lobbies exist in Barbuda and Tobago, they are more subtle than the movement in Nevis. In 1998, a Nevis referendum vote to separate from St. Kitts fell short of the two-thirds majority needed, although many supporters of independence for tiny Nevis continue their support of these efforts. While many see Trinidad and Tobago as a model of successful twin-island states, others suggest that the system is highly flawed and unbalanced.

The French West Indies

The French West Indies consist of Guadeloupe, Martinique, St. Barts (Barthelemy) and Saint Martin—all important tourist destinations. Saint Martin is unique in that it shares a small island with its Dutch counterpart, Sint Maarten. While the French-Dutch border is well marked and is part of the tourist appeal of the island, there are no customs or passport controls between the two entities, and there never have been.

However, in the mid-1980s, when France began requiring visas of nationalities that had not previously needed them (e.g., the United States of America and Canada), the French government considered erecting border checkpoints on the small island. Local Dutch and French authorities, cruise companies and airlines fought the measure because of the damage it would have done to tourism. The primary international gateway to French Saint Martin is via Dutch Sint Maarten's Princess Juliana International Airport. Given the need to acquire a French visa, tourists to the island would inevitably choose to stay in the south. Eventually the proposal was defeated.

The French side of the island is nearly twice the size of the Dutch portion and is conspicuously French in its language, religious landscape, architecture, foodways, music, relaxed atmosphere and more conservative values. Beaches and resorts are the primary basis of tourism. St. Barts is an upscale destination with white sand beaches, lush green hills and a yacht harbor. Tourism in Martinique and Guadeloupe thrives on beachfront development, shopping, cruises, sailing, golf, diving, and, increasingly, heritage tourism. All of the islands in the French West Indies are markedly French, which gives them a unique array of cuisines and French imported products that are popular among tourists.

The French West Indies is an especially popular destination for French tourists, because they can travel somewhere tropical and far away and speak French the entire time. Guadeloupe, Martinique and Saint Martin are known as EU 'outermost regions', meaning they are part of the European Union because they are an integral part of France. However, although they are part of France, they are not part of the Schengen Agreement, and they lie outside the EU VAT (Value Added Tax) Area. This means that passengers flying from France (even EU citizens) to the French West Indies are required to travel with a valid passport or approved national identity card, and being outside the VAT zone provides some tax advantages not available in continental France on alcohol, tobacco and other products. The British and Dutch Caribbean territories are in a similar situation, but they have fewer rights because they are not part of the European Union as the French West Indies are.

Aruba and the Former Netherlands Antilles

Aruba belonged to the Netherlands Antilles until 1986 when, as part of its preparations for full independence, it separated from the Antilles and became an autonomous part of the Kingdom of the Netherlands. The island has stopped short of being fully independent, but future plans still include separation from the Netherlands. Aside from tourism, Aruba has a thriving oil refining industry and some offshore banking. Aruba is one of the most popular destinations in the region, although its tourism was noticeably affected by the terror attacks of September 11, 2001. Likewise, the island experienced a tourism blow in May 2005, when an American teenager went missing on the island and was feared dead. The news of her disappearance hit hard as Americans canceled or rerouted their travel plans. The young woman was never found, which has not boded well for an island that depends on tourism for its economic survival.

As noted earlier, the island of St. Martin is shared by two countries: France and the Netherlands. The Dutch portion (south), Sint Maarten, is a popular cruise destination and shopping haven, because it is, like many Caribbean islands, a duty-free port. Gambling is allowed only on the Dutch side of the island, so casinos and bordello bars have become popular attractions for tourists staying on both sides of the island. Nude beaches and luxury resorts dot the coastline of Sint Maarten, and slave- and plantation-related historic sites are becoming fashionable attractions on both sides of the border.

Bonaire and Curacao

Bonaire and Curacao are both known for their petroleum refineries and transshipping facilities, primarily with oil connections to Venezuela. Tourism is a very significant part of their economies as well, as it is

ISSUES AND INSIGHTS World Cricket's Premier Event Fails to Attract Fans

Organizers of international cricket's tourist destinations were left facing many empty seats inside stadiums as well as poor business with vacant hotel rooms outside them. The seven-week Cricket World Cup in the West Indies in 2007, which was eventually won by Australia, was promoted as the perfect sports tourism package. It offered sun-drenched beaches, Caribbean hospitality and the chance to watch the sport's elite contest at its premier tournament. However, organizers, governments and the tourism industry were extremely disappointed after only a fraction of the tens of thousands of fans expected actually took up the expensive trip. The nine Caribbean nations that staged the World Cup matches and warm-up games for the event were Jamaica, St. Kitts and Nevis, Antigua and Barbuda, St. Vincent and the Grenadines, Saint Lucia, Barbados, Grenada, Trinidad and Tobago and Guyana.

A number of reasons have been forwarded for the problems associated with the event. In sporting terms, the shocking first-round elimination of India and Pakistan, with the Indian team having the world's biggest fan base, meant that Indian supporters went home early. The poor form of the West Indies' host team did not help either, as none of them qualified for the semi-finals. In addition, the event was regarded by many as being too long, as it can be difficult for fans to secure seven weeks of holidays. Furthermore, the nature of the event meant that although fans knew the locations of matches in later stages of the competition, they did not know what teams would be competing. Therefore, fans had to second guess the results of games in order to pre-book flights, accommodation and tickets. The unpredictable nature of sport meant that teams did not perform as expected. In some instances, fans had tickets to games but could not get flights between the islands to see them.

Visa requirements were altered for the event. A special CARICOM visa was introduced for some visitors to Dominica and the nine host countries for the period February 1 to May 15, 2007. This was supposedly for security and ease of movement, as visitors with the special visa were able to move freely between the host countries without further immigration formalities. Unfortunately, the visa was introduced too late in the lead-up to the event, after some fans had already purchased flights to it. However, even more confusing was the fact that nationals from some countries did not need a CARICOM special visa, including Canada, France and its overseas territories, Germany, Ireland, Italy, Japan, the Netherlands and its overseas territories, the United Kingdom and its dependent territories. South Africa, Spain, the United States and its dependent territories, and nationals and residents of CARICOM Member States (Except Haiti). On the other hand, nationals of some of the potentially biggest markets for the event did require such visas, including Australia, Bangladesh, India, Pakistan, New Zealand and Sri Lanka.

Other issues included the relative high cost of accommodation for nationals of most of the counties that were participating, as well as the high entrance prices. The situation contrasted with major sports events in other parts of the world, such as the soccer or rugby world cups, which even budget tourists can usually afford to attend. Ticket prices were high, in part, because regional governments wanted to recover the large sums spent on infrastructure. For example, the Jamaican Government spent US$81 million via a loan from China to refurbish the Sabina Park ground and constructing a new multi-purpose facility in Trelawny. Another US$20 million was budgeted for off-the-pitch expenses, putting the total at more than US$100 million, or JM$7 billion.

The total amount of money spent on all the stadia was estimated at over US$301 million. Unfortunately, while there were substantial pre-tournament ticket sales, the actual attendance was approximately 40 percent less than what would have been expected. This was primarily the result of the loss of Pakistan and India from later rounds and the difficulties fans had getting to venues via the Caribbean's airline system. This meant that returns to the hotel sector and to food and beverage suppliers were well below what was expected. Furthermore, it appeared that some hotels actually had rooms available during what would normally be the peak holiday season leading up to Easter. This was due to some potential visitors not coming to the Caribbean because of fears of event related congestion. The World Cup therefore highlights the difficulties for developing countries with a limited financial and infrastructure base to run such large events.

for St. Eustatius and Saba, the two smallest islands of the Dutch Caribbean. Saba is a well-known honeymoon destination owing to its small physical size and relatively secluded location. St. Eustatius is a trendy destination for hikers and divers. On October 10, 2010, the Netherlands Antilles, which included Sint Maarten, Bonaire, Saba, Curacao and Saint Eustatius, was dissolved and the Dutch Caribbean territories were reorganized. Sint Maarten and Curacao joined Aruba and the Netherlands to become the four autonomous "constituent countries" within the Kingdom of the Netherlands. In the same process, Bonaire, Saba and Saint Eustatius became "special municipalities" within the Netherlands, and they are now legally an integral part of the Netherlands. This is very similar to the British model, wherein Northern Ireland, Scotland, England and Wales are "constituent countries" within the United Kingdom of Great Britain and

Northern Ireland. From a tourism perspective, this means that Bonaire, Saba and Saint Eustatius are governed and promoted directly from Amsterdam, although they continue to maintain their own regional tourism offices. As self-governing constituent countries within the kingdom, Aruba, Sint Maarten and Curacao have more autonomy in their development goals and programs.

5.3 | SOUTH AMERICA

Current South America Issues and Resources (http://wrgeography.com/southamerica.html)

South America is the fourth largest continent in the world, yet it is considerably smaller than Asia, Africa and North America. It comprises 12 percent of the earth's land area, but only 6 percent of the planet's population. The Andes Mountain range and the vast Amazon Basin rainforest limit the potential settlement areas of the continent. Along with North America, South America is part of the Western Hemisphere. The Western Hemisphere is also referred to as the New World, in comparison to all of the other continents that comprise the Old World. The term the Americas is also used to refer to both North and South America together. The name, "America" comes from the Italian navigator Amerigo Vespucci, who was one of the first Europeans to sail and map the coastline of northeastern South America (in the early 1500s). He was also one of the first to propose that the Americas were not a part of Asia but a separate continent (a new world). Using information from his sailings, by 1507 Spanish and German cartographers had applied the name America to South America. Soon thereafter it was applied to North America, as well.

Physical Geography of South America

South America shares some common features with Africa, to which it was attached some 245 million years ago. It is a very compact continent with few large peninsulas, seas or bays. There are also very few islands associated with South America. The islands that do exist are mostly at the very southern tip of the continent, including the Falkland Islands (known in Argentina as the *Malvinas*) and Tierra del Fuego. The country of Trinidad and Tobago, which are islands off the north coast of South America, and the Galapagos Islands off the coast of Ecuador are the two other significant island groupings of South America.

Highland Massifs

Like Africa, South America's geologic structure is mostly comprised of large high land massifs separated by major drainage basins. Excluding the Andes Mountain system, there are two major highland regions that were once part of the core of the ancient continent of Gondwana, along with much of the continent of Africa. These are the *Guiana Highlands* to the north of the Amazon River, and the *Brazilian Highlands* to the south of the Amazon. The Guiana Highlands are much smaller and have an average elevation of about 1,000 ft (305 m), though their highest peak, Maverick Rock on *Mount Roraima*, is 9,220 ft (2,810 m). Mount Roraima is located at the border of Venezuela, Guyana and Brazil and is a high sandstone plateau surrounded by steep cliffs. These have kept it isolated from surrounding areas and allowed for flora and fauna to develop that are unique only to the plateau. Early descriptions of Mount Roraima by European explorers are believed to have influenced Arthur Conan Doyle's novel, *The Lost World* (1912).

The Brazilian Highlands are also known as the *Brazilian Plateau*, and they have an average elevation of about 3,000 ft (914 m), with their highest peaks being over 9,000 ft (2,750 m). These highlands cover about half of the country of Brazil, and most of the country's population resides on their Atlantic coastal edge. Geologically, it was created by massive basalt lava flows in central Gondwana, though that is hard

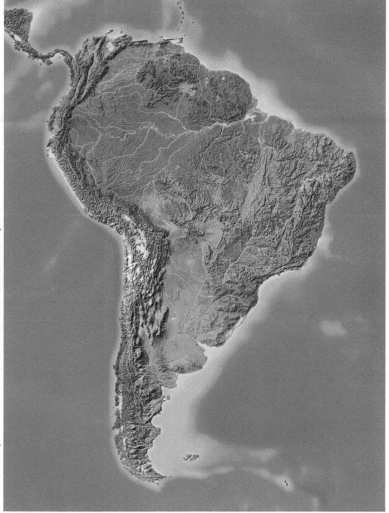

Physiographic map of the South American continent showing the pattern of major surface features and water bodies.

to see today. *Patagonia* is a smaller third tableland area of ancient rock located in the very southern part of Argentina, south of the Negro River. Large agricultural farms are found throughout both the Brazilian Highlands and Patagonia.

Basins

The major basins of South America include the Orinoco River Basin, the Amazon Basin (or Amazon Depression), the Paraguay River Basin and the Pampas of Argentina. The *Orinoco River Basin* is north of the Guiana Highlands and southeast of the Andes Mountain Range. The river drains the interior of the countries of Venezuela and Colombia and creates a grassland region also known as *the Llanos*. The *Paraguay River Basin* is situated between the Andes Mountains and the southern part of the Brazilian Highlands. The Paraguay River, South America's second largest river, enters the Atlantic Ocean at the city of the Buenos Aires, Argentina. South of Buenos Aires is *the Pampas*, which is a vast outwash plain of the Andes Mountains (see Map, page 413).

The Amazon

The *Amazon Depression* (also known as the *Amazon Basin*) is the world's largest river basin. The Guiana Highlands form its northern boundary, the Andes Mountains form its western boundary, and the

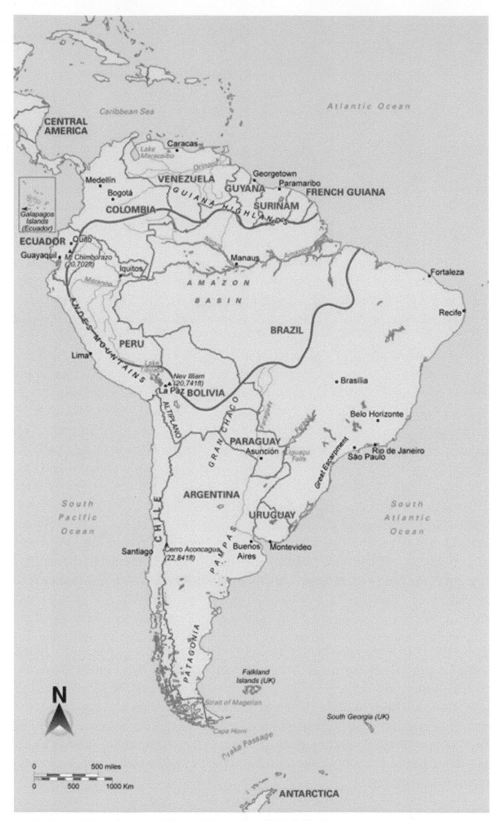

▲ *Countries, Major Cities, and Physical Geography of South America*

Brazilian Highlands form its southern boundary. It is very flat and reaches an elevation of only 385 ft (117 m) in Iquitos, Peru, on the eastern slopes of the Andes Mountains. The elevation of Manaus, Brazil, near the center of the Amazon basin, is only 144 ft (44 m).

The Amazon River is about 4,000 miles (6,400 km) long. Only the Nile River in Africa and the combined Missouri and Mississippi Rivers in North America are longer. (Claims in 2007 that the Amazon is actually the longest river in the world have not yet been widely accepted as of 2015.) The Amazon, however, contains 20 percent of all of the river waters of the world, and its outflow is ten times greater than that of the Mississippi/Missouri River. In addition, it has more than 1,000 tributaries, several of which are over 1,000 miles (1,609 m) long, and it can be as much as 20 miles wide in some parts during the rainy season (in the southern hemisphere summer).

The Amazon rainforest is the largest rainforest area in the world, and it may contain the earth's oldest forested areas. Some of its trees are more than 300 ft (91 m) high, hovering over an incredible diversity of plants and animals. Unfortunately, human activity has been transforming the rainforest, which as recently as 1980 was only occupied by small Indian tribes and wild animals. Today, hundreds of square miles of tropical rainforests are cut annually to make way for large-scale agriculture, ranches and timber production. Road construction by the Brazilian government, designed to open the Amazon for settlement and development, is leading to the widespread destruction of its natural environment and the loss of its biodiversity.

Biodiversity

South America has more animal and plant species than any other major region on the planet. Although several theories exist as to why this is the case, the most plausible seems to suggest that it was the rise of the central Andes Mountains, about 10 to 12 million years ago, that contributed to this biological diversity. Prior to the rise of these mountains, South America was divided into several unconnected land masses, separated by large seas. Plants and animals evolved separately in these areas. The rise of the central Andes provided a land bridge that connected northern and southern parts of South America, allowing the spreading and intermixing of flora and fauna, especially out of the older northern Andes and across modern-day South America, Central America and the Caribbean.

Overall, scientists believe that Brazil contains 20 percent of the global species biodiversity, with over 50,000 known plant species and 10 percent of known animal species (excluding those in the oceans). Beyond Brazil, however, the Yasuni National Park in eastern Ecuador, which is part of the Amazon Basin, is believed to have the greatest biological diversity per hectare on the planet. In that location, researchers have documented 596 bird species, 150 amphibian species (especially frogs and toads), 1,100 tree species (with up to 655 species in a single hectare), over 200 mammal species and some 100,000 insect species. All of these are the highest numbers recorded anywhere in the world. Unfortunately, multinational oil companies have been proposing major developments in the Yasuni National Park area that may threaten both this biodiversity and the native tribal groups that live in the area.

The Andes Mountain Range

One major geologic difference between South America and Africa is the Andes Mountain Range, which forms a spine along the entire continent on its western edge and is the longest continuous mountain range in the world. The Andes were formed by the movement of the South American continental plate over the Pacific Ocean plate. The Pacific Ocean plate slides under the South American plate in a process known as *subduction* (the Pacific plate is being subducted under the South America plate). Material from the Pacific plate, as well as sediments scraped from the South American plate, become heated as they slide deeper into the earth's crust. The more volatile of these materials melt and seep upward, pushing the land area above into high mountains and creating earthquakes and volcanic activity. The entire western coastline of South America is dotted with active volcanoes and is subject to frequent earthquakes.

The Andes are actually a very young mountain range, with their oldest geological faults (in central Colombia) dating back to about 25 million years ago. The ranges further to the south are estimated to be only 10 million years old. The Andes are about 4,300 miles (7,000 km) long and second only to the Himalayas in their average elevation (13,000 ft or 4,000 m). Several large rift valleys exist within the Andes. These valleys are formed by faulting and the spreading apart of mountains, rather than by erosion. One of these large rift valleys is the Magdalena River Valley in Colombia.

Altiplano

The most famous of the Andean valleys, however, is the Altiplano (high plain) of Bolivia and southern Peru (on the border with nothernmost Chile). This valley has a base elevation of 12,000 to 15,000 ft (3658 to 4572 m) and is as wide as 125 miles (201 km). Only in Tibet are similar conditions found on the planet.

CHILE EARTHQUAKE 27.0.2010 (8.8 magnitude)

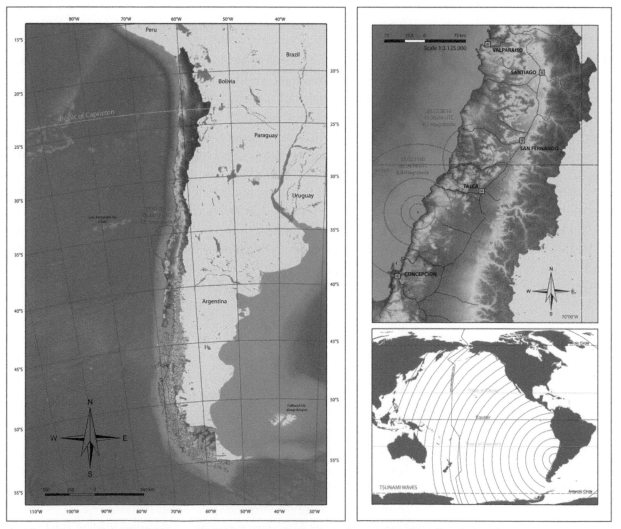

▲ *The Chilean earthquake of February 27, 2010 was one of several very large quakes that affect the Pacific Ring of Fire every decade.*

Several internal drainage basins are scattered across the Altiplano that contain marshes and lakes. The largest of these is *Lake Titicaca* on the border of Peru and Bolivia.

Volcanic fields of lava flows, cinder cones and stratovolcanoes are found in four separate zones in the Andes. One of these zones lies along the northern border between Chile and Argentina (south of Bolivia and the Altiplano), where the world's highest active volcano, *Nevado Ojos del Salado* ("Eyes of the Salty Summit" 22,608 ft (6891 m)), is situated. The highest peak in South America is *Mount Aconcagua* in Argentina on the border with Chile, a little north of Santiago. At 22,831 ft (6959 m), it is the highest peak in the Western Hemisphere. (However, when measured from the distance out from the very center of the earth, the highest peak in the world is actually *Mount Chimborazo* 20,702 ft (6310 m) in Ecuador, which sits close to the equator.) In the southern portion of the Andes, the mountain valleys are submerged below sea level, and the peaks form a vast region of islands off the Pacific coast of southern Chile. Because of the high precipitation in this area, glaciers and fjords are also widespread.

Climate of South America

The equator runs just north of the Amazon River through the Amazon Basin. As already noted, the *intertropical convergence zone* (ITCZ) extends roughly along the equator around the entire earth and directly through the Amazon region. The ITCZ is an area where air from the northern and southern hemispheres comes together and rises into the upper atmosphere. In doing so, it creates a band of low pressure and brings moisture from ground level to higher altitudes where condensation occurs, resulting in rain. Along the ITCZ, rain falls all year long and on a frequent basis. Temperatures are hot and the air is humid. Tropical rainforests are the dominant vegetation under such conditions and are the natural vegetation throughout the Amazon Basin.

North and south of the ITCZ are regions dominated by high pressure. South America has a large high-pressure air system located most of the year off the coast of Chile, as well as high-pressure systems in the North Atlantic and South Atlantic. High-pressure systems are characterized by air descending from the upper atmosphere that is dry and cloud free. The biggest impact of these high-pressure systems is on the west coast of South America, in the area from northern Chile through Peru where desert climate conditions exist. This is where the *Atacama Desert* is located, which is possibly the driest place on Earth. In addition to high-pressure, the Atacama is affected by a cold off-shore current (which limits ocean air humidity), and the Andes Mountains block humid tropical air from penetrating from the Amazon Basin, putting the Atacama in a rain shadow.

Deserts

Desert-like conditions also extend into Patagonia in southern Argentina, which lies in the rain shadow of the southern Andes Mountains. Unlike in the Atacama Desert, the wind and moisture here come from the Pacific Ocean to the west, and the dry rain shadow is below the eastern slopes of the Andes. In addition, the southern Andes extend further south toward Antarctica than any other landmass on the planet. This places them in the high southern latitudes and into the *Southern Ocean* where an intense *Marine West Coast* climate brings heavy rain and snow to their west coast side. This is also home to the town of Puerto Toro, Chile, which is the most southern permanently inhabited community in the world.

The high-pressure systems and ITCZ low-pressure belt migrate north and south following the sun through the year. January is the southern hemisphere summer when these air masses migrate southward. In July, the ITCZ moves northward, as do the high-pressure systems off the coasts of South America. The low-pressure Marine West Coast storms in the far south also move northward at this time.

Climate researchers have found that the ITCZ has been gradually migrating northward from the equator over the last 300 years. It was located very close to the equator during the earth's little ice age (1400 to 1850 CE), and although its location varies from year to year and in different parts of the planet, it

Photo: Alan A. Lew

▲ *Late summer snow can still be seen above the ski resort town of Farallones, located at 8200 ft (2500 m) elevation in the Andes mountains above Santiago, Chile. Relatively young, the Andes have not had sufficient time for erosion to diminish their elevation and relief as has happened in other parts of the world. The Andes have played a critical role in the development of society in South America and have become an important part of the region's tourism appeal.*

is mostly north of the equator today. This, along with higher sea-surface temperatures, may have contributed to the increasing intensity of droughts in the Amazon Basin since about 2000. In 2005, the Amazon experienced a one-in-one hundred year drought, and a lack of rain between July and September 2010 resulted in rivers being at their lowest levels ever recorded and in widespread and uncontrolled fires. The droughts of 2005 and 2010 both caused the Amazon to expel more carbon dioxide (CO_2) greenhouse gasses than it absorbed. This is a major problem, because the Amazon rainforest is considered the "lungs of the world" due to its normal ability to absorb CO_2 from anthropogenic (human) activities and to expel oxygen in return. Therefore the Amazon rainforest acts as a major carbon sink which if lost would mean a substantial transfer of CO_2 into the atmosphere.

Human Geography of South America

There are basically four major groups that have shaped the human geography of South America. These include American Indians, the Spanish and Portuguese who conquered the Indians starting in the 1500s, black Africans who were mostly brought to South America as slaves, and immigrants from Europe and Asia who came to South America after its countries gained their independence from colonial rule. Each of these groups established a distinct cultural pattern of settlements and livelihoods, portions of which are still strong on the contemporary landscapes of the continent.

Earliest Migrations

The earliest humans to settle in South America most likely arrived from North America after crossing from the Asian continent sometime between 45,000 and 12,000 years ago. It is quite certain that human groups lived in Chile 11,000 years ago. By 3,000 years ago, a village-based civilization existed in the valleys and coastal areas of Ecuador and Colombia where they built temple mound complexes, created fine ceramics and cultivated many crops, including *maize* (known as *corn* in North America). Their culture spread to Peru and Chile where new technologies and innovations appeared.

The Chavin culture (900 BCE to 300 CE) built communities as high as 9,842 ft (3,177 m) in the Andes of Peru. Cultures in the central Andes built some of the most elaborate stone-walled terraces and irrigation canals of their time. Numerous advanced societies developed in the northern and central Andes starting about 100 BCE, occasionally conquering and incorporating one another. Growing populations, especially in the Andes, led to the development of early cities and the first empires of South America. The best known of these empires was that of the *Inca* (Inka). The Inca state expanded out of northern Peru to the north and south through the Andes Mountains starting about 1100 CE, though it never encompassed all of the advanced agricultural societies of its time. Its expansion was only stopped by the arrival of the Spanish in 1532.

Inca Collapse

At the height of the Inca period (1438 to 1533 CE), it is estimated that there were 10 to 20 million human inhabitants in South America, half of whom lived in the central and northern Andes. As in North and Central America, the arrival of the Spanish in the 16th and 17th centuries brought diseases that the people of South America had no resistance to. In some areas, up to 95 percent of the indigenous population died of these new illnesses in what is known as a *demographic collapse*. This impact may have been the greatest in the Amazon Basin (see Issues and Insights).

Today the major Indian regions of South America are the Amazon and Orinoco basins, Paraguay, north and central Chile, and the Andean Highlands of Venezuela, Colombia, Ecuador, Peru and Bolivia. These were all areas of major Indian societies prior to European contact. The diversity of languages traditionally spoken among different Indian groups was immense. More than a hundred different linguistic families once existed in South America. Many were very localized and only spoken by a small group of people, and many have become extinct in modern times due to Indian acculturation into European languages.

European Colonization

The two most powerful countries of Europe in the early 1500s were Spain and Portugal, who were at the vanguard of European overseas colonial expansion. In 1494, the two countries signed the *Treaty of Tordesillas*, which divided the world in half between them, based on a line drawn down the middle of the Atlantic Ocean. Spain received the western half of that line and Portugal was granted everything to the east of that line. This line was intended to give all of the New World to Spain and all of the uncolonized Old World to Portugal. However the line ended up giving a large chunk of South America to Portugal, which became the Portuguese colony of Brazil.

Spain and Portugal both had policies to keep out all other Europeans from their territories. It is estimated that fewer than 150,000 Spaniards actually migrated to South America. Up to one million Portuguese migrated to Brazil, drawn by the gold rush there in the 1700s. Many of these migrants eventually returned to their homeland. Although their numbers were not large, Spanish and Portuguese political and religious domination of the continent had a significant impact on the culture that has developed in both South and Central America in modern times.

Roman Catholicism

A major impact of the Iberian invaders of South America was the introduction of *Roman Catholicism*. Well over 80 percent of the people in South America consider themselves Catholics. Their Catholicism,

ISSUES AND INSIGHTS   Terra Preta and the Lost Civilization of the Amazon

The *demographic transition* (or collapse) of the 16th and 17th centuries, caused by the introduction of new diseases to North and South America by Europeans, may have been the greatest in the Amazon Basin. Early Spanish explorers in the mid-sixteenth century reported the presence of a large, agriculture-based civilization along the Rio Negro, a major tributary of the Amazon River, extending from the Llanos in Colombia. Later explorers considered these tales a hoax because they found only scattered small and nomadic tribal groups, which typify the area today, and no visible signs of a more advanced civilization in any of the inner Amazon Basin. In addition, it was believed that the highly acidic tropical rainforest soils in the Amazon could not produce enough food to support a large, advanced civilization.

Modern day soil scientists, however, have found a large number of what they called *terra preta* ("black earth" in Portuguese) throughout some portions of the Amazon Basin.

The origins of this soil are unknown, though it is used for small scale farming in the region today. This soil is extremely nutrient rich and able to support highly-intensive agriculture. It is composed of charcoal, pottery shards, plant and animal residues, fish and animal bones, as well as a large number of minerals and microorganisms. In addition, it regenerates and by itself grows new nutrients on an annual basis.

Scientists originally thought that terra pretra was created by volcanic ash from the Andes or was the remains of sediment that was once at the bottom of lakes. However, they now believe that humans created it between 450 BCE and 950 CE, and that it may have enabled the rise of an extensive agricultural civilization in the Amazon Basin. Because there are no rocks in the central Amazon, large plant-based (such as wood) structures would have quickly succumbed to the natural rainforest, leaving few other signs of what might once have existed there.

however, is often combined with more traditional and local religious beliefs. Shamanic practices are common among Indian populations, while African traditions are widespread among black populations. Protestant Christian evangelism has increased considerably since the late 20th century.

African Slavery

As was discussed in the Sub-Saharan Africa section, the Portuguese were the first Europeans to import African slaves to work in their colonies in the New World. Starting in the early 1600s, large numbers of African slaves were brought to both the Portuguese and Spanish areas of South America. It is estimated that some 4 million Africans were brought to Brazil, and 3 million were brought to Spanish South America (which includes the modern day countries of Venezuela, Colombia, coastal Ecuador and Peru, and northern Argentina). Large numbers of Africans were also brought to British Guiana (now Guyana), Dutch Guiana (now Suriname) and French Guiana. The major black regions of South America today include the coastal areas of the three Guianas, Venezuela, Colombia, Ecuador, Peru, most of eastern Brazil and northwestern Argentina. Africans have contributed much to the culture and cuisine of South America, especially in these regions.

Although the social customs that South America inherited from Spain and Portugal emphasized a separation of race and class, much intermixing among the indigenous people and the immigrants to South America took place. Important designations have been developed to describe these interracial offspring, even if they are very simplistic. As mentioned previously, Mestizo refers to those who have a mix of European and Indian blood; *caboclo* is the term used in Brazil for those who have a mix of Portuguese and Indian blood; *mulatto* refers to those of mixed European and African blood; *zambo* is the term for a mix of African and Indian blood; and *cholo* has the same meaning as mestizo, though it also refers to those of Indian blood who aspire to the higher social status of the dominant European social classes.

Colonial Independence

Most of the countries in South America gained their independence from colonial rule in the early 1800s. Independence allowed the countries of South America to be more open to immigration from other regions of the world besides the Iberian Peninsula. Large-scale immigration started around 1850, reaching its peak in the late 1800s, and decreasing considerably in the 1930s. (This period of migration roughly paralleled the era of large European migration to the United States.) During this time, some 11 to 12 million people migrated to South America, over half of whom went to Argentina and about one-third to Brazil. The largest numbers were from Italy, Spain and Portugal. Chinese laborers came in the 1800s to help build railroads and created Chinatowns in South America, much as they did in North America. South Asians and Indonesians migrated to the three Guianas, where they were specifically invited as laborers.

After World War II, large numbers of Koreans were invited to migrate to Argentina, while many Japanese settled in Brazil, Bolivia and Argentina. Together, these three countries have the largest Japanese population outside of Japan. Today the region of South America most dominated by Europeans and other relatively recent immigrants extends from southern Chile and Argentina to southeastern Brazil. In these areas, large-scale commercial agriculture predominates, including wheat and dairy farms, and cattle and sheep ranches.

Countries and Tourism of South America

Tourism is an important part of the social and economic environment of several South American countries, although the region as a whole is one of the least visited (along with Sub-Saharan Africa) in terms of arrivals from outside the region. Recent data, however, suggest that South America is experiencing a renewed growth in intra- and extra-regional tourist arrivals.

The continent of South America has a varied image, ranging from beautiful beaches, Amazon jungles and snow-peaked mountains on the pleasant end of the spectrum, to dangers associated with corruption, criminal violence and narcotics-related guerilla warfare on the other end. Several countries in the region appear regularly in the global media as hotspots for violent crime and drug trafficking. Travel warnings issued by many tourist-generating countries, such as the U.S., Australia, Canada and the U.K., have featured countries in South America.

Culturally and ecologically, South America is very diverse. The continent has a rich variety of native peoples, whose cultures are prominently on display for tourists, and an extensive colonial heritage of Spanish, Portuguese, French, British and Dutch influences. It is a large region extending from the equatorial tropics in the north and central areas to a cold, polar climate in the far southern reaches of Chile and Argentina, with temperate climatic zones in between. Tourism is unevenly spread across the continent, with Brazil, Ecuador, Peru, Chile, and Argentina (and in previous years Venezuela) being the most visited countries in the region. Colombia, the Guianas (French Guiana, Suriname and Guyana), Bolivia, Paraguay and Uruguay are secondary destinations.

The Guianas

The Guianas is a generic term that refers to the two countries of Guyana and Suriname, and the French department of Guiana. The Guianas demonstrate the diversity of colonial influences in South America aside from Spanish and Portuguese rule. Guyana was a British colony, formerly known as "British Guiana." Suriname was a former Dutch colony known as "Dutch Guiana," and French Guiana is still a part of France. In addition to these three Guianas, the country of Venezuela was formerly known as "Spanish Guiana," and the adjacent Brazilian state of Amapa used to be called "Portuguese Guiana." The colonial language continues to be the official language in each of these Guiana territories.

French Guiana (officially known simply as *Guyane*) is the most eastern of the three Guianas. The country of France has 101 departments, of which five are overseas (the rest are in European France). These five include French Guiana, the islands of Guadeloupe and Martinique (in the Caribbean), and the islands of Reunion and Mayotte (in the Indian Ocean). Like the rest of France, all of these use the euro as their currency. The French colonized French Guiana in the 17th and 18th centuries, and used it in part as a penal colony for prisoners from France. Large numbers of French settlers died from tropical diseases in the early years of settlement. Since the 1960s, Guyane has served as an important base for French, and later European, space launches.

As an integral part of France, Guyane has a high level of social welfare that is funded from Europe, including some EU development funds (though it is not legally part of the EU). Because of this, French Guiana is the most developed and has the highest incomes of the three Giuanas. The territory's 250,000 people include many immigrants from France and the French West Indies, as well as from French-speaking Haiti. *Creoles*, usually defined as having a mix of African and French ancestry, comprise about 65 percent of the population.

Suriname

Suriname is the middle of the three Guianas and is the smallest of the countries on the South American continent in both land area and population (573,000 in 2014; French Guiana has fewer people, but it is not a country). Its name probably derives from the name given to the local natives by the English who first attempted to settle the territory in the early 1600s. Disputes between the Netherlands and England resulted in the Dutch deciding to take Suriname while giving their small colony of New Amsterdam to the English. (New Amsterdam is better known today as the U.S. city of New York.) Independence from the Netherlands was negotiated and granted in 1975. Nearly a third of the population of Suriname has migrated to the Netherlands seeking greater economic and political security in the years immediately prior to independence.

Guyana

Guyana (formerly British Guiana) is the westernmost of the three Guianas. It was at one time a Dutch colony, but the U.K. ruled it for 200 years until its independence in 1970. Today, it is the only South American country that is part of the *British Commonwealth of Nations*. Its population of 735,000 (2014) is much larger than that of Suriname, but their ethnic make-up is similar. The largest ethnic group in both is descended from contract workers brought to the former colony from different parts of India. They are known as *Hindoestanens* in Suriname, where they comprise 37 percent of the population; in Guyana they consist of 43.5 percent. The second largest group is Creoles (about 30 percent in each country), who are part West African (whose ancestors came as slaves) and mixed European ancestry (mostly Dutch in Suriname and British in Guyana). *Amerindians* and *Maroons*, a mix of African and Amerindian, comprise about 15 percent of each country's population.

Although they have some of the most dramatic natural scenery in all of South America, Guyana, Suriname and French Guiana comprise one of the least visited corners of the continent. The tourism industry in all three lands ranks below gold and bauxite (aluminum) mining, timber, fishing and petroleum in economic importance. Tourist offerings in all three countries are very similar, focusing on trekking in rainforest nature preserves, wildlife viewing, riverboat tours, waterfalls (Guyana has some 300 waterfalls), capital cities and some Amerindian culture. Most of the region is comprised of uninhabited rainforests, rivers and swamps, with few interior roads to facilitate travel and the spread of tourism, which tends to be primarily located in coastal cities and towns.

Border Disputes in the Guianas

Travel in the Guianas has been significantly affected by ongoing border disputes between Guyana and its neighbors, Venezuela and Suriname. Although relations have improved since the mid-1990s, territorial disagreements still hinder normalized and neighborly relations. More than 50 percent of Guyana's territory is claimed by Venezuela, and the far southeastern corner of the country is claimed by Suriname.

Guyana's maritime boundaries with the two countries are also contested in a region that has considerable offshore petroleum potential. These conflicts have contributed to a lack of cross-border travel, shopping and tourism between Guyana and its neighbors. In fact, there are no official border crossings (or bridges) between Venezuela and Guyana, although indigenous people commonly cross by canoe for trade and to partake in illicit drugs and gold mining. Even air travel between the two countries must be done via a third country such as Brazil or Trinidad and Tobago.

There has been talk since 2005 of building road links between Venezuela, Guyana and Suriname, with official border-crossing points, but this has yet to come to fruition. Suriname has a road link with Guyana, which facilitates some cross-border travel, but strained relations between the two countries and poor infrastructure still prevent the development of larger-scale economic cooperation and tourism. There is also a car-carrying riverboat between Suriname and French Guiana. Political relations aside, building roads and bridges between the countries in this region would create a logistical challenge, because driving in Venezuela and French Guiana is on the right, while in Guyana and Suriname it is on the left. This is one of only a few borderlands in the world that face this situation; the Takutu River Bridge between Brazil and Guyana was completed in 2009. There, traffic has to change driving directions through a series of roundabouts and crossover lanes.

Brazil

Brazil is, by far, the largest country in South America, comprising 48 percent of the continent's land area, and with 203 million people, it makes up almost half of its population. Worldwide, Brazil is the fifth largest country in both land area and population. It is the world's largest *lusophone* (Portugese-speaking)

Photo: Dallen J. Timothy

▲ *Copacabana Beach in Rio de Janeiro is one of the most famous beaches in the world. It is extremely popular for tourists and local residents and is surrounded by urban services and a dramatic natural environment.*

country, and the only one in the Americas. As mentioned above, it became a Portuguese colony in 1500, and Rio de Janeiro was even made the capital of the *Portuguese Empire* in 1808 when Napoleon invaded Portugal and the royal family fled to Brazil. Portugal expanded its control over modern-day Brazil through successful battles with France (in coastal areas) and Spain (moving into the Amazon interior from the Andes). Sugar plantations dominated the colony's economy until the discovery of gold in the 1690s in the Matto Grosso area of the interior Brazilian Highlands. The independent Empire of Brazil was established in 1822, which became the Federative Republic of Brazil in 1889.

Brazil's population is highly concentrated along the eastern coastline, south of the Amazon River. The two largest metropolitan areas (Sao Paulo at 19.5 million; Rio de Janeiro at 11.1 million) are both on the southeast coast. According to the Brazilian census, 48 percent of the population defines itself as white, 44 percent as brown (referring to multiracial), and 7 percent as black. In addition to Portuguese, there are some 180 native Amerindians languages spoken in remote areas, although Amerindians comprise only .28 percent of the country's population.

Brazil's economy is the largest in South America, and although tourism is an important sector, it falls far behind agriculture and heavy industry in relation to total GDP. Major sectors of Brazil's exports include timber products, coffee, sugarcane, beef, wheat, cocoa, shoe and textile manufacturing, mining, steel, airplanes and automobiles. It also has offshore oil reserves and is a world leader in the production and use of ethanol as an alternative to fossil fuel based petroleum.

Transportation in Brazil

Brazil has an extensive transportation infrastructure with good air and land links throughout the entire country. The primary exception to this is the jungle interior region, where most travel is undertaken by boat on the Amazon River and its tributaries. The Amazon Basin, with its capital, Manaus, and numerous national parks, is at the center of the country's *ecotourism* activities, where people travel by boat, visit

Photo: Dallen J. Timothy

▲ *Favelas, such as this one in Rio de Janeiro, are becoming the focus of new tourism promotional efforts. There is considerable interest among tourists in seeing how the world's poor live and work, and how such shanty towns function. Local governments now see the economic potential of "slum tours" as part of the urban tourism product and have begun devoting money and effort to highlight a part of the city that for years was ignored and even hidden from tourists' view.*

indigenous villages, and hike in the rainforest. Some of the country's wealthy class have begun investing in boats to offer public transportation and tours through the rivers and jungles. Dozens of eco-lodges have been built and are managed by natives with only cursory interference from state and national governments, and academic researchers have begun examining the role of Brazilian natives as vital stakeholders in rural tourism development. Part of Brazil's recent promotional efforts to cultivate rural tourism has been an emphasis on *agritourism*, where domestic urbanites and foreigners visit sugarcane, cocoa, tropical fruit and coffee plantations and spend time on cattle ranches in the country's interior. Of all the countries in South America, Brazil has by far the most diverse tourism industry, with nearly every sector and activity represented.

Rio de Janeiro

Rio de Janeiro is perhaps Brazil's best-known destination owing to its bounteous beaches, lively culture, active nightlife, accessible location, international gateway status and world-renowned media image, which received a significant boost in 2014 when the city hosted the FIFA World Cup. Carnival is Brazil's most notorious festival and is celebrated all over the country, although Rio's event tends to draw the most international visitors. While the city's Copacabana Beach is one of the best-known beaches in the world, Rio de Janeiro is also well known for its *favelas*, or shantytowns, which originally grew as people from the countryside moved into the city in search of work. Without work or a place to live, the urban immigrants ended up squatting on empty land. The process has continued since the 1960s and 1970s, so that over the years many favelas have developed into very large neighborhoods, with more than 100,000 residents. They have a reputation of being lawless areas where drug activities and crime run rampant and where local law enforcement has no control.

Slum Tourism

Rocinha is the best-known shantytown in Rio, with some estimates placing its population at around a quarter of a million people. Although there were early attempts to remove the favelas, many of them have now been formalized into recognized urban communities with government money being spent to upgrade them and improve the infrastructure. Many favelas now have running water, sewage treatment systems and electricity. In recent years, the idea of slum tourism, which includes visits to slum areas and squatter settlements to view poverty and to become immersed in a unique aspect of local culture and society, have become an important part of Brazil's tourism product. The commodification of these previously off-limits places in tourism development, are being studied by tourism scholars as a form of *pro-poor* tourism, wherein residents can become more involved in the economic and social benefits of tourism rather than just being gazed upon and photographed by the tourists.

Sao Paolo

Sao Paulo is one of the largest cities in the world and is best known for its cosmopolitan atmosphere and famous art galleries and shows. The city is known among homosexual travelers for hosting what is said to be the world's largest annual gay gathering, *Sao Paulo Pride*. Other important urban coastal destinations include Recife, Porto Alegre, Salvador de Bahia and Fortaleza. Salvador de Bahia is Brazil's center for African culture, and is a former center of the Portuguese sugarcane slave trade. The city is known for its Afro-Brazilian music, food and culture.

Fortaleza and Recife are well established beach and urban destinations with many historic buildings, famous nightlife and a wide variety of entertainment and recreational options. Brazil is also a hub for heritage tourism, and the country has more UNESCO-designated *World Heritage Sites* than any other country in Latin America, outside of Mexico. Most of the heritage product derives from historic cities and towns (e.g., Salvador de Bahia, Ouro Preto, Olinda and Goias) with an emphasis on the colonial past. As already noted, indigenous cultural heritage is also an important part of tourism, but it does not receive the same level of attention by government officials, tourism developers, or tourists that the coastal and interior historic towns receive because they are more conducive to 'branding' with the internationally-recognized UNESCO label.

Another important part of the European heritage of Brazil is found in the southern reaches, in and near the city of Porto Alegre. Beginning in the late 1800s, the southernmost states of Brazil saw heavy waves of immigrants from Germany, Italy, Poland and other parts of Europe and Asia. Many of their original settlements still resemble transplanted European villages. The German architecture, food, festivals and handicrafts, and the physical characteristics of blue eyes and blonde hair among the Itajai Valley population, for example, appeal to Brazilian travelers as a kind of European getaway. The region is also trendy for travelers from Germany who visit to see how their *diaspora* kin have fared across the Atlantic.

The Southern Cone

The region of South America that encompasses the four southernmost countries of Argentina, Chile, Paraguay and Uruguay is known as the *Southern Cone*. These four countries are all situated south of the *Tropic of Capricorn* and together comprise the most prosperous region in South America, sharing high standards of living (comparable to Eastern Europe), high levels of economic development, and long life expectancies. (Southernmost Brazil, from Sao Paolo southward, is sometimes included in the Southern Cone. However, since Brazil was covered above, this discussion does not cover that area. Paraguay is sometimes not included as a Southern Cone country, though it is included here.)

All of the Southern Cone countries are overwhelmingly Catholic and Spanish speaking. There are some variations, for instance in Paraguay where the Amerindian *Guarani* language is an official language (along with Spanish) that is understood by 90 percent of the population. There is also *Rioplatense Spanish*, which is highly influenced by Italian and spoken in Argentina and Uruguay. Another Amerindian language, Araucanian, is the predominant language in the southernmost part of Patagonia in Argentina and in adjacent areas of southern Chile.

Native Americans, however, only make up about 3 percent of the population of the Southern Cone, and they are mostly found in northern Paraguay and southern Patagonia and nearby Chile. Whites, originating from Europe, comprise 80 percent of the population of this region, followed by Mestizos (European and Ameridian mixed) at 16 percent. Unlike large areas in other parts of South America, blacks and mulattos comprise only .2 percent of the people in the Southern Cone. Paraguay is the one exception, where an estimated 95 percent of the population is mestizo.

Spanish Settlement

The countries of the Southern Cone had all been under Spanish colonial rule since the early to mid-1500s, and along with more northern countries in the Andes, they fell under the colonial era Viceroyalty of Peru. One of the earliest settlements in the Southern Cone was where Buenos Aires, Argentina, is located. It was originally settled by the Spanish in 1536, but native tribes destroyed it. Modern-day Buenos Aires was founded in 1580 and became the region's capital in 1776. Most of Spanish South America achieved independence from Spain in the 1800s, including Argentina and Colombia in 1810, Paraguay and Venezuela in 1811, Chile in 1818, Peru in 1821, Ecuador in 1822, Bolivia in 1825 and Uruguay in 1828. For most of these countries, the path to independence was long; it took different forms over a decade or more, and it was hard fought against Spanish military forces and between Spanish loyalists and independence supporters.

The Southern Cone contains a variety of mountain, flatland and coastal scenery, with small towns and arctic climates in the south, and larger colonial-era cities in the north. Unlike southeastern Brazil, the Atlantic Coast of the Southern Cone has a wide continental shelf, with broad beaches, sand dunes and rich off-shore fisheries, all of which attract recreational tourists. Southern Chile and Argentina's Patagonia region feature high mountains with permanent glaciers and alpine scenery in dozens of national parks and preserves. Most *Antarctic cruises* depart from Ushuaia, Argentina, or Punta Arenas, Chile, both on the island of Tierra del Fuego. The island, in its own right, is a significant tourist destination with glaciers, parks and rural villages.

Iguazu Falls

One of the Southern Cone's most spectacular attractions is the world-renowned Iguazu Falls, which lies on the border of Argentina and Brazil only a few kilometers from their tri-point border with Paraguay. Iguazu is one of the world's most spectacular falls, descending 230 ft (70 m) over a span of 1.7 miles (2.7 km). The falls, with its surrounding Iguazú National Park (Argentina) and Iguaçu National Park (Brazil), has been inscribed on UNESCO's list of World Heritage Sites and is an important international tourist destination for both countries.

Argentina

Argentina is the second largest country in South America, and the eighth largest in land area in the world. Its population of 43 million is 86 percent of European descent, mostly Spanish and Italian. Archaeologists have found evidence of human settlements in Patagonia that are 11,000 years old. Most of the European population of Argentina arrived in the 19th and early 20th centuries, and the country is second only to the U.S. in the number of European immigrants it received during that time period.

Photo: Dallen J. Timothy

◀ *Iguazu Falls is among the most important natural attractions in both Argentina and Brazil. This view on the Argentinian side illustrates the immensity and grandeur of the falls. It is a UNESCO World Heritage Site and is usually listed on several different lists of the Seven Natural Wonders of the World.*

Argentina's Route 40 is a favored corridor for visitors. It runs the length of the country from north to south and passes through some of Argentina's most vivid cultural and natural landscapes. Argentinian and Chilean wines have become favorites among wine connoisseurs in recent years. *Viticulture* has grown considerably in response, and several wine routes have been developed in both countries, which include visits to vineyards, wineries and sales shops. The Patagonian region saw large numbers of immigrants arriving from Europe in the early and mid-twentieth century, particularly from Wales and Germany. The communities they established bear a strong resemblance to their European landscapes, and family names and traditions reflect a strong European heritage. The urban heritage of Buenos Aires and other large cities is typical of Latin American colonial pasts in terms of architecture, layout and population. Health, sport and meetings and conventions are now a primary thrust of the Tourism Secretariat to complement the more traditional culture- and nature-based activities that have long defined Argentinian tourism.

Uruguay

Uruguay has a population of only 3.3 million, about half of whom live in the capital city of Montevideo. Uruguay shares a lot in common with its larger neighbor, Argentina. It is almost entirely of white European ancestry (95 percent; mostly Spanish and Italian, like Argentina), though about 10 percent have some African ancestry and 5.5 percent have some Amerindian ancestors. It has a very successful agricultural export economy, trading mostly with Brazil, Argentina, the U.S. and China. In addition, Uruguay has one of the most liberal societies in the Americas. It was the first in South America to legalize same-sex civil unions (in 2009), the first country in the Americas to legalize the possession of cannabis (in 2011), and the first country in the world to give every child a free laptop and Internet access (again in 2009). On a global scale, Uruguay is one of the least tourism-oriented economies in South America, although it

Photo: Dallen J. Timothy

▲ *Outside of Asuncion, these ruins of the Jesuit Missions La Santisima Trinidad de Parana are Paraguay's most significant heritage attraction for international visitors. These mission ruins reveal an interesting history of religious settlement and Spanish colonization in South America. This site was inscribed on UNESCO's World Heritage List in 1993.*

has been experiencing considerable growth in recent years. Like its neighbors, Uruguay's tourism focus is cultural heritage, rural tourism and ecotourism, although recent promotional efforts have spotlighted beach resorts and thermal springs and spas.

Chile

Chile occupies the narrow western slopes of the Andes Mountains, across from Argentina and Bolivia. Its shape is distinct, extending 2,880 miles (4,600 km) from north to south, but only an average of 109 miles (175 km) east to west (265 miles (430 km) at its widest). As discussed above, it includes the extremely dry Atacama Desert in the north and heavy rain and glaciers in its south. It is also part of the *Pacific Ring of Fire*, with both volcanoes and earthquakes. Chile also stands out as having the highest living standards and the most successful economy in South America, and it is often held up as a role model for other developing countries in the New World. About 60 percent of the country's 17 million people are majority European ancestry, about 4.5 percent are mostly Amerindian, and the rest are mestizo mixes of those two. Santiago, Chile, is one of South America's key international gateway cities, with many flights fanning out to other cities around the globe. It is a major center of commerce for the continent, and as a result Santiago hosts many business travelers each year.

The country's largest cities are located in the central region, and the long, slender shape of the country means that one can travel from the coast to the upper reaches of the snow-capped Andes within a short distance. Glaciers and ski resorts appeal to mountain enthusiasts, while the country also is home to beach resorts. Fishing, golf, hiking, birding, sea kayaking and other outdoor adventures are heavily promoted by the Chilean Tourism Promotion Corporation.

Chile's Easter Island is one of the earth's most remote islands—more than 2,000 miles (3,200 km) from South America—and one of the most mysterious. Nobody is quite sure about the origins of the island's *Moai* sculptures, but they are the foundation of tourism to the island. The sculptures and the entire island are protected as *Rapa Nui National Park*, which has become a center for *New Age* travelers who believe in the mystical earth powers of the island and its earlier inhabitants.

Paraguay

As mentioned above, Paraguay is the most indigenous of all the countries of the Southern Cone, with almost all (95 percent) of the country's 6.7 million people being mestizo mixes of Spanish and local *Guarani*. It is also the least developed of all the Southern Cone countries, with a high poverty rate (estimated at 30 to 50 percent) and a large informal economy. However, it was also one of the fastest growing economies of South America in the 2000s, and is one of the largest soybean producers in the world.

Paraguay is possibly the least tourism-oriented country in all of South America. It has an adequate transportation system and accommodations, but it lacks an appealing image abroad due to a long history of internal political turmoil and unrest. Most tourism in Paraguay occurs in the southern portion of the country in the capital, Asuncion, with its colonial architecture and parks along the Paraguay River. Secondary foci are the central city of Concepcion, the Jesuit Missions of La Santisima Trinidad de Parana and Jesiis de Tavarangue, which together are a UNESCO World Heritage Site, and a handful of national parks and nature preserves scattered around the country. Brazilians and Argentines also cross into Paraguay to purchase electronic goods that are available at lower prices due to low taxes.

The Andes: Peru and Bolivia

Bolivia and Peru have historically been home to some of the most important early civilizations of South America. The *Norte Chico* civilization was the oldest known major civilization on the continent, occupying the coast of Peru between 3,000 and 1,800 BCE. It was a complex, pre-ceramic culture that is best known for its large architectural structures. The present-day *Ayamara* people of Bolivia (about 25 percent of Bolivia's population) trace their ancestry to the *Tiwanaku* civilization of coastal Peru and Bolivia. They grew from a village-based society in 1,500 BCE to a major empire in 1,000 CE, at which time they disappeared, possibly due to a dramatic change in climate.

Many other groups also arose in the coastal and mountain areas of Peru and Bolivia before the rise to dominance of the *Incas*, who controlled the entire region from Ecuador to northern Chile in the 15th century until the arrival of the Spanish. In 1532, *Francisco Pizarro* defeated the Inca Emperor *Atahualpa*, and within ten years the Spanish had established the Viceroyalty of Peru, which eventually governed almost all of colonial Spanish South America. Because of its history, Peru was the strongest royalist (loyal to the king of Spain) holdout against 19th century independence movements that swept Spanish South America.

Bolivia's population of 11 million (2010) is 30 percent *Quechua*-speaking and 25 percent Ayamara speakers. These are two languages among 34 other native languages, plus Spanish, that are officially recognized for use in Bolivia. Approximately 55 percent of Bolivia is Amerindian, 15 percent is European white, and the remaining 30 percent are mestizo. This makes Bolivia the most important purely indigenous country in South America (Paraguay, by contrast, is almost entirely mestizo). Peru's population of 29.5 million is similar to Bolivia in its diversity, with Native Indian groups at an estimated 45 percent, mestizos at 40 percent, European whites at 10 percent and blacks at 5 percent. However, only Spanish and Quechua (the Incan language) are officially recognized in Peru.

◄ *Ruins of the Incan city of Machu Picchu in the Andes Mountains of Peru, The ancient Incas lived in highly advanced societies and were key in domesticating several of the food staples eaten throughout the world today, including potatoes, tomatoes, corn (maize), peppers and various squashes. Machu Picchu is an iconic tourist attraction and the most popular destination in Peru.*

Land-Locked

Bolivia and Paraguay are the only two countries in South America without access to the sea. Some observers believe this to be one of the reasons they have lagged behind other countries on the continent in terms of economic development and prosperity. Most tourist activity in Bolivia concentrates on indigenous culture, which is appropriate for a country where approximately half the population still practices traditional lifestyles and cultural customs.

Bolivia has several of unparalleled distinctions, including being the highest and most inaccessible country on the continent. This adds to its tourist appeal and is one of the reasons for its ability to preserve native cultures. It is sometimes referred to as the other Tibet. It is also one of the few countries on Earth that has two capital cities: La Paz and Sucre. In addition, Bolivia shares the world's highest navigable lake, *Titicaca*, with Peru, and La Paz is one of the highest cities in the world at 11,750 ft (3,580 m). The Lake Titicaca region is among the most visited parts of Bolivia, often viewed on side trips from Puno, Peru. In addition to being twin capitals, La Paz and Sucre are attractive towns that feature traditional architecture and unique urban landscapes.

Tourism in Peru

Much of Peru's tourism industry is based on ancient remnants of the Incas, as well as on the living cultures of their descendants today. *Machu Picchu*, near Cusco in southern Peru, is a UNESCO World Heritage Site (since 1983) and one of the most recognizable heritage icons in the world. In 2007, the New7Wonders organization elected Machu Picchu as one of the new *Seven Wonders of the World* because it embodies global heritage at its best. The location is visited by thousands of tourists each year and is growing in popularity among New Age and other nature spiritualists as a worship site owing to what some believe was the site's possible role as an ancient Inca place of worship.

Although cultural heritage dominates Peru's tourism image, the country also promotes itself as a nature and adventure destination with river rafting, bird watching and mountain climbing. Volunteer tourism has gained a strong foothold in Peru and in other parts of South America, as people travel on short-term health service missions from the United States and Canada to work among the poor of Peru. In the southeastern corner of the nation, Lake Titicaca has become a mythical destination for people who have read about the mystique of the floating islands and the interesting cultures. Lake Titicaca, the earth's highest navigable lake, is home to the unique culture of the Uro Indians, who build homes on floating islands they construct from reeds and other plant materials.

Unfortunately, Peru's tourism industry has suffered during the past quarter century from media-led images of Shining Path communist guerilla group's kidnappings of Peruvians and foreign aid workers in the 1990s, crimes against tourists in Lima and the harassment of tourists by narcotics gangs. The northeast border region is considered dangerous because of possible incursions by Colombian drug gangs and armed insurgents. While most of Peru is relatively safe, suffering from the same ailments that other countries in the region face, these persistent images keep many less-adventuresome tourists from visiting. However, despite these representations, tourism has become an important economic sector in Peru.

Gran Colombia: Venezuela, Colombia and Ecuador—and the Galapagos Islands

The three northern countries of Ecuador, Colombia and Venezuela (as well as Panama) once comprised the country of *Gran Colombia*, which was created following Simon Bolivar's independence victory over the Spanish in 1821. Bolivar became the republic's first and only president until 1831 when Gran Colombia (officially the Republic of Colombia) was disbanded. Prior to Gran Colombia, these three countries were part of the *Viceroyalty of New Granada*, which also extended northward to the Caribbean coast of

Nicaragua and eastward into the Guianas. The modern flags of the three core countries all have stripes of red (bottom), blue (middle) and yellow (top), reflecting their shared history.

In addition to history, these three countries share a lot in common. All have high population densities in their urban and coastal areas, with very small populations in their interiors (the Amazon and Orinoco river basins). All three countries are renowned for their biological diversity, though high rates of poverty (ranging from 30 to 50 percent of their populations) and a long history of political corruption and instability have also threatened those special resources. About 65 percent of the overall population of this region are mestizos. Colombia and Ecuador have a significant minority of Quechua speakers, reflecting their indigenous ties to the Incas, though their populations are not nearly as large as in Peru and Bolivia.

Colombia

Colombia, with 49 million people, is the second most populous country in South America, and the world's third largest Spanish speaking country behind Mexico and Spain. It has suffered from political instability, internal guerrilla movements, drug smuggling and high crime rates, much of which reflects its great income inequality. The U.S. State Department and the governments of other countries regularly warn their citizens about travel to the country. In addition to the Quechua speakers living in the mountains, there are some 101 other Amerindian languages spoken in Colombia, besides Spanish.

Through much of the 1990s and 2000s, Colombia had major challenges developing its tourism owing to unsafe conditions for foreign travelers. Several Colombian terrorist organizations involved in illicit drug trafficking have waged attacks against the country's infrastructure, public places, modes of transportation and civilian targets. Foreign aid workers, missionaries and tourists were kidnapped and murdered, and rebel occupation of national parks and protected areas has prevented domestic and international tourists from visiting them. Several parks under guerrilla control continued charging visitors to enter, keeping the funds to support their efforts. By the late 2000s, many of these issues began to subside, and adventurous tourists visiting the country reported wonderful experiences due, in part, to the lack of mass tourism found there.

One of the strongest and most notable forms of tourism over the past few decades has been *VFR* (visiting friends and relatives). Colombians living throughout the world often travel to their homeland to visit family members and friends. In addition to their remittances sent home to Colombia from overseas work, these visits are important to the economy. Ecotourism and heritage, urban and beach tourism all have great potential in Colombia, and recent guidebooks suggest that it is relatively easy to avoid the country's trouble areas.

Venezuela

Venezuela has 29 million people (2010) and is one of the world's leading oil exporters. It also has natural gas, and its subsidized petroleum prices result in the cheapest gas in the world at about USD $0.12 per gallon (U.S. 3.2 cents per liter) as of 2014. Despite its oil riches, poverty and social instability continue to hinder the country's development. While approximately 90 percent of Venezuela's export earnings come from oil, tourism and agriculture have long been important parts of the economy as well. Margarita Island, off of Venezuela's Caribbean coast, is a popular cruise port of call and an otherwise desirable sun, sea and sand destination. Together with Lima in Peru, Venezuela's capital city of Caracas has become one of the least expensive airline gateways to all of South America, not only to Venezuela.

Venezuela has many national parks, beaches and Amazon rainforests that attract many tourists each year. *Angel Falls*, for example, located in Venezuela's Guiana Highlands, is the world's tallest waterfall, dropping 3,212 ft (979 m). Unfortunately, violent crime, including armed robbery, murders, and kidnappings have grown quickly in Venezuela since 2011, leading many tourist-generating countries (such as the United States and Canada) to warn citizens against inessential travel there.

Ecuador

Ecuador is named after the equator, which it straddles, and its official name translates into English as the *Republic of the Equator*. Its population is 15.6 million (2014), but it also has a much smaller land area than the other Gran Colombia countries. For tourism promotion purposes, Ecuador, which is slightly smaller than the U.S. state of Nevada, is divided into five geographic regions: the Amazon, the northern Andes, the southern Andes, the Coast and the Galapagos Islands.

Tourism in the hot and humid Amazon region centers on ecotourism, adventure travel and Amazon natives. The Northern Andes region is home to the capital, Quito, and the town of Otavalo, where indigenous people continue to live in much the same way that they have for centuries. Quito is located in an Andean valley at 9,350 ft (2,850 m) above sea level, which gives the town a cooler climate than it would have in lower elevations so close to the equator. Old Quito itself has been listed by UNESCO as a World Heritage Site for its colonial architecture. The city is a major trading crossroads for Ecuador's many indigenous people. As a result, it has many lively markets, cultural attractions and handicraft centers. The Northern Andes is home to adventure and sport tourism trips out of Quito, with a focus on kayaking, hiking, trekking, windsurfing and horseback riding. Nature-based/ecotourism, agritourism, archeological sites, and indigenous markets and culture are the foundations of tourism in the south Andes area. The coast is known for beach resorts, ecotourism, whale watching, and cultural arts and crafts.

The Galapagos

The Galapagos Islands are almost 620 miles (1,000 km) off the coast of South America, but they are still one of Ecuador's most sought-after tourist destinations. The islands are especially well known as the place that inspired *Charles Darwin* to conceptualize the theory of *evolution*. While there is some beach-based and intra-island cruise tourism, the Galapagos are ranked by avid nature enthusiasts worldwide as the ultimate *ecotourism* destination, not only for their diverse nature, but also because of their connections to Darwin.

The islands are somewhat protected as a national park, but researchers have consistently voiced concern that the masses of tourists traveling to the Galapagos in the name of ecotourism have begun to exceed their carrying capacity, resulting in negative impacts. These islands are one of the few places in the world promoted by the government ministry (Ministry of Tourism of Ecuador) as a *scientific tourism* destination. Scientists from around the world work in the Galapagos on issues related to preserving reptile, bird and plant species, developing water-based agriculture, preserving marine life and developing educational programs.

The Falkland Islands

Since 1833, these islands off the coast of Argentina have been controlled by the British as an overseas territory, and Falkland Islanders are United Kingdom citizens. However, Argentina also lays claim to the islands, which it refers to as *Islas Malvinas*. On April 2, 1982, the Argentinian military dictatorship then in power asserted its claim over the Falklands and invaded them, in what it argued was a re-establishment of its legal rights over the islands. The timing of the invasion appeared to be for domestic political reasons, as the dictatorship was becoming increasingly unpopular and the invasion was a potentially populist measure. The Argentinian military did not believe that Britain would seek to retain control of the islands. The United Kingdom, however, saw this as an act of aggression on its sovereign territory and responded with military force, resulting in the Argentinians surrendering on June 14, 1982.

Argentina still has not given up its claim to the islands, however, as they regularly appear part of Argentina on some of the maps produced there. In March 2013, the Falkland Islands held a referendum on its political status, and 99.8 percent of voters favored remaining under British rule. However, this was dismissed by the Argentine government which does not recognize the Falklands government as a party in negotiations. The relationship with Argentina has stymied tourism development on the Falklands. Nevertheless, the islands have become an important part of cruise ship itineraries to the Antarctic territories and sub-Antarctic islands such as South Georgia and the South Sandwich Islands that are also British overseas territories.

While the battle was devastating, and many lives were lost on both sides, it put the Falkland Islands on the tourist map. Most of the world's population had never heard of the Falklands until the altercation, and once the hostilities ended, tourism began to boom as people from Europe, the United Kingdom and North America began to visit. Today, fishing and sheep farming are the primary industries (95 percent), with tourism becoming a more important sector since the 1980s. Much of the islands' tourism product focuses on ecotourism and environmental history. It was visited by Charles Darwin as part of the voyage of the Beagle and played a crucial role along with the Galapagos Islands and southwestern Australia in the development of his evolutionary theories. War history with battlefield tours is also among its more popular activities.

ISSUES AND INSIGHTS ◖ Whale Watching

Since the growth of the international environmental conservation movement in the 1960s and the Save the Whale movements of the early 1970s, whales have been regarded as representative of the success of marine conservation policies in many countries. Whales have become an international conservation icon species, also referred to as *charismatic mega fauna*. Other examples include elephants, pandas and harp seal pups, all which possess considerable emotive potential. The public appeal of whales, particularly in the Western world, has been sustained through a proliferation of books, magazines, films, television shows, music and works of art.

Related to the media interest in whales has been a growing fascination with observing whales and dolphins in their natural environment. There has been a shift from their consumptive use to the non-consumptive utilization of whales since the worldwide ban on commercial whaling by all countries (generally starting in 1986) except Japan, Norway and Iceland. This has resulted in financially viable businesses based on taking tourists to watch whales in the open sea and a rapid worldwide growth in marine mammal-based tourism.

Whale watching has become an important economic, educational and recreational activity. Organized whale watching appears to have begun in the United States in the 1950s on small, isolated scales in Hawaii and California. It grew slowly at first, and the concept did not reach Europe, Australia and New Zealand until the mid-1980s. The whale watching industry is a significant contributor to many coastal communities, particularly in more peripheral coastal and maritime areas. Nevertheless, the overwhelming growth in the whale watching industry has stirred some serious concerns over planning and management issues.

Whale watching can provide both benefits and costs to animals and humans. Much work is needed to develop techniques to quantify both the biological risks and the ethical values of whale watching. Management must confront both human demands and the impact it has on the animals. Indeed, management officials often find it difficult to come up with ways to ensure both tourist satisfaction and the continued success of marine mammal programs. In relation to whales and dolphins, management concerns range from maintenance of the whale's habitat, harassment of the whales and dolphins by commercial and private vessels, and the definition and enforcement of the policies that affect both tourists and whales.

One of the objectives of ecotourism is to change the tourists' behavior and lifestyle so that their actions become more environmentally responsible both during the tourism experience and over the longer term. Opportunities to view whales may therefore provide an important platform for public education about the natural environment. It has even been argued that any negative impacts on populations of marine mammals that are constantly pressured by tourist boats and viewers may be counterbalanced by the education and awareness that whale watchers receive. If presented clearly and concisely, this education may increase a whale watcher's overall environmental awareness, and benefit all cetacean species. However, if inadequate education is provided, and the operator is instead simply exploiting a natural resource, then such operations may need to be extensively reviewed.

Name _____

Human Mobilities

1. Identify and describe two examples of human mobility that shaped the historical development of one or more of the subregions of North America and South America.

   _____

   _____

   _____

   _____

2. Identify and describe the role of human mobility in the contemporary societies of one or more of the subregions of North America and South America.

   _____

   _____

   _____

   _____

Tourism Destinations

3. Describe the contemporary international image of one or more of the subregions of North America and South America. How do these images impact the potential of the regions as tourist destinations?

   _____

   _____

   _____

   _____

4. Select one or more of the subregions of North America and South America and describe how the countries in the sub-regions might use their cultural and physical resources to improve their global image and attractiveness.

   _____

   _____

   _____

   _____

Sustainable Environments

5. Identify and describe tine major environmental challenges of one or more of the subregions of North America and South America. What challenges do the countries in the region face in addressing these issues?

_____

_____

_____

_____

6. Identify and describe the major social and economic development challenges of one or more of the subregions of North America and South America. What challenges do the countries in the region face in addressing these issues?

_____

_____

_____

_____

Instructor Questions (answer any additional questions from your course instructor in the space below)

_____

_____

_____

_____

Conclusions: Global Opportunities and Futures

CHAPTER 6 OVERVIEW QUESTIONS

HUMAN MOBILITIES

- How does high degree of modern human mobility today contribute to the economic and cultural globalization of almost every corner of the planet?

TOURISM DESTINATIONS

- In what ways will the landscape of tourism in destinations change in response to the global social and environmental changes that are likely to occur in coming years?

SUSTAINABLE ENVIRONMENTS

- What can individuals, like yourself, do to affect, shape or ameliorate the possible coming impacts of climate change?

INTRODUCTION

This chapter returns the reader to some of the major issues facing the world today and that set the context for understanding our planet's geography. The first section revisits the issue of globalization, which we had previously discussed in Chapter 1, and which provides the impetus for the continued growth in international and domestic travel, specifically, and world and regional development in general. This is because globalization is bound up with technological developments in transport and communication as well as with financial deregulation. However, as the chapter points out, such developments also have a "dark side" in that they potentially expose countries and places to a new range of economic and environmental risks. Like any form of development tourism can therefore have both positive and negative impacts.

The second section outlines some of the issues that are combining with globalization trends to set the framework for the future of our planet. Some of the major trends identified include demographic change, particularly with respect to an aging population, and various dimensions of global environmental change, the most significant of which is concern over the world's climate. The chapter also includes a discussion of geography and tourism related careers, then concludes with some final thoughts as to the value of a geographic understanding as we live, work in and travel our world.

6.1 | GLOBALIZATION AND TOURISM GEOGRAPHY

Current Globalization Issues and Resources (http://wrgeography.com/globalization.html)

We are living in a time of unprecedented change. Society, the environment and the economy are shifting and changing at rates that humans have never seen before, and which the planet itself has seldom experienced. To a great extent, these changes are due to processes of *globalization*. Globalization is a complex concept that is both a process and a state or condition. In its simplest form, globalization can be understood as the increased interdependence of social and economic systems across time and space. In one sense, globalization has always existed. For example, we know that there was trade between ancient Rome and China over 2000 years ago, while the Silk Road through Central Asia has been significant for trade and travel for over 3000 years. But what is different today is the extent to which relations and *connectivities* across space and time have speeded up to the point where, in some cases, they are virtually instantaneous.

Space-Time Distantiation and Compression

Although globalization processes are not evenly spread across the planet, because some places are clearly more connected than others, all places have experienced what is termed *space-time convergence*. This occurs through two processes: "space-time distantiation" and "space-time compression" (Hall 2005). Space-time distantiation refers to the "stretching" of social and economic relations over time and space, such as through the utilization of communications technology (including the Internet) and by advances in transport technology. Space-time compression refers to the increased velocity and speed of material and non-material flows over a given distance. Again, this is related to technological change, including communications, transport and social media technologies. All of this has practical implications. For example, on average, people continue to travel further than ever each year for work and leisure.

Globalization Influences

Four major shifts have occurred in globalization in recent years that provide a framework for understanding global change (Harvey 1990): financial deregulation, technological change and innovation, media and communications, and the cost and time of moving commodities. All of these factors impact global development and tourism.

Financial Regulation

Since the early 1970s, the global financial system has become increasingly deregulated. This has meant that the international financial system has become more decentralized and has operated through the functions of the market. However, this has also provided for substantially increased volatility, as seen for example in the global financial crisis that started in 2008 and the Asian financial crisis of the late 1990s (Visser and Ferreira 2013). Within the current global financial system, economic capital has the ability to move rapidly around the world in order to take advantage of local economic conditions, such as exchange rate differences, labor costs and tax and regulatory regimes. The drive for financial deregulation has also been felt in movements towards free trade and the lessening of foreign investment restrictions under the negotiations of the World Trade Organization (WTO). These have affected foreign investments and the

◄ *Our Shrinking Planet. Advances in transportation technologies have made the furthest corners of the planet easier to get to today than at any time in history.*

formation and development of transnational firms, and they have had a broader impact on the regulation of the mobility of travel and labor.

Technology Change and Product Innovation

The recent rate of technological change, the types of change and the rapid diffusion of innovation have been taking place at a pace never seen before (Hall and Williams 2008). Transport, and aviation technology in particular, has rapidly advanced in a manner that has revolutionized the nature of mass tourism. In the 19th century, the revolution was the train and the bicycle, and in the 20th century it was the automobile and the plane. In the 21st century, it may be space. Changes in transport as well as in communications technology have profound effects both on how we travel around the world and also how we see it. For example, when the Jules Verne novel *Around the World in Eighty Days* was first published in 1873, it described an almost impossible feat of rapid travel, but by 1959 it had become possible to circumnavigate the globe in 611 hours on scheduled transport routes. By the early 1980s, the same journey could be made on scheduled air routes in 32 hours, while the space shuttle circled the globe once every 90 minutes.

Media and Communications System

The "*information revolution*" has changed the way in which we receive information and how we see the world. The personal computer and computer tablet, mobile phones and the Internet have all led to the development of '*knowledge-based*' economies. However, the centrality of such technology also divides the world, with only just over half of the world's population having access to a mobile phone. The development of *information and communication technologies (ICT)* and their increasing use by consumers has radically changed the communication, knowledge and economic relationships between the companies and their customer base. In the case of tourism, access to travel information and bookings has become much more consumer-driven. Traditional face-to-face travel agents are now catering to those who do not have or do not want to use new communications technologies, or they are forced to develop new roles as knowledge brokers. In addition, consumers also can gain knowledge of destinations through a wider range of media than ever before, including social networks.

Cost and Time of Moving Commodities, Services and People

The cost and time of moving commodities, services and people has been reduced dramatically in recent years. The real cost of travelling internationally has fallen sharply as has the time it takes to travel long distances. For example, in the 1970s jet aircraft between eastern Australia and the west coast of North America had to travel via Fiji and Hawaii in order to refuel. In the 1980s, improved aviation technology and fuel efficiency meant that aircraft only had to refuel in Hawaii. In the 1990s, further developments, particularly the addition of winglets to the wings of long-haul aircraft, meant that you could fly non-stop between Sydney and Los Angeles for the first time. Non-stop commercial passenger flights are now available between Singapore and the United States, stretching distances even further. Although the promise of supersonic commercial travel becoming widely available to the public has not yet been delivered, more people are able to fly further than ever before. Similarly, trains are now designed to be faster than ever, while cruise ships have gotten increasingly larger.

▶ *A Connected World. Telecommunication and information technologies have actually reduced the necessity to travel, as the colleague on the other side of the globe is now as easy to talk to as the neighbor next door. This has not, however, diminished the desire to travel, and actually it may have increased it, as we learn about ever more places that peak our interest.*

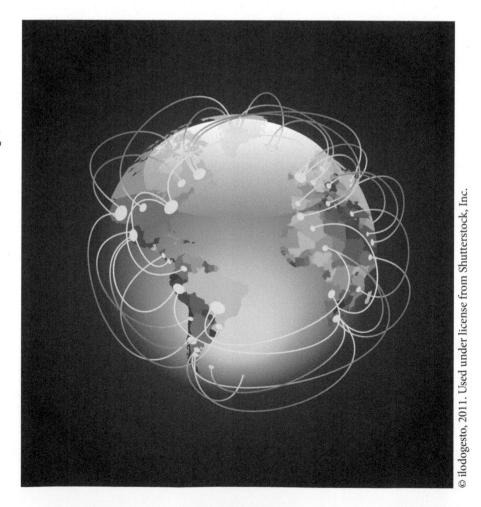

© ilodogesto, 2011. Used under license from Shutterstock, Inc.

Globalization has meant that an international outlook is increasingly important for graduates. The labor market, particularly for high-skilled jobs for the well-educated, is best described now as being global in scope rather than local or even national. Whether you're a high school student researching colleges, a current tourism or geography major or a recent graduate navigating the job market, knowing the career opportunities available to degree holders can help you make educational choices, seek out internships or find a promising entry-level position. In the past, many who graduated with a geography degree went into teaching. Nowadays the spread is far wider, with research, planning and management careers in government, the commercial sector and nongovernment organizations (NGOs) providing many positions.

The skills acquired in any general college or university degree (being literate and articulate, well organized, clear-headed, able to define problems and solve them, and have initiative) are important for all employers. The particular skills of a tourism geographer (the ability to synthesize disparate types of information, to design and execute research projects, and to have some numerical and analytical capacity) are invaluable factors in the professional job market. Significantly, such skills are valuable in the longer term. The workplace is increasingly structured in ways that encourage or require high flexibility in employment, given that jobs are often on a contract or short-term basis. Government jobs for instance are no longer "for life" in a particular agency, but instead people move through various research and/or policy positions in a variety of organizations. It is also notable that many tourism geographers shift back and forth between the public and private sectors. In many cases, there are two recurring underlying expectations that employers have.

The first is knowing one's subject. This means having a good degree (i.e., a good grounding in geography, as well as in a specific sub-discipline or disciplines, such as tourism), and the ability to show that one has applied his or her knowledge through university coursework as well as in any vacation or part-time work or volunteering while still studying. Also, one should be prepared to learn on the job, and to pick up extra qualifications or skills when required and as opportunity presents. A second aspect is good communication skills. Being able to write and speak clearly are essential attributes for a graduate or for any employee. That's why students write essays, present seminars and work collaboratively undertaking fieldwork or when in the laboratory. If you're interested in the human and physical environment and how people work with it, studying tourism geography offers career opportunities involving economics, people, resource extraction and management, environmental planning and more.

Human geographers focus on how people create places, use them and behave in geographic space. *Physical geographers* study the physical environment of planet Earth, including land formation, climate patterns, plant and animal distributions and water models and issues. Human and physical geography are integrated through the practice of environmental resource management and community and regional planning,

both of which are significantly grounded in sustainable development. In the tourism area, human geographers may work in community and destination planning, interpretation and tour guiding, visitor management, real estate development or urban and regional planning. They may design people-centered spaces with functional and environmental requirements in mind. Many geographers also find work in the conservation and management of natural and cultural heritage sites, including national parks, as well as other visitor attractions. Physical geographers advise governments and business about the environmental effects of their activities. They are often involved in the development of environmental impact assessments, studying climate and climate change, natural hazards research and risk assessment, national park interpretation and trail design and the scientific dimensions of natural resource management.

Those students who have studied both human and physical geography in substantial depth and who work in natural resource management are often highly sought after by government agencies, consultants and firms in areas related to conservation, nature-based tourism, waste and pollution management, landscape planning and assessment, and wilderness management. An additional career path for tourism geographers exists for those who use spatially-based information and communication technology and software. Some work as analysts for *geographic information systems* (GIS), *geographical positioning systems* (GPS), and in the field of *remote sensing*, reviewing database information about an area's features to help with surveying and with urban and regional planning. Others focus more on cartography, field surveying and computer-mapping software development. Geographers with GIS skills are also increasingly being employed in more market-oriented applications related to retail locations and tracking the spatial patterns of consumers.

According to the *Association of American Geographers*, geography graduates find work with businesses, governments, research institutes, universities and consulting agencies. Possible government employers in the United States include the Bureau of Land Management, the Forestry Service, state and national environmental and planning agencies, the National Climate Data Center and the US Geological Survey. Mining and resource extraction companies hire geographers to help find deposits and assess environmental impact. Builders and municipal governments seek out geographers to help plan major construction and development projects, as well as to undertake planning for growth management, resort development and transport. Environmental agencies, such as the World Wildlife Fund, the United Nations Environmental Program and UNESCO also hire geography graduates.

Read More

Earthworks Jobs: http://www.earthworks-jobs.com/index.shtml
Geography Jobs: What Are You Going to Do With a Degree in Geography?: http://geography.about.com/od/careersingeography/a/jobsgeography.htm
What Kind of Job Can You Get With a B.S. in Geography? leHow.com http://www.ehow.eom/info_8034535_kind-job-can-bs-geography. html#ixzzlKNDYyl52

. . .

The Challenge of Globalization

The effects of globalization have had profound and, in many cases, quite unexpected consequences. For example, the food that we eat now often travels thousands of miles from where it is grown to where it is consumed. A visit to the supermarket is an exercise in globalization. The growth in world trade has led to massive economic changes with heavy industries moving from some regions only to be replaced by *service industries*, including tourism. Indeed, some former industrial sites have now become locations for heritage tourism. The world is also more connected than ever before. No longer bound by what we can see on television, radio and newspapers, it is now possible to gain access to a range of alternative news sources that can provide profoundly different ways of viewing the world. At the same time, such connectivity also raises new concerns.

Borders

Many countries are voicing the difficulties they have in controlling *illegal immigration* and cross-border criminal activity. As a result of an influx of illegal migrants seeking to escape the social and political unrest in Africa and the Middle East since about 2011, some European countries are looking for ways to reinstate national passport controls among EU states. At the end of 2014, the border-free *Schengen region* embraced more than 400 million people in 22 EU countries, as well as in Switzerland, Liechtenstein, Norway and Iceland. It extended from Portugal to Russia's borders on the Baltic, and from Reykjavik to Turkey's border with Greece. And even the few EU members that were not part of the Schengen Area in 2014 (which included Bulgaria, Croatia, Cyprus and Romania) are legally obliged to join the area within a reasonable period of time. Given the extent to which ease in cross-border travel is critical to international tourism, any changes in Europe's border politics to curb illegal migration will also have profound implications for tourism.

Despite declarations from the World Tourism Organization (UNWTO) and the United Nations as to the Right to Tourism and the liberty of tourist movements, nowhere in international law is there enshrined a right to cross international boundaries and enter foreign spaces. Even the advisory *Universal Declaration of Human Rights* only gives a right of exit and entry to one's own country and freedom of mobility within a citizen's own country. Article 13 states: "(1) Everyone has the right to freedom of movement and residence within the borders of each state. (2) Everyone has the right to leave any country, including his own, and to return to his country."

Invasive Species

Another concern associated with globalization has been the spread of disease and biologically *invasive species*. Since the 17th century, invasive alien species have contributed to nearly 40 percent of all animal extinctions for which the cause is known. It is also estimated that over half a million species have been accidentally or deliberately introduced to locations that lie beyond the natural limits of their geographic range. Of course, many of these species have been extremely beneficial for humankind, for example, with respect to food production. In some cases, introduced species have served to replace the native equivalent. On the Great Plains of North America, cattle and to a lesser extent, sheep, replaced the bison, while wheat and other cereal crops replaced the native grasslands. However, the introduction of alien species has also led to major economic losses. In the United States alone, it has been estimated that invading alien species cause losses adding up to almost US$120 billion per year. Significantly, there is an extremely positive correlation in statistical terms between international trade, including tourism, and the introduction of invasive species.

Diseases

Closely related to the transport of invasive species is the transport of disease, with the global air transport network being responsible for the global patterns of emerging diseases. In fact, tourism is both affected by diseases and is a major vector (or "path") for their distribution. The international spread of *SARS* (Severe Acute Respiratory Syndrome) in 2003 was closely associated with the speed at which international aviation moves travelers from one transport hub and major population center to another. SARS spread in a matter of weeks to 30 countries, infecting a reported number of over 8,400 people after eight months. It is believed to have caused the death of more than 900 people by the time the pandemic was finished. The economic loss from SARS was substantial and was estimated to have cost the Asian region nearly

---

ISSUES AND INSIGHTS **Welcome to the Anthropocene**

The world's climate and environment have remained remarkably stable in the last 11,700 years, an epoch referred to by many physical geographers and geologists as the Holocene. This has been particularly important in helping the human race develop agriculture, towns and cities, technology and many of the other things that characterize our cultures and economies. However, some believe that human activity has so irrevocably altered our planet that we have entered a new geological age that has come to be described as the *Anthropocene*. The term Anthropocene was coined by Nobel Laureate chemist Paul Crutzen to describe the rapid anthropogenically-induced environmental changes that affect not only the climate but also biodiversity loss, biotic transfer, deforestation, desertification, urbanization and the production of waste—all of which are factors that change not only geomorphology but that also appear in the future geological record.

There are three main ideas about when the Anthropocene began:

1. Thousands of years ago with the rise of agriculture
2. Around 1800, which was the year that human population hit one billion and the amount of carbon dioxide started to rise significantly as a result of the switch from burning wood to burning coal in the Industrial Revolution
3. After the Second World War when the nuclear age began. Sediments deposited worldwide that year contain the radioactive signature from the first atom bomb tests in the United States and from the attacks on Hiroshima and Nagasaki that ended the war. This is significant because in order to formally define a new geological epoch, it must be recognized in the layers of sediment that will eventually form rocks.

The choice of 1945 is also significant as it coincides with an event that Professor Will Steffen of the Australian National University describes as the "*Great Acceleration.*" Since 1945, human population has more than doubled to 6.9 billion while the global economy has increased tenfold over the same period. At the same time it is estimated that the number of wild animals has been reduced by almost half. This means that humans are consuming exponentially more resources.

Similarly, from a preindustrial value of 270–275 ppm, atmospheric carbon dioxide had risen to about 310 ppm by 1950. Since then, atmospheric CO_2 concentration has risen from 310 to 380 ppm, with about half of the total rise since the preindustrial era occurring in just the 30 years from the mid 1970s to the mid 2000s (Steffen et al. 2007). However, there are a number of other markers. The present *mass extinction* of biodiversity as a result of human activity will leave a new series of fossils. There will also be new sediment layers consisting of human-made concrete, bricks and asphalt from roads, cities and infrastructure, as well as the polluted muds of estuaries.

Such *terra forming*, as a result of rapid urbanization (more than half of the world's population now lives in urban areas) has led to about an enormous increase in the long-term rate of erosion and sedimentation. Changes are likely to be compounded by sea-level rise as a result of climate change, although the *Intergovernmental Panel on Climate Change* (IPCC) predicts less than a meter of sea level rise over the next 90 years. However, more than five meters of sea level rise are possible over the coming centuries as the Greenland and West Antarctic ice caps melt. In New Orleans, large areas of the city are already below sea level. The combination of a rising sea level and the subsidence of the Mississippi Delta on which the city is built suggest that it will succumb at some point in the future. This rapid environmental change will lead to a remarkable, though perhaps short-lived, sedimentary signal in the geological record.

Sources

Crutzen, P J. 2002. Geology of mankind: The anthropocene. Nature, 415:23.
Steffen, W. 2008. Looking back to the future. AMBIO: A Journal of the Human Environment, 37(spl4): 507–513.
Steffen, W., P Crutzen, and J. McNeill. 2007. The Anthropocene: Are humans now overwhelming the great forces of nature. AMBIO: A Journal of the Human Environment 36(8): 614–621.
Steffen, W., J. Grinevald, P Crutzen, and J. McNeill. 2011. The Anthropocene: Conceptual and historical perspectives. Philosophical Transactions of the Royal Society A: Mathematical, Physical and Engineering Sciences, 369(1938), 842–867.

$15 billion. As of the end of 2014, increasing border surveillance measures are being put in place to curb the spread of the Ebola virus from West Africa. A number of airlines also stopped flying routes to the affected countries while some countries, such as Kenya, banned flights from the most affected region.

The tourism industry was one of the most affected sectors as a result of SARS; it suffered many cancellations and a substantial drop in traffic. This occurred not only because of negative media coverage but also because many countries issued "travel advisories" that recommended against travel to affected areas. Unfortunately for tourism, the traditional way to manage the spread of disease is to restrict access and the movement of people, which therefore means that destinations not affected by the disease may still be impacted by the measures taken to control the disease (Hall and Lew 2009).

The increased movement of people has therefore created a whole new series of challenges for those interested in the geography of tourism. As this book has stressed, from a tourism geography perspective, graduates need to be aware of both the cultural and natural aspects of a country or place, as both dimensions affect the landscapes and attractions that draw tourists. But, just as importantly, it makes us all the more aware of the nature of change and how such changes are gradually impinging on the future of tourism.

The Changing Nature of Destination Competitiveness

Changes in the global economy can also have major impacts on the relative competitiveness of destinations (measured in terms of the numbers of tourists attracted), as well as upon the emergence of new source areas. China and India, along with a number of other newly emerging economies of Southeast Asia and South America, are clearly poised to be major players in the world tourism market. These economies are significant because of their strength, the emerging middle-classes and their desire for consumer goods and services, and their relative resilience to economic and financial crises. Indeed, one of the ironies of globalization is that the more interconnected they are, the more vulnerable they can be when national economies face major problems due to unsustainable levels of debt, real estate speculation, irresponsible banks or even natural disasters and civil wars (Hall 2010, 2013).

Financial Crises

The economic and financial crisis that affected the global economy from 2008 on had considerable effect on the demand for tourism services (Visser and Ferreira 2013). For example, although they continued to travel, Europeans adapted their behavior to their new economic circumstances, primarily by travelling to less distant destinations and reducing their length of stay and their spending. Tourist activity in Europe therefore fell overall by approximately 5.6 percent in 2009. However, this figure hides wide disparities with some regions, especially in eastern or northern Europe, being particularly affected and recording a net fall of up to 8 percent in the number of tourists.

In addition, European financial crises in the early 2010s, and their effect on tourism hit some economies such as Greece, Portugal and Spain especially hard. This was in great part because the tourism sector is interrelated with the construction industry. Importantly, the global financial crisis also provided further signals that future growth in tourism was likely to be centered more on the emerging economies than on the old mature ones, which include Europe, Japan and North America. Therefore, although UNWTO estimates foresee a rise in international tourist arrivals in the future, it is suggested that such growth will be slower in Europe than in other regions of the world, such as Asia, which is increasing its market share at the expense of Europe and North America (UNWTO 2011).

At a more local scale competition is even more intense. The construction of new or the expansion of existing airports combined with the growth of budget airlines has seen a growth of tourism in hitherto

◀ *A World of Destinations. The growing middle classes, especially in developing countries, will ensure an increasing demand for travel and tourism for decades to come. How destinations compete for tourist dollars has become a fundamental aspect of today's global economy.*

© Terry Chan, 2011. Used under license from Shutterstock, Inc.

relatively undeveloped destinations, not all of which is successful. In some coastal areas, the competition between airports has been repeated with respect to the construction of areas that cruise ships can visit. Such measures place even more pressure on existing destinations to revitalize themselves. This has led to many towns and cities building convention and exhibition centers, stadia, heritage precincts, and museums and art galleries, often at significant public expense in order to try and attract visitors. Many such developments would be regarded as uneconomic if they were conducted by the private sector.

Branding

An additional issue faced in destination competitiveness is that if there are easily *substitutable* attractions, then destinations can be played off against each other by airlines or cruise ship companies to provide the lowest costs. In some cases, destinations may, in effect, subsidize, large operators to come to the destination. Faced with this competition, countries and destinations are developing a range of new strategies to try and differentiate themselves that often focus on some of the more intangible elements of a destination such as its culture and its "*brand*." However, other elements such as hallmark events, goods and wine tourism, gaming and nature-based tourism also appear significant.

Critical to the future competitiveness of destinations will be their relative cost, which is determined not only by the costs that exist at the destination itself but also by the expense that travelers experience getting there. These expenses depend strongly on the cost of fuel for transport, but they are also starting to include various taxes and other measures that are slowly being brought in to try and cover the true environmental costs of travel as well as the provision of transport infrastructure. Therefore, the final section of the chapter outlines some of the major issues that will affect the future world geography of tourism.

6.2 | FUTURES

Current Globalization Issues and Resources (http://wrgeography.com/futures.html)

By definition, the future is unknown. Nevertheless, there are a number of trends that are clearly influencing tourism patterns and flows at present and that will continue to do so for the foreseeable future. This section will discuss several of these, including demographic change, energy and resource supply, and various dimensions of global environmental change. These are all significant themes that have been touched on as we have gone through the book.

Demographic Changes

The major demographic factor that will affect the future of tourism is the aging of the world's population (Hall 2005; UNWTO 2010). This substantial demographic change has occurred because of improvements in healthcare and a declining birth rate in most of the developed world. At the end of the 20th century, 11 percent of the world's population was aged 60 and above. The United Nations estimates that 20 percent will be 60 years or older by 2050; and that by 2150, approximately one-third will be 60 years or older. Just as importantly, the older population itself is also aging. An eight to tenfold increase in the number of very old people (aged 80+ years) is expected between 1950 and 2050. Substantial regional differences in the aged population are also expected. For example, by 2020 some 20 percent of Europeans are expected to be 65 years or older, but only 5 percent of Africans are 60 years or older. However, in some developed countries today, the proportion of older persons is close to 20 percent. According to the United Nations, by 2050 that proportion will reach one in four and, in some countries, as high as one in two.

Given that the vast majority of the world's tourists come from the developed countries, an aging population will clearly have substantial implications for the international tourism industry in terms of tourist behaviors and expectations. Not only may particular types of tourism favored by older travelers continue to grow in popularity, such as cruising, but second homes and retirement homes and the provision of health facilities for retirees are likely to become increasingly important in destination development. Areas of Mediterranean Europe and the Iberian Peninsula (Spain and Portugal), as well as the Southwest United States and Florida in North America are already subject to substantial seasonal and permanent retirement migration.

Retirement Tourism

However, a critical factor in the tourism, leisure and second home patterns of retirees is not only their time budget but also their level of income. The growth of retirement tourism requires continued economic growth in order to maintain the pensions and savings of retirees. Ironically, this means that in many countries people may have to work longer. In some cases, this will be because of legislated changes to the retirement age, while many retirees will have to continue to work either out of choice or of necessity because of the inadequacy of retirement savings. A further implication of an aging population, at least in part, is that there will also be a growing number of people with impaired mobility who have specific needs and must be integrated into the tourist supply and service structure.

Population Booms and Busts

Another influential demographic factor with respect to tourism is that, although the world's developed countries are expected to grow in total population by about 4 percent to over 1.2 billion by 2050, population in developing countries is predicted to grow by up to 55 percent to more than 8 billion. It is predicted that among the major developed countries only the United States is estimated to have significant population growth by 2050. The United States is expected to have reached a population of 420 million by 2050.

But Europe is expected to have 60 million fewer people than today, with some countries, such as Bulgaria, potentially losing more than a third of their populations. Romania could have 27 percent fewer and Russia 25 million fewer people. Germany and Italy are expected to shrink by approximately 10 percent. Japan, which currently has only 14 percent of the population under 15, may have shrunk in size to approximately 100 million people by 2050. In contrast, developing countries are predicted to expand dramatically in population. Under this scenario, Western Asian nations are expected to gain about 186 million people by 2050, and sub-Saharan African countries more than one billion people. By 2050, it is expected that India will be the largest country in the world by population, having long surpassed China.

Ethnic Diasporas

Although the aging of the world's population will be the dominant demographic factor in tourism trends in the future, other factors will also be significant. For example, the increase in single parenting and the fact that many people will choose never to have children, or have them later in life, not only influences demographic characteristics, but also affects leisure travel. The growth of large migrant populations in many developed countries, particularly as the attraction of new labor sources may be required to sustain economic growth within an aging population, is likely to contribute to the development of new diaspora tourism patterns as migrant groups seek to retain relationships with their "home" countries.

Even in a domestic context, research has shown that people of different ethnic or national origins often seek different experiences during their vacation times, including visiting friends and relatives, vacationing at amusement parks, camping in the wilderness and visiting national parks, or staying close to home for *staycation*-types of activities. As we have seen throughout this book, migration is an extremely important driver of domestic and international travel flows.

The implications of these predicted demographic changes for tourism are incredibly complex and need to be seen in relation to other factors, particularly to global environmental change and to the availability of resources and energy. Even assuming the world's environmental systems can sustain continued population growth, the economic and political instability that may result from resource insecurity has the potential to severely damage tourism growth because of its impact on tourism-generating regions. Indeed, a major constraint on both population growth and tourism may be the future cost of energy.

Changing Energy Sources

The cost of energy and, in particular, the cost of fuel has a major impact on tourism (Hall and Lew 2009). This is primarily in the area of transport, but increased energy costs affect all aspects of tourism. Of major concern has been the concept of "*peak oil*," which refers to the point in time when the maximum rate of petroleum production is reached in any area under consideration (e.g., from a specific site to the global scale). After this point, the rate of production goes into decline, often becoming progressively more expensive to produce. (Also see Chapter 1.)

Oil supply is a significant issue for the supply of liquid fuels, with almost 60 percent of the world's oil consumed in the transportation sector. In aviation, for example, although there is experimentation with biofuels, air fleets are almost completely dependent on oil for fuel supply. The lack of potential substitutes for oil in many forms of transport means that transport demand for oil is relatively price-inelastic, although there is also a degree of elasticity and substitution with respect to consumer demand for different modes of transport. However, demand for oil is not only being fuelled by growth in the aviation sector but also by car purchases in such newly developed economies as China and India.

One alternative for oil production is the development of so-called "*unconventional*" supplies, including oil sands, shale and coal sources. However, such sources require significant energy inputs for the production of oil and also have high *greenhouse gas* emissions. *Biofuels* may be a substitute in some circumstances, but they also potentially create new issues because the production of biofuel can lead to the loss of biodiversity if previously unfarmed land is used. It can also lead to competition for land with food production if existing agricultural land is utilized. In the case of the latter, this may also serve to increase the cost of food production and transport. In itself, it is also highly dependent on oil and other energy inputs (Gössling et al. 2012a).

Water Use

Another area of concern with respect to environmental change is water, as a result of both climate change and competition between users. *Water stress* is a function of renewable fresh water availability, abstraction rates and the share of consumptive use (Gössling et al. 2012b, 2015). Global water use is increasing due to population and economic growth, changes in lifestyles, technologies, international trade and the expansion of water supply systems. Tourism is both dependent on fresh water resources and an important factor in its use. Tourists need and consume water when washing or using the toilet, when participating in activities such as ski and winter tourism (snowmaking), and when using spas and swimming pools. Fresh water is also needed to maintain hotel and resort gardens and golf courses, and it is embodied in tourism infrastructure development and in food and fuel production.

Although tourism does increase global water consumption, direct tourism-related water use is considerably less than 1 percent of world consumption. It is not likely to become significant at the global scale in the fore seeable future even if the sector continues to grow at anticipated rates of around 4 percent in international tourist arrivals per year. This issue has had significant negative repercussions in recent years in parts of the U.S. Southwest (e.g., Las Vegas and Southern California) where, owing to drought, residents are heavily restricted from using water for household purposes, yet tourism services such as resorts and golf courses are able to continue using water because of tourism's economic importance in the region.

Arid Regions

However, the situation is different at the regional level of analysis because tourism concentrates traveler flows in time and space, as well as often in dry regions where renewable water reserves are limited. Many island resorts, for example, have a high degree of water stress, as do some of the resort areas in Mexico, the Mediterranean and the Middle East. Many of them either have to ship or freight water in or have had to develop desalinization plants. The future of water supply for many tourist areas is likely to be complicated by the potential effects of climate change. Regional modelling demonstrates that already *water-scarce* or *water-sensitive* regions, such as the Caribbean, will be particularly affected by declining rainfall. Major concerns with respect to decreased precipitation, rising temperatures and changes in runoff patterns also exist in northern Africa, the Mediterranean and the Middle East. The Middle East is already the world's most water-stressed region and pressures continue to increase as populations increase and climate changes (Gössling and Hall 2006; Gössling et al. 2015).

Climate Change

Climate change also sets new challenges for tourism in other ways. Tourism is considered a highly climate-sensitive economic sector similar to agriculture, insurance, energy and transportation. Climate defines the length and quality of tourism seasons and plays a major role in destination choice and visitor spending patterns. Climate affects a wide range of the environmental resources that are critical attractions for tourism in many destinations, such as wildlife and *biodiversity*, water levels and quality, snow conditions and glacier extent. Climate also has an important influence on environmental conditions that can deter tourists, including wildfires, insect or water-born pests, and extreme weather events such as floods, droughts, storms and tropical hurricanes (cyclones) (Gössling et al. 2012a; Herring et al. 2014).

In the future, climate change could determine a restructuring of travel models and seasonality and affect many destinations. Declining snow cover in alpine and high latitude regions will cause a decline in winter tourism activities such as skiing. Even though artificial snow making can extend the lifespan of some ski resorts, artificial snow still requires the availability of water and low enough temperatures to make snow. At the same time, rising sea levels are bringing changes for tourism in coastal areas, with some parts of the United States, such as Florida, already affected by combinations of sea level rise and high tides. Clearly, the extent to which a nation or a specific region is economically dependent on tourism, and the relative cost-distance from markets, will have implications for the overall vulnerability of that destination to climate change.

Islands and Reefs

Island states, which have some of the economies that are most dependent on tourism in the world, are also the most vulnerable to climate change. In the case of island states, not only are they vulnerable to sea-level rise and high-magnitude weather events, those in the tropical and sub-tropical areas are also susceptible to *coral reef* loss. This is important as both a tourist attraction and to reduce the effects of storms.

Recent assessments of the world's coral reefs suggest that three-quarters of them are at risk due to over-fishing, pollution, climate change and other factors. The reef-building coral polyps live in partnership with algae that give corals their color and provide nutrition. When seawater gets too warm, the polyps expel the algae, and the coral turns white. This event is known as coral bleaching. Although reefs can recover, the more often it happens, the harder it is for them to do so. Climate change predictions suggest that by 2030 roughly half of the world's reefs will experience bleaching in most years-rising to 95 percent by the 2050s.

To make matters even worse, the slow decrease in the *pH level* of seawater as it absorbs more carbon dioxide, usually referred to as ocean acidification, reduces the availability of carbonate ions in the water, which many creatures use to build shells and skeletons out of calcium carbonate. The decrease in available carbonate ions means that organisms, such as coral, mollusks and plankton, will struggle to build or maintain their hard protective or supportive structures. Although some organisms, such as sea grass, may grow better in waters rich with carbon, this does not bring some of the benefits that can come from coral reefs.

Sensitive Environments

Other highly significant vulnerabilities include the concentration of tourism in coastal regions and estuarine systems, and in alpine regions, as well as having environmentally-based attractions and a substantial proportion of the market based on *long-haul travel*. A significant risk to the tourism industry is the increased likelihood of *extreme high-magnitude weather events* and the attendant infrastructure damage, tourism interruption and negative effects on destination image. Sensitive coastal and mountain tourism infrastructure, in particular, is likely to be vulnerable.

Assessments of the integrated effects of climate change (both shifts in climatic means and extremes), climate-induced environmental change and climate-related societal change suggest that they will have far-reaching impacts on tourism destinations. Such impacts are already becoming evident at destinations around the world at such high latitude destinations as the Arctic and the Antarctic Peninsula. The regional manifestations of climate change will generate both negative and positive impacts in the tourism sector, and these impacts will vary substantially by market segment and geographic region. However, from a market perspective, one of the notable impacts includes a gradual shift in preferred destinations and tourist spending to *higher-latitude*, temperate nations and higher elevation mountainous areas, as well as the development of new seasonal patterns of visitation to current destinations in order to avoid uncomfortably high temperatures.

Destination Choice

Tourists have the greatest capacity to adapt to the effects of climate change, with relative freedom to avoid destinations impacted by climate change. Climate, the natural environment, personal safety and travel cost are four primary factors in destination choice and consumer demand, with global climate change anticipated to have an impact on all of these. All tourism businesses and destinations will also need to adapt to climate change in order to minimize associated risks and to capitalize on new opportunities. However, it appears that industry knowledge of the capacity of current climate adaptations to respond successfully to future climate change is rudimentary.

Climate change is slowly entering into the decision-making processes of a range of tourism stakeholders (e.g., entrepreneurs, investors, insurance companies, tourism corporations, governments, nongovernment organizations, and national park and conservation agencies). In general, *sustainability issues* are huge right now throughout the various sectors of the tourism economy. Accommodations, transportation and tour companies are all trying to incorporate green and sustainable processes and products. This is especially true for industry associations and for the larger companies in these sectors. It is less noticeable among smaller companies with fewer resources. There are also quite a few small, *niche companies* that are trying to differentiate themselves through sustainability. In addition, there are quite a few government and non-profit (NGOs) and community development organizations that are involved in sustainable development in general, with tourism being one of their areas of focus.

Nevertheless, given that the large information requirements, policy changes and investments required for effective adaptation by tourism destinations may require decades to implement, the process

▶ *The North Pole. Most of the Arctic Ocean has been covered by ice throughout human history. With recent global warming, however, most of us will likely see an ice-free Arctic Ocean, at least in the summer months, within our lifetime.*

of adaptation must commence as soon as possible for those destinations likely to be impacted by mid-century. This is well within the life spans of those reading this book. This is especially important for many of the less wealthy countries in the tropics, and in particular for island states, that depend on international tourism for a large share of their GDP and foreign exchange earnings. Such destinations will need to develop a greater understanding of not only the implications of climate change for tourism products and environmental services (e.g., coastal zones, coral reefs, water supply, heritage assets), but also for the implications of emerging climate policy regimes for their relative cost and accessibility. Most importantly, in many cases governments will need to decide whether international tourism will be the best development alternative (Gössling et al. 2009).

ISSUES AND INSIGHTS Arctic Tourism: Last Chance to See?

The polar regions are regarded by many commentators as the proverbial "canaries on the coal mine" with respect to global environmental change, and to climate change in particular. The Arctic is manifesting the early stages of a human-induced greenhouse signature. Surface air temperatures in the Arctic have warmed at approximately twice the global rate, with a figure of 1-2°C representing the really averaged warming north of 60°N since there was a temperature minimum in the 1960s and 1970s. From 1980, the warming of much of the Arctic has been even higher (about 1°C/decade) in winter and spring. It is strongest over the interior portions of northern Asia and of northwestern North America. The extent of recent warming is such that it has been recognized as the warmest period in the Arctic for the last 2000 years, with four of the five warmest decades in that period occurring in the past 50 years. Precipitation in the Arctic has also increased at about one percent per decade over the past century, although the trends are spatially highly variable and very uncertain because of deficiencies in the meteorological record.

Consistent with the observed increases in Arctic surface air temperatures is also widespread evidence of reductions of Arctic sea ice and glaciers, reductions in the duration of river and lake ice in much of the sub-Arctic, and, since the 1980s, warming and melting of permafrost. Arctic sea ice has also shown a substantial reduction since the 1950s. Sea ice in the Arctic shrank to its smallest size on record in September 2007, when it extended over an area of just 4.13 million km^2 (1.59 million mi^2), beating the previous low of 5.32 million sq km measured in 2005. Using previously classified submarine data, researchers found that the average thickness at the end of the melt season has decreased by 1.6 m, or some 53 percent from 1958 to 2008. The peak winter thickness of 3.64 m in 1980 (from submarine data) decreased to 1.89 m by the winter of 2008 (in satellite data), a net decrease of 1.75 m or 48 percent in thickness.

The lowest average Arctic Sea ice extent in the satellite record was recorded in September 2007. In 2014 the seasonal daily minimum of 5.02 million km^2 (1.94 million mi^2) that was set on September 17, 2014 (6th lowest in the satellite record), Arctic sea ice extent averaged for the month of

September 2014 was 5.28 million km^2 (2.04 million mi^2), also the 6th lowest in the satellite record. This is 1.24 million km^2 (479,000 mi^2) below the 1981 to 2010 average extent, and 1.65 million km^2 (637,000 mi^2) above the record low monthly average for September that occurred in 2012.

Less ice means more open water exposed to short wave solar radiation that is absorbed and transformed into heat. This provides a strong positive feedback that further accelerates the melting of Arctic sea ice. (Sea ice areas covered with snow have a high albedo that reflects 80 percent of the incoming solar radiation back into space. In contrast, the open ocean has a low albedo that reflects only 20 percent of solar radiation, absorbing the other 80 percent.) The loss of Arctic sea ice is significant because it is not just the ocean that is subject to change, but also the coastline and the hinterland. The larger heat transfer from the ocean to the atmosphere (the maritime effect) will also help moderate autumn and winter cold temperatures on land.

The decline in Arctic sea ice extent is inseparable from the increased interest in maritime access to the polar seas from countries such as Canada, Denmark, Norway, Russia and the United States. This interest focuses upon fishing, mineral exploration and exploitation, and trade and tourism, as well as on concerns over political sovereignty. Given the number of visitors to the Arctic, it should therefore be of no surprise that tourism is regarded as such a key component of the economy, along with fisheries and mineral and energy development.

For example, from 2006 to 2008, just prior to the global economic and financial crisis, cruise passenger numbers to Greenland jumped by 30 percent from 22,051 in 2006 to 28,891; and in nine years the total number of travelers to Greenland more than doubled from 26,410 in 1999 to 57,223 in 2008. As with many coastal polar areas, much of the tourism growth is coming from cruise tourism. The number of cruise ships operating in Greenland waters has grown from 13 in 2003 to 39 in 2008, with the number of port arrivals increasing from 164 to 375 in the same period. Cruise numbers are now increasing again.

These growth figures are matched from elsewhere in the Arctic. The number of visitors to Svalbard, Norway, increased

continued

by almost a third between 2000 and 2006, while a similar growth rate also occurred for Alaska. The extent of Arctic change, and the possible loss, or at least significant changes in the distribution, of such iconic species as polar bears, musk ox and reindeer, have given rise to the phenomenon of "*last chance*" tourism. This refers to the notion that it might be the last chance to see the various attractions before they disappear (also sometimes referred to as "*doom tourism*").

One travel magazine reported: "The world has never traveled to the Arctic like now. Aided by global warming— that's opening up areas never before visited—but tinged by a quiet urgency, it's here the world gets a live demonstration of how our world is changing" (Destinations of the World News 2008: 2). Round's (2008, 46) observation that, "The plight of the region has become such a part of our contemporary back ground that it's no wonder demand for the region has become so high. The message is quite clear: 'come quickly or you'll miss it', provides an interesting ethical dilemma for tourists, should you go now and see it knowing that you are possibly further contributing to its demise by seeing it, or should you stay home and potentially miss it forever?'"

In some areas of the Arctic, climate change, rather than having a negative impact on the regional economy, is often regarded as being beneficial along with maritime transport in general as access to many northern areas is improved. While the old Arctic is rapidly disappearing, a new geography of the far northern hemisphere is emerging.

Geographer Laurence Smith has referred to the area of the globe north or 45° N. Latitude as the *New North*. This is a region dominated by the eight countries that have territorial claims to the Arctic Ocean (also known as the Northern Rim countries): Canada, Greenland/Denmark, Finland, Iceland, Norway, Russia, Sweden and the United States. As the world's climate warms, each of these is staking its claims to natural resources that had previously been protected by harsh environmental conditions. Aboriginal populations, military installations and some fossil fuel extraction activities have traditionally sparsely settled the *Northern Rim* region. Today, however, populations are growing rapidly and following new roads and sea routes, along with facing growing territorial issues as each country seeks to secure or expand its claims to the Arctic Ocean.

Sources

Hall, C. M. 2010. Tourism and environmental change in polar regions: Impacts, climate change and biological invasion. In *Tourism and change in polar regions: Climate, environments and experiences*, ed. C. M. Hall and J. Saarinen, 42–70. London: Routledge.

Hall, C. M., and J. Saarinen. 2010. Polar tourism: Definitions and dimensions. *Scandinavian Journal of Hospitality and Tourism* 10(4): 448–467.

Hall, C. M., and J. Saarinen. 2010. Last chance to see? Future issues for polar tourism and change. In *Tourism and change in polar regions: Climate, environments and experiences*, ed. C. M. Hall and J. Saarinen, 301–310. London: Routledge.

Round, A. 2008. Paradise lost. *Destinations of the World News*, issue 21(March): 44–51.

Smith, L. C. 2011. Agents of change in the new north. *Eurasian Geography and Economics* 52(1): 30–55. Available online at: http://www.sscnet.ucla.edu/geog/downloads/297/393.pdf

Arctic Sea Ice News and Analysis, National Snow and Ice Data Center: http://nsidc.org/arcticseaicenews/

6.3 | CONCLUSION

What can you do with the knowledge that you have gained from this book?

Geographers are infatuated with everything about planet Earth, from its physical landscapes to its cultural diversity. That is what this book is all about. It has introduced you to the planet from the perspective of geography. The goal of this book was to make you a better inhabitant and traveler of the planet by increasing your knowledge and appreciation of it. As the planet becomes smaller, through easier, faster and cheaper transportation and communication, being a better citizen of planet Earth is essential. The geographic skills that you have been exposed to, and that you can further build upon through future study and career development, include:

- landforms, landscapes and land features
- human cultures, with a focus on settlement histories, patterns and landscapes
- the use of maps and other spatial information in understanding physical and human features
- environmental and community development issues and processes
- globalization and the global economy

Each of these geographic concepts is personally *scalable*. You can apply them to the community you live in, to understand better how and why the community looks and works the way it does. They can be applied to the region or nation that you live in, as you travel to neighboring communities and regional attractions, to understand what you are seeing from the window of your car, train or plane. And they can be applied on an international scale to understand better the global news that you hear and read about, and to plan your personal travels better, whether for work or for play.

Most importantly, we also hope that by being interested in the diverse cultures, environments and landscapes of our planet, you will want to find ways to care more for it. It is, after all, the only one we have. As the book has highlighted, tourism can be either a positive or a negative influence on destinations, depending on perspectives and what tourism development was hoping to achieve. But regardless of perspective, tourism is undoubtedly a major force. We hope that as both a *geographer* and as a *tourist* you will act as a force for good.

REFERENCES CITED

Gössling, S., and C. M. Hall, eds., 2006. *Tourism and global environmental change*. London: Routledge.

Gössling, S., C. M. Hall, and D. Scott, 2009. The challenges of tourism as a development strategy in an era of global climate change. In *Rethinking Development in a Carbon-Constrained World: Development Cooperation and Climate Change*, ed., E. Palosou, pp. 100–119, Helsinki: Ministry of Foreign Affairs.

Gössling, S., C. M. Hall, and D. Scott, 2012a. *Tourism and climate change: Impacts, adaptation and mitigation*. London: Routledge.

Gössling, S., C. M. Hall, and D. Scott, 2015. *Tourism and Water*. Bristol: Channelview.

Gössling, S., P. Peeters, C. M. Hall, J-P. Ceron, G. Dubois, L. V. Lehmann, and D. Scott, 2012b. Tourism and water use: Supply, demand, and security-An international review. *Tourism Management*, 33: 1–15.

Hall, C. M. 2005. *Tourism: Rethinking the social science of mobility*. Harlow, UK: Prentice-Hall.

Hall, C. M. 2010. Crisis events in tourism: Subjects of crisis in tourism. *Current Issues in Tourism* 13(5): 401–417.

Hall, C.M. 2013. Financial crises in tourism and beyond: Connecting economic, resource and environmental securities. In *Tourism and Crises*, Eds. G. Visser and S. Ferreira, pp. 12–34. London: Routledge.

Hall, C. M. and A. Lew, 2009. *Understanding and managing tourism impacts: An integrated approach*. London: Routledge.

Hall, C. M. and A. Williams, 2008. *Tourism and innovation*. London: Routledge.

Harvey, D. 1990. Between space and time: Reflections on the geographical imagination. *Annals of the Association of American Geographers* 80: 418–34.

Herring, S. C., M. P. Hoerling, T. C. Peterson, and P. A. Stott, Eds., 2014: Explaining Extreme Events of 2013 from a Climate Perspective. *Bulletin of the American Meteorological Society*, 95 (9), S1–S96.

UNWTO 2010. *Demographic change and tourism*. Madrid: World Tourism Organization and the European Travel Commission.

UNWTO 2011. *Tourism towards 2030 global overview*. Madrid: UNWTO.

Visser, G., and S. Ferreira Eds. 2013. *Tourism and Crises*. London: Routledge.

Name _____

Human Mobilities

1. Identify three examples of human mobility that shape the community you live in. Describe the impact of each of these on your community.

_____

_____

_____

_____

2. Identify and describe three examples of human mobility that have global impacts. Describe the impact of each of these on your community.

_____

_____

_____

_____

Tourism Destinations

3. In what ways is the community that you live in shaped by the global tourism economy? How do globalization and tourism influence (or not) the image of your community?

_____

_____

_____

_____

4. How is a global perspective useful in understanding the geographic landscapes of the world and the places in it? Cite three specific examples or concepts in geography (from this chapter) that could help you, personally, to enhance your understanding of the world.

_____

_____

_____

_____

Sustainable Environments

5. In what ways are a geographic understanding of our planet and its environments and communities necessary in order to create a more sustainable world?

_____

_____

_____

_____

6. What is the relationship between economic and cultural globalization and local sustainability? How might globalization help local sustainability efforts? How might globalization hurt local sustainability efforts?

_____

_____

_____

_____

Instructor Questions (answer any additional questions from your course instructor in the space below)

_____

_____

_____

_____